Class. Quantum Grav. 11 (1994) L121–L123. Printed in the UK

LETTER TO THE EDITOR

A note on the symmetries of the gravitational field of a massless particle

Peter C Aichelburg†§‖ and Herbert Balasin‡¶+

† Institut für Theoretische Physik, Universität Wien, Boltzmanngasse 5, A-1090 Wien, Austria

‡ Institut für Theoretische Physik, Technische Universität Wien, Wiedner Hauptstraße 8–10, A-1040 Wien, Austria

Received 28 July 1994

Abstract. It is shown that the metric of a massless particle obtained from boosting the Schwarzschild metric to the velocity of light, has four Killing vectors corresponding to an $E(2) \times \mathbb{R}$ symmetry group. This is in agreement with the expectations based on flat-space kinematics but is in contrast to previous statements in the literature (Schücker). Moreover, it also goes beyond the general Jordan–Ehlers–Kundt (JEK) classification of gravitational pp-waves as given by Ehlers and Kundt.

PACS numbers: 0420, 0250, 9760L

Several years ago Sexl and one of the authors (PCA) obtained a gravitational impulsive plane-fronted wave with parallel rays (pp-wave) by boosting the Schwarzschild metric to the singular limit [3]. Since then this metric* has been widely used to discuss ultra-relativistic scattering processes in both the classical and the semiclassical context [4, 5]. Moreover, the method of boosting was also applied to other configurations like the Kerr or the Reissner–Nordstrøm metric [6, 7]. In flat spacetime the trajectory of a free particle of zero rest mass is described by its initial position and 4-momentum which is null. The subgroup of the Poincaré group that leaves the trajectory invariant is isomorphic to $E(2) \times \mathbb{R}$. Its generators act in the null-hyperplane orthogonal to the given vector. On physical grounds one would therefore expect that the gravitational field generated by a massless particle shares the same symmetries. As early as in 1962, in a now classical paper [2], Ehlers and Kundt gave a complete classification of symmetries for pp-waves together with their standard form for each symmetry. A comparison of this classification with the metric of a massless particle tells one that the metric should have only two instead of the expected four Killing vectors, namely the translation along the given lightlike propagation direction and the corresponding space rotation. This contradiction was emphasized in a paper by Schücker [1] who argued that the full symmetry implies the presence of infinite mass density which general relativity does not allow for. In this note we show that the symmetry of the field in question indeed admits the expected four Killing vectors. The above-mentioned contradiction is resolved by

§ E-mail address: pcaich@pap.univie.ac.at

‖ This work was also supported in part by the FUNDACION FEDERICO

¶ E-mail address: hbalasin@ecxph.tuwien.ac.at

+ Supported by the APART-program of the Austrian Academy of Sciences

* denoted by the AS-metric by some authors.

0264-9381/94/100121+03$19.50

noting that the metric, which is an impulsive gravitational wave, has in its standard form a distributional metric coefficient. Thus, allowing for distributions, a careful analysis of the Killing equations shows the presence of the full symmetry. It is also straightforward to see that the JEK classification is based on regular functions only.

Our approach to deriving the Killing vectors of the AS-geometry will start from its unboosted counterpart, the Schwarzschild geometry. Since both geometries, Schwarzschild and AS, belong to the Kerr–Schild class they admit a decomposition

$$g_{ab} = \eta_{ab} + f k_a k_b$$

where η denotes the flat part of the decomposition, k a geodetic null vector-field and f a scalar function. The existence of the flat background part in the decomposition gives a natural meaning to the concept of boosts as its isometries. Moreover, this geometrical aspect will turn out to be of great calculational advantage. Choosing Kerr–Schild coordinates x^a for the Schwarzschild metric brings η into its standard form and f and k become respectively

$$f = \frac{2m}{r} \qquad k^a = \begin{pmatrix} 1 \\ x^i/r \end{pmatrix} \tag{1}$$

where x^i denotes spatial cartesian coordinates according to the (flat) Minkowski metric and r its corresponding radial distance. With respect to these coordinates the Killing vectors of (1) corresponding to staticity and spherical symmetry can be written as

$$\begin{aligned} &\xi^a_t = (\partial_t)^a \\ &\xi^a_{\hat\omega} = \hat\omega^a{}_b x^b \qquad \hat\omega^a{}_b \xi^b_t = 0 \qquad \hat\omega^{ab} = -\hat\omega^{ba} \end{aligned} \tag{2}$$

where $\hat\omega$ denotes an infinitesimal rotation matrix. The index-raising in the second line of (2) was performed with the help of η, which will be implicitly assumed in the following unless stated otherwise. Interpreting the Kerr–Schild coordinates as referring to a Lorentz-frame associated with an observer that is at rest asymptotically, allows us to find the form of the Killing vectors with respect to a boosted asymptotic observer:

$$\begin{aligned} \xi^a_t &= \frac{1}{m} P^a \\ \xi^a_{\tilde\omega} &= \tilde\omega^a{}_b x^b \\ \xi^a_{\tilde a} &= \frac{1}{m}((\tilde a x) Q^a - (Qx)\tilde a^a) \\ \hat\omega^a{}_b &= \tilde\omega^a{}_b + \frac{1}{m}(Q^a \tilde a_b - \tilde a^a Q_b) \end{aligned} \tag{3}$$

where P and Q denote the timelike and spacelike vectors spanning the boost 2-plane and vectors in the orthogonal 2-plane are tagged by a tilde (i.e. $P^a \tilde a_a = Q^a \tilde a_a = 0$). P and Q are conveniently normalized to $\pm m^2$ and tend to a null vector p in the limit $m \to 0$. To take $m \to 0$, we multiply ξ_t and $\xi_{\tilde a}$ with m thereby obtaining the limit

$$\begin{aligned} &(m\xi^a_t) \to p^a =: \zeta^a_t \\ &\xi^a_{\tilde\omega} \to \tilde\omega^a{}_b x^b =: \zeta^a_{\tilde\omega} \\ &(m\xi^a_{\tilde a}) \to (\tilde a x) p^a - (px)\tilde a^a =: \zeta^a_{\tilde a} \,. \end{aligned} \tag{4}$$

From their mutual commutation relations

$$[\zeta_t, \zeta_{\tilde\omega}] = 0 \qquad [\zeta_t, \zeta_{\tilde a}] = 0 \qquad [\zeta_{\tilde a}, \zeta_{\tilde a'}] = 0 \qquad [\zeta_{\tilde\omega}, \zeta_{\tilde a}] = \zeta_{\tilde\omega \tilde a}$$

one sees that the ζ form a representation of the Lie-algebra of $E(2) \times \mathbb{R}$.

The corresponding limit of the Schwarzschild geometry

$$g_{ab} = \eta_{ab} - 8\delta(px)\log\rho\ p_a p_b \tag{5}$$

may either be obtained by employing a 'singular' boost [3] or by solving the Einstein equations with a boosted source [8]. The result is a pp-wave with profile function $f = -8\delta(px)\log\rho$, where ρ denotes the radial distance in the orthogonal 2-plane. It is now easy to verify that all the ζ of (4) have the Killing property

$$\begin{aligned} L_\zeta g_{ab} &= (\zeta\partial)f\ p_a p_b + \partial_a\zeta_b + \partial_b\zeta_a + f(p_a\partial_b(p\zeta) + p_b\partial_a(p\zeta)) \\ &= (\zeta\partial)f\ p_a p_b + \partial_a\zeta_b + \partial_b\zeta_a = 0 \end{aligned} \tag{6}$$

where ∂ is the covariant derivative with respect to η. However, for the sake of explicitness, let us exhibit the only non-trivial term of (6) relative to $\zeta_{\tilde{a}}$

$$(\zeta_{\tilde{a}}\partial)f = -8(\tilde{a}x)p^2\delta'(px) + 8(px)\,\delta(px)\frac{1}{\rho}e^a_\rho\tilde{a}_a$$

which vanishes due to the product $px\,\delta(px)$ and the fact that p is null.

This shows that the Killing vectors of the Schwarzschild metric survive the ultrarelativistic boost in the sense that the resulting vector fields are Killing with respect to the boosted metric. Comparing (5) with the general classification of Killing vectors for pp-waves [2] one would only expect the presence of ζ_t and $\zeta_{\tilde{\omega}}$. The fact that the two-parameter family of vector fields $\zeta_{\tilde{a}}$ is also Killing depends crucially upon the presence of the $\delta(px)$-factor in f, which also explains the discrepancy with [2], where no distributional profile functions were considered. The implications of this aspect for the general classification of pp-waves will be investigated in a forthcoming publication.

The authors wish to thank H Urbantke for a helpful comment. One of the authors (PCA) thanks A Ashtekar for a discussion on this problem during a meeting in Cocoyoc, Mexico in 1990.

References

[1] Schücker T 1989 *Phys. Lett.* **136A** 200
[2] Ehlers J and Kundt W 1962 *Gravitation, an Introduction to Current Research* ed L Witten (New York: Wiley) p 85
[3] Aichelburg P and Sexl R 1971 *Gen. Rel. Grav.* **2** 303
[4] 't Hooft G 1987 *Phys. Lett.* **198B** 61
[5] Verlinde E and Verlinde H 1992 *Nucl. Phys.* B **371** 246
[6] Loustó C O and Sánchez N 1990 *Int. J. Mod. Phys.* A **5** 915; 1992 *Nucl. Phys.* B **383** 377
[7] Ferrari V and Pendenza P 1990 *Gen. Rel. Grav.* **22** 1105
[8] Balasin H and Nachbagauer H 1994 The ultrarelativistic Kerr-geometry and its energy–momentum tensor *Preprint* gr-qc 9405053

Class. Quantum Grav. **11** (1994) 2395–2410. Printed in the UK

On loop states in quantum gravity and supergravity

Hans-Jürgen Matschull†
II. Institut für Theoretische Physik, Universität Hamburg, Luruper Chaussee 149, 22761 Hamburg, Germany

Received 21 March 1994, in final form 23 June 1994

Abstract. The Wilson loop functionals in terms of Ashtekar's variables were the first (formal) solutions to the quantized Hamiltonian constraint of canonical gravity. Here it is shown that the same functionals also solve the supergravity constraints, and some evidence is presented that they are not states of a quantum theory whose classical limit is general relativity. For $2+1$ dimensional gravity and supergravity it is shown explicitly how some crucial properties of Einstein gravity are lost when using 'densitized' constraints.

PACS numbers: 0460D, 0465

1. Introduction

In this paper some features of the so called loop states of canonically quantized gravity are discussed. In the first part we will focus on $N = 1$ supergravity in four dimensions. For this theory the canonical treatment 'au Dirac' [1] in the metric (or vierbein) representation leads to first class constraints, which in the quantized version become inhomogeneous second order differential equations with non-analytic coefficients [2], whereas in the connection representation, using Ashtekar's variables [3], they become polynomials in the canonically conjugate variables and homogeneous in the momentum variables [4]. However, when using the connection representation one has to deal with complex phase space coordinates obeying certain reality conditions. In [4] the quantum versions of these conditions are not analysed completely, thus we will briefly discuss them here.

As for the connection representation of pure gravity, one has to multiply some constraints by the determinant of the spatial metric to obtain polynomial expressions. However, in supergravity this multiplication not only produces a densitized Wheeler–DeWitt operator but also the diffeomorphism constraint becomes a density of weight 2, because both are generated by the commutator of the supersymmetry constraint with its conjugate.

We will see that the densitized supersymmetry constraints are formally solved by the *same* Wilson loop functionals which are known to solve the Wheeler–DeWitt constraint of pure gravity [5]. In contrast to pure gravity, however, a single loop functional now solves *all* constraints without considering functionals that depend on knot classes only, i.e. the state functionals are no longer invariant under spatial diffeomorphisms.

Though this result comes out when dealing with supergravity, it can be reproduced for pure gravity, too, just by dropping all the fermionic quantities from the action and the constraints, which was shown in [6]. When defining the polynomial constraints in this way, a single loop functional again becomes a solution to *all* constraints. Thus multiplying

† Work supported by DFG.

the diffeomorphism constraint with extra factors destroys the invariance of the theory under spatial coordinate transformations. In fact, for the classical theory we will explicitly see that the invariance under *spacetime* diffeomorphisms is lost, if we use the densitized Hamiltonian constraint and allow the metric to be singular.

Another property of the loop functionals in supergravity is that they are purely bosonic states, i.e. they do not depend on the fermionic variables. There has been a discussion about the existence of such states [7–9]. For the metric representation it was shown in [9] that purely bosonic states do not exist, because they cannot fulfill one of the supersymmetry constraints. This suggests that the Wilson loop states are nothing but solutions which are artificially generated by multiplying the constraints with the metric determinant, as the solutions themselves are annihilated by this determinant.

To make all these arguments more precise, in the second part we will discuss $N = 2$ supergravity in $2 + 1$ dimensions, which is completely soluble both classically and canonically quantized. To define the connection representation for this model properly, no extra factors of the dreibein are needed to make the constraints polynomial.

But, if one starts from the four-dimensional densitized constraints and reduces them to three dimensions, or, equivalently, if one multiplies the Hamiltonian constraint by the metric determinant in the same way as for the $3 + 1$ dimensional theory, one obtains a different set of constraints, the 'densitized connection representation'. In this representation one again finds the loop states, and the quantum theory becomes completely different. In particular, there are now infinitely many states even if the spatial topology is trivial and one requires the state to be invariant under diffeomorphisms (which does not follow from the constraints!), whereas in the 'correct' quantum theory there are only finitely many classical degrees of freedom, i.e the wave functional depends on finitely many variables only, and there is only one state for trivial topology.

This will lead us to the conclusion that the theories described by the densitized constraints are, in some very crucial properties, different from general relativity, and the loop states might not represent states of a quantum theory whose classical limit is given by a full diffeomorphism-invariant theory like Einstein gravity.

2. $N = 1$ supergravity in four dimensions

The canonical formalism for $N = 1$ supergravity has been worked out in the metric representation [2] and in terms of Ashtekar's new variables in [4]. To set up a slightly different notation and to draw attention to a fact that has not been discussed in [4], we quickly derive the constraints from the classical action.

The first order action for $N = 1$ supergravity [10, 11] is usually written as

$$I'[E, \Omega, \psi, \bar{\psi}] = \int \mathrm{d}^4x\, L_{\mathrm{EH}} + L_{\mathrm{RS}} \tag{2.1}$$

where

$$\begin{aligned} L'_{\mathrm{EH}} &= \tfrac{1}{2} E E_A{}^M E_B{}^N R_{MN}{}^{AB}[\Omega] \\ L'_{\mathrm{RS}} &= \mathrm{i}\varepsilon^{MNPQ}\big(\bar{\psi}_M \sigma_N D_P \psi_Q - D_M \bar{\psi}_N \sigma_P \psi_Q\big) \end{aligned} \tag{2.2}$$

are the Einstein–Hilbert and Rarita–Schwinger action, respectively. The notation is as follows. Indices from the beginning of the alphabet always denote flat tangent space vectors, those from the middle of the alphabet curved spacetime indices. The vierbein $E_A{}^M$ (or its inverse $E_M{}^A$) thus has the flat index A running from 0 to 3, raised and lowered by the Lorentzian metric $\eta_{AB} = \mathrm{diag}(-,+,+,+)$, and the curved index M, taking the values t, x, y, z. E is the determinant of $E_M{}^A$ and ε^{MNPQ} the Levi-Cività tensor density with $\varepsilon^{txyz} = 1$.

The spin connection Ω_{MAB} defines the Lorentz covariant derivative of a tangent space vector with flat index

$$D_M V_A = \partial_M V_A + \Omega_{MAB} V^B \qquad \left[D_M, D_N\right] V_A = R_{MNAB} V^B. \tag{2.3}$$

We introduce Ashtekar's variables by mapping the $4 \cdot 6$ components of the real $so(3,1)$ spin connection Ω_{MAB} onto $4 \cdot 3$ complex components of the $so(3,\mathbb{C})$ connection A_{Ma}, $a = 1, 2, 3$. The mapping is given by

$$A_{Ma} = J_{aAB} \Omega_M{}^{AB} \qquad A^*_{Ma} = J^*_{aAB} \Omega_M{}^{AB} \tag{2.4}$$

where the coefficients J_{aAB} and J^*_{aAB} form a basis of self-dual and antiself-dual antisymmetric tensors, i.e.

$$J_{aAB} = -\tfrac{1}{2}\mathrm{i}\varepsilon_{AB}{}^{CD} J_{aCD} \qquad J^*_{aAB} = \tfrac{1}{2}\mathrm{i}\varepsilon_{AB}{}^{CD} J^*_{aCD} \tag{2.5}$$

and ε^{ABCD} is the flat Levi-Civita tensor with $\varepsilon^{0123} = -\varepsilon_{0123} = 1$. The J-symbols are complete and orthonormal

$$\begin{aligned} &J_{aAB} J_a{}^{CD} + J^*_{aAB} J^{*CD}_a = \delta^{[C}_{[A} \delta^{D]}_{B]} = \tfrac{1}{2}\delta^C_A \delta^D_B - \tfrac{1}{2}\delta^D_A \delta^C_B \\ &J_{aAB} J_b{}^{AB} = J^*_{aAB} J^{*AB}_b = \eta_{ab} \qquad J_{aAB} J^{*AB}_b = 0 \end{aligned} \tag{2.6}$$

where $\eta_{ab} = \delta_{ab}$ is the spatial component of the flat metric (remember that $a = 1, 2, 3$ whereas $A = 0, 1, 2, 3$). Thus the relation (2.4) can be inverted to give

$$\Omega_{MAB} = J_{aAB} A_{Ma} + J^*_{aAB} A^*_{Ma}. \tag{2.7}$$

An explicit representation for J is

$$J_{aAB} = \tfrac{1}{2}\mathrm{i}\eta_{aA}\delta^0_B - \tfrac{1}{2}\mathrm{i}\eta_{aB}\delta^0_A - \tfrac{1}{2}\varepsilon^0{}_{aAB}. \tag{2.8}$$

Here we have used the fact that that the range of the small indices is just a subset of that of the big indices. Another useful property of J is that it provides a four-dimensional representation of $so(3)$, which commutes with its conjugate

$$\begin{aligned} J_{aA}{}^B J_{bB}{}^C &= -\tfrac{1}{4}\eta_{ab}\delta^C_A + \tfrac{1}{2}\varepsilon_{abc} J_{cA}{}^C \\ J^*_{aA}{}^B J^*_{bB}{}^C &= -\tfrac{1}{4}\eta_{ab}\delta^C_A + \tfrac{1}{2}\varepsilon_{abc} J^*_{cA}{}^C \\ J^*_{aA}{}^B J_{bB}{}^C &= J_{bA}{}^B J^*_{aB}{}^C. \end{aligned} \tag{2.9}$$

Thus A_{Ma} is the $so(3,\mathbb{C})$ representation of the spin connection acting on self-dual tensors $T_{AB} = J_{aAB} T_a$ as

$$D_M T_a = \partial_M T_a + \varepsilon_{abc} A_{Mb} T_c \tag{2.10}$$

and its field strength is given by

$$F_{MNa}[A] = \partial_M A_{Na} - \partial_N A_{Ma} + \varepsilon_{abc} A_{Mb} A_{Nc}. \tag{2.11}$$

It follows from (2.4) that this field strength is related to the curvature R_{MNAB} by

$$F_{MNa} = J_{aAB} R_{MN}{}^{AB} \qquad F^*_{MNa} = J^*_{aAB} R_{MN}{}^{AB} \tag{2.12}$$

or

$$R_{MNAB} = J_{aAB} F_{MNa} + J^*_{aAB} F^*_{MNa}. \tag{2.13}$$

With the last formula the Einstein–Hilbert action splits into a part holomorphic in A and its conjugate. So after some algebra, which uses the self-duality of J, L'_{EH} becomes the real part of

$$L_{\mathrm{EH}} = -\tfrac{1}{2}\mathrm{i}\varepsilon^{MNPQ} E_M{}^A E_N{}^B J_{aAB} F_{PQa}[A]. \tag{2.14}$$

As in [2, 4] we represent the gravitinos as complex 2-component Grassmann valued spinors, transforming in the self-dual complexified $su(2)$ representation of the Lorentz group. But we will not write out the spinor indices explicitly. Instead, we introduce the four Hermitian 2×2 matrices σ_A with $\sigma_0 = 1$ and σ_a the usual Pauli matrices. Matrices with curved indices are obtained by multiplication with a vierbein: $\sigma_M = E_M{}^A \sigma_A$. Note that the algebra of these matrices is given by the J symbols

$$\sigma_A \sigma_a = 2\mathrm{i} J_{aAB} \sigma^B \qquad \sigma_a \sigma_A = -2\mathrm{i} J^*_{aAB} \sigma^B. \tag{2.15}$$

The covariant derivatives of a spinor χ and its conjugate $\bar{\chi} = \chi^\dagger(\mathrm{i}\sigma_0) = \mathrm{i}\chi^\dagger$ read

$$D_M \chi = \partial_M \chi - \tfrac{1}{2}\mathrm{i} A_{Ma} \sigma_a \chi \qquad D_M \bar{\chi} = \partial_M \bar{\chi} + \tfrac{1}{2}\mathrm{i} A^*_{Ma} \bar{\chi} \sigma_a\,. \tag{2.16}$$

Using the properties of J it is straightforward to show that $\mathrm{i}\bar{\chi}\sigma_A\chi$ is real and transforms as a vector. Note that $\bar{\chi}\chi$ is not a scalar but the 0-component of a vector.

With these definitions we see immediately, also, that the Rarita–Schwinger action splits into a part holomorphic in A and its conjugate, thus L'_{RS} is the real part of

$$L_{\mathrm{RS}} = 2\mathrm{i}\varepsilon^{MNPQ} \bar{\psi}_M \sigma_N D_P \psi_Q. \tag{2.17}$$

The total action I' is given as the real part of $I[E, A, \psi, \bar{\psi}] = \int \mathrm{d}^4x\, L_{\mathrm{EH}} + L_{\mathrm{RS}}$. As this is a holomorphic function of A, the equations of motion for A are the same as those for Ω in $I'[E, \Omega, \psi, \bar{\psi}]$, which directly follows from the Cauchy–Riemann differential equations for holomorphic functionals. As shown in [4], the complex action becomes real (up to a total derivative) if the equations of motion for A_{Ma} are satisfied and thus I and I' imply the same equation for all fields. Therefore we can take the complex action as starting point for the canonical quantization:

$$L = \mathrm{i}\varepsilon^{MNPQ}\big(-\tfrac{1}{2} E_M{}^A E_N{}^B J_{aAB} F_{PQa} + 2\bar{\psi}_M \sigma_N D_P \psi_Q\big). \tag{2.18}$$

The independent variables are $E_M{}^A$, A_{Ma}, $\bar{\psi}_M$ and ψ_M. They naturally split into configuration and momentum variables and Lagrange multipliers if we split the spacetime spanned by the coordinates $M = t, x, y, z$ into a spatial hypersurface spanned by $m = x, y, z$

and a canonical time coordinate t. The momenta of the configuration variables A_{ma} and ψ_m are

$$\frac{\delta L}{\delta(\partial_t A_{ma})} = -\mathrm{i}\varepsilon^{mnp} E_n{}^A E_p{}^B J_{aAB} = \mathrm{i}\tilde{e}_a{}^m$$
$$\frac{\delta L}{\delta(\partial_t \psi_m)} = -2\mathrm{i}\varepsilon^{mnp}\bar{\psi}_n\sigma_p = \bar{\pi}^m \tag{2.19}$$

leading to the Poisson brackets

$$\{\tilde{e}_a{}^m, A_{nb}\} = \mathrm{i}\eta_{ab}\delta^m_n \qquad \{\psi_m, \bar{\pi}^n\} = -\delta^n_m\,\mathbf{1}. \tag{2.20}$$

Thus we should use $\tilde{e}_a{}^m$ and $\bar{\pi}^n$ as phase space variables rather than $E_m{}^A$ and $\bar{\psi}_m$. However, a problem arises when we use $\tilde{e}_a{}^m$ (and its conjugate A_{ma}) as the basic operators in the quantum theory. As can be read off from its definition, the 'densitized dreibein' $\tilde{e}_a{}^m$ is complex. But it obeys some reality conditions, as the 12 real components of $E_m{}^A$ are mapped to the nine complex components of $\tilde{e}_a{}^m$. Thus there must be six reality conditions, which are found to be

$$\tilde{e}_a{}^m\tilde{e}_a{}^n = \varepsilon^{mpq}\varepsilon^{nrs}G_{pr}G_{qs} = \tilde{g}^{mn} \in \mathbb{R}. \tag{2.21}$$

The square of the complex dreibein yields the real densitized spatial metric. The A_{ma} also obey six reality constraints which are computed in [4]. For the classical theory we only have to check that the reality constraints are consistent, in the sense that they are conserved under the time evolution given by the Hamiltonian. In fact, the reality constraints for A_{ma} are just obtained by requiring $\tilde{g}^{mn}$ to remain real under the time evolution.

In quantum theory, however, a more serious problem arises: if we want to represent the wavefunction by a functional $\Psi[A_{ma}]$ (let us drop the fermions for a moment), we have to specify properties of the functional such that it carries the right number of degrees of freedom. In the original approach to Ashtekar's formulation of pure gravity A_{ma} is obtained by a canonical transformation from a set of real coordinates, thus each component of the connection represents exactly one real degree of freedom, and there are nine reality conditions on A_{ma} as well as on $\tilde{e}_a{}^m$ (simply requiring them to be real).

If a complex configuration variable represents exactly one real degree of freedom it is easy to define a wavefunctional with the right properties. One just has to consider holomorphic wavefunctionals only, and the basic operators can be represented by the usual differential operators acting on holomorphic functions [12]. On the other hand, if a complex coordinate represents exactly two real degrees of freedom, again there is no problem in defining the quantum operators. Just take Ψ to be an arbitrary functional, and define the operators by the usual holomorphic and anti-holomorphic differential operators.

But for the case given above we have nine complex coordinates $\tilde{e}_a{}^m$ representing 12 real degrees of freedom, and it is not clear how to transfer these properties to the quantum theory. Note that this problem only arises when using the Ashtekar representation. In the metric representation it can be solved by defining the wavefunctional as a function $\Psi[E_m{}^A]$ of the real fields $E_m{}^A$ instead of $\Psi[\tilde{e}_a{}^m]$. In fact, if one evaluates the theory further in the metric formalism, one finds that parts of the Lorentz constraints imply that the wavefunctional depends on the real fields $E_m{}^A$ via a holomorphic function of $\tilde{e}_a{}^m$ only.

For the connection represention it is not straightforward to see how this problem can be solved. In [4] it is assumed that the wavefunctional is holomorphic, but there is no argument why. If it is not, the meaning of $\delta/\delta A_{ma}$ is not defined. To avoid these difficulties, we will fix the Lorentz gauge in the same way as one usually does in pure gravity, namely by requiring the three spacelike vierbein vectors to point along the spatial hypersurface, or equivalently $E_m{}^0 = 0$. The densitized dreibein then becomes $\tilde{e}_a{}^m = -\varepsilon^{mnp} E_n{}^A E_p{}^B J_{aAB} = \frac{1}{2}\varepsilon^{mnp}\varepsilon_{abc}e_{nb}e_{pc}$, where $e_{ma} = E_m{}^a$ is the spatial part of the vierbein. Thus $\tilde{e}_a{}^m$ is real. As a consequence, also the components A_{ma} carry one real degree of freedom only, because their real part is now given as the three-dimensional spin connection as in pure gravity, plus a bilinear term in the gravitinos. Thus with this gauge fixing the wavefunction can be given as a holomorphic functional of A_{ma}.

The constraints are now obtained by differentiating L with respect to the Lagrange multipliers $E_t{}^A$, A_{ta}, $\bar{\psi}_t$ and ψ_t. They read

$$
\begin{aligned}
\mathcal{H}_A &= -\mathrm{i}\varepsilon^{mnp} E_m{}^B J_{aAB} F_{npa} - 2\mathrm{i}\varepsilon^{mnp}\,\bar{\psi}_m \sigma_A D_n \psi_p \\
\mathcal{L}_a &= \mathrm{i} D_m \tilde{e}_a{}^m + \tfrac{1}{2}\mathrm{i}\bar{\pi}^m \sigma_a \psi_m \\
\mathcal{S} &= 2\mathrm{i}\varepsilon^{mnp}\sigma_m D_n \psi_p \\
\overline{\mathcal{S}} &= -\,D_m \bar{\pi}^m .
\end{aligned}
\qquad (2.22)
$$

The constraints $\mathcal{H}_A$ combine the Hamiltonian constraint

$$
\mathcal{H}_0 = -\tfrac{1}{2}\varepsilon^{mnp} e_{ma} F_{npa} - 2\mathrm{i}\varepsilon^{mnp}\bar{\psi}_m D_n \psi_p \qquad (2.23)
$$

and the diffeomorphism constraints

$$
\mathcal{H}_a = -\tfrac{1}{2}\mathrm{i}\varepsilon^{mnp}\varepsilon_{abc} e_{mb} F_{npc} - 2\mathrm{i}\varepsilon^{mnp}\,\bar{\psi}_m \sigma_a D_n \psi_p . \qquad (2.24)
$$

In Ashtekar's representation the variables A_{ma} appear as multiplication operators and $\tilde{e}_a{}^m$ is replaced by a differential operator. To quantize the constraints it is therefore necessary to write them as polynomials in $\tilde{e}_a{}^m$. $\mathcal{L}'_a = \mathcal{L}_a$ and $\overline{\mathcal{S}}' = \overline{\mathcal{S}}$ are polynomials already, so only $\mathcal{S}$ has to be multiplied by a suitable function of e_{ma} to obtain a polynomial $\mathcal{S}'$. The Hamiltonian constraint is then obtained by taking the bracket of $\overline{\mathcal{S}}'$ with $\mathcal{S}'$, which, of course, again gives a polynomial. It turns out that we have to multiply $\mathcal{S}$ by the determinant $e = \frac{1}{6}\varepsilon^{mnp}\varepsilon_{abc}e_{ma}e_{nb}e_{pc}$.

With the help of the polynomial $\mathcal{S}' = e\mathcal{S}$ we then define a new combination of the Hamiltonian and diffeomorphism constraints as a matrix $\mathcal{K}' = \{\mathcal{S}', \overline{\mathcal{S}}'\}$. It is equal to $e\mathcal{H}_A\sigma^A$ plus a term proportional to $\mathcal{S}$. So the complete set of constraints is

$$
\begin{aligned}
\mathcal{K}' &= \tfrac{1}{2}\mathrm{i}F_{mna}\,\tilde{\sigma}^m\tilde{\sigma}^n\sigma_a - 2\mathrm{i}\varepsilon_{abc}\,\sigma_a D_{[m}\psi_{n]}\,\tilde{e}_b{}^m\bar{\pi}^n\sigma_c \\
\mathcal{L}'_a &= \mathrm{i}D_m\tilde{e}_a{}^m + \tfrac{1}{2}\mathrm{i}\bar{\pi}^m\sigma_a\psi_m \\
\mathcal{S}' &= -2\tilde{\sigma}^m\tilde{\sigma}^n D_{[m}\psi_{n]} \\
\overline{\mathcal{S}}' &= -D_m\bar{\pi}^m
\end{aligned}
\qquad (2.25)
$$

where $\tilde{\sigma}^m = \tilde{e}_a{}^m\sigma_a$.

To solve the quantized constraints, we only need to solve $\mathcal{L}'_a$, $\mathcal{S}'$ and $\overline{\mathcal{S}}'$. Choosing the representation

$$
\hat{\tilde{e}}_a{}^m = \hbar\,\frac{\delta}{\delta A_{ma}} \qquad \hat{\bar{\pi}}^p = \mathrm{i}\hbar\,\frac{\delta}{\delta\psi_m} \qquad (2.26)
$$

they become those given in [4]:

$$\widehat{\mathcal{L}}'_a = \mathrm{i}\hbar D_m \left(\frac{\delta}{\delta A_{ma}}\right) + \tfrac{1}{2}\hbar(\sigma_a \psi_m)_\alpha \frac{\delta}{\delta \psi_{m\alpha}}$$

$$\widehat{\mathcal{S}}' = -2\mathrm{i}\hbar^2 \varepsilon_{abc}\sigma_c D_m \psi_n \frac{\delta}{\delta A_{ma}} \frac{\delta}{\delta A_{nb}} \tag{2.27}$$

$$\widehat{\overline{\mathcal{S}}}' = -\mathrm{i}\hbar D_m \left(\frac{\delta}{\delta \psi_m}\right).$$

On a wavefunctional $\Psi[A_{ma}, \psi_m]$ the constraints $\mathcal{L}'_a$ and $\overline{\mathcal{S}}'$ just generate local Lorentz and supersymmetry transformations, respectively. Thus they require Ψ to be invariant under $\delta A_{ma} = D_m \lambda_a$, $\delta\psi_m = \frac{1}{2}\mathrm{i}\lambda^a \sigma_a \psi_m$ and under the chiral supersymmetry transformation $\delta\psi_m = D_m \epsilon$, which does not act on A_{ma}.

A formal solution to all constraints can now be given as *the same* functional that was found to solve the Wheeler–DeWitt constraint of pure gravity [5], namely the Wilson loop functional

$$T_\eta = \mathrm{Tr}\ \mathcal{P}\mathrm{exp}\tfrac{1}{2}\mathrm{i} \oint_\eta \mathrm{d}s\ \dot{\eta}^m(s) A_{ma}(\eta(s))\sigma_a \tag{2.28}$$

where $\eta(s)$ is a smooth, non-intersecting loop in the spatial hypersurface, and $\dot{\eta}^m(s)$ denotes its tangent vector. As is well known, when computing the second derivative of T_η with respect to A_{ma} and A_{nb} one obtains an expression symmetric in m, n. Thus T_η is a solution to all the constraints, because the second derivative in $\mathcal{S}'$ is antisymmetrized, and T_η also solves $\mathcal{L}'_a$ and $\overline{\mathcal{S}}'$, because it is Lorentz invariant and does not depend on ψ_m.

There are some questions arising here. As shown in [9], purely bosonic states (in the sense that the wavefunction does not depend on the gravitino) do not exist in the metric representation of supergravity, described by the unprimed constraints. This is reasonable because the wavefunctional should be supersymmetric and therefore cannot depend on the bosonic variables only. The loop solution, however, is purely bosonic in this sense. This property is obviously lost when introducing the primed constraints.

As was realized shortly after the discovery of the loop solutions for pure gravity, they are annihilated by the determinant of the spatial metric and therefore somehow represent states with singular metric [13]. So far this is not a serious problem, as Ashtekar's action and the constraints are polynomials in the canonical variables, and thus they are able to handle singular metrics. But if the solutions represent states with singular metric, another problem arises: as we multiplied the constraints by e, it seems that these solutions are not solutions to the original constraints, but they are simply annihilated by this factor. Note that the factors of $\tilde{e}_a{}^m$ are ordered *to the right* in $\mathcal{S}'$, thus the extra factors act on the wavefunctional first. It is often argued that this problem can be solved by considering new state functionals in terms of special loops which are no longer annihilated by $\widehat{e}$, but we will see below that the singular metric problem is in fact more involved.

The third problem is that these solutions also solve the constraints for pure gravity. If we drop all fermions from the constraints we are left with $\mathcal{L}'_a$ and $\mathcal{K}'$ only, and $\mathcal{K}'$ again is proportional to the antisymmetrized second derivative with respect to A_{ma}. The solutions do not 'see' the factor F_{mna} in $\mathcal{K}'$, because they are already annihilated by $\tilde{e}_a{}^{[m}\tilde{e}_b{}^{n]}$. Thus we can replace the curvature by the Rarita–Schwinger field strength $D_{[m}\psi_{n]}$ without changing the solutions, and this replaces $\mathcal{K}'$ by $\mathcal{S}'$, thus switching from pure gravity to supergravity.

This problem is somehow related to that arising when constructing loop solutions for matter coupled gravity or when adding a cosmological constant; e.g. there are extended Wilson loops solving the Wheeler–DeWitt constraint of gravity coupled to scalar fields [14], which do not 'see' the mass of these fields, nor do the loop states depend on the cosmological constant. Of course, this also implies that the loop solutions are generated by multiplying the constraints by the determinant of the vierbein and do not correspond to any solution of the Wheeler–DeWitt equation in its original metric representation.

There is still another crucial feature of the loop solutions for supergravity. In the usual connection representation of pure gravity the loops are annihilated by the Wheeler–DeWitt constraint but not by the diffeomorphism constraints, simply because they are not invariant under spatial diffeomorphism. One usually solves this problem by defining a loop representation and requiring the wavefunctional to depend on the knot class of the loop only, which is invariant under diffeomorphisms. But for supergravity we saw that T_η solves all constraints, though it is not invariant under diffeomorphisms. To make this problem more explicit, observe that the usual Wheeler–DeWitt and diffeomorphism constraints are both obtained from $\mathcal{H}_A$: the diffeomorphism constraint is given by $\mathcal{D}'_m = e_{ma}\mathcal{H}_a$ and the Wheeler–DeWitt constraint is $\mathcal{H}' = e\mathcal{H}_0$.

On the other hand, if we are allowed to obtain the Wheeler–DeWitt operator by multiplying $\mathcal{H}_0$ by a third-order function of the dreibein there is no reason why this should not be allowed for the diffeomorphism constraint. This is exactly what we have done when using (2.27) as the constraint algebra for supergravity. Doing the same for pure gravity, the loop functionals become solutions to *all* constraints, too: there is no need to consider functionals on knot classes only in the loop representation, as every arbitrary function which has support on smooth loops only is a solution. Thus densitizing the diffeomorphism constraint somehow destroys the invariance of the solutions under diffeomorphisms. How this works in detail for the loop states is discussed in [6].

To summarize, there are the following problems when considering the loop states as physical states of supergravity:

- the states are purely bosonic, which is not possible in the metric formulation;
- the solutions seem to be artificially generated by multiplying the constraints by an extra factor;
- they do not see the difference between gravity and supergravity, nor the value of the cosmological constant;
- the invariance under diffeomorphisms is lost if one also multiplies the diffeomorphism constraint with an extra factor.

As we do not know the 'correct' version of quantum gravity, we cannot argue that these problems imply that the theory described by the primed constraints is 'wrong'. But if we consider the corresponding $2+1$ dimensional theory below, which can be solved completely, we are able to show explicitly the differences between the theory described by the original constraints and the densitized constraints. We will find that the theory given by the primed constraints is, in a certain sense, not *general relativity*, and at the classical level a similar argument will hold for the $3+1$ theory.

3. $N = 2$ supergravity in three dimensions

The bosonic variables for $N = 2$ supergravity are the dreibein $e_\mu{}^a$, $a = 0, 1, 2$ the flat and $\mu = t, x, y$ the curved index, and the spin connection $A_{\mu a}$ with curvature

$$F_{\mu\nu a} = \partial_\mu A_{\nu a} - \partial_\nu A_{\mu a} - \varepsilon_{abc} A_\mu{}^b A_\nu{}^c. \tag{3.1}$$

We use the same notation as in [8], i.e. the flat indices are raised by the Lorentzian metric $\eta_{ab} = \text{diag}(-,+,+)$, and we have $\varepsilon^{012} = -\varepsilon_{012} = 1$. The gravitinos are again represented by 2-component complex spinors ψ_μ. As the Lorentz group $SO(2,1)$ has a real spinor representation $SL(2,\mathbb{R})$, we choose real gamma matrices

$$\gamma_0 = \begin{pmatrix} 0 & 1 \\ -1 & 0 \end{pmatrix} \qquad \gamma_1 = \begin{pmatrix} 0 & 1 \\ 1 & 0 \end{pmatrix} \qquad \gamma_2 = \begin{pmatrix} 1 & 0 \\ 0 & -1 \end{pmatrix} \tag{3.2}$$

obeying

$$\gamma_a \gamma_b = \eta_{ab} \mathbf{1} - \varepsilon_{abc} \gamma^c . \tag{3.3}$$

The Lorentz covariant derivative of a vector v_a, a spinor χ and its conjugate $\bar{\chi} = \chi^\dagger (\mathrm{i}\gamma_0)$ reads

$$\begin{aligned} D_\mu v_a &= \partial_\mu v_a - \varepsilon_{abc} A_\mu{}^b v^c \\ D_\mu \chi &= \partial_\mu \chi + \tfrac{1}{2} A_\mu{}^a \, \gamma_a \chi \\ D_\mu \bar{\chi} &= \partial_\mu \bar{\chi} - \tfrac{1}{2} A_\mu{}^a \, \bar{\chi} \gamma_a . \end{aligned} \tag{3.4}$$

The Lagrangian is

$$L = \tfrac{1}{2} \varepsilon^{\mu\nu\rho} e_\mu{}^a F_{\nu\rho a} + 2 \varepsilon^{\mu\nu\rho} \, \bar{\psi}_\mu D_\nu \psi_\rho . \tag{3.5}$$

Splitting spacetime into space and time, i.e. splitting the index μ into t and $i = x, y$, the configuration variables become A_{ia} and ψ_i and their momenta are

$$\frac{\delta L}{\delta(\partial_t A_{ia})} = \varepsilon^{ij} e_j{}^a \qquad \frac{\delta L}{\delta(\partial_t \psi_i)} = -\, 2 \varepsilon^{ij} \bar{\psi}_j \tag{3.6}$$

where $\varepsilon^{xy} = \varepsilon_{xy} = 1$. The Poisson brackets read

$$\left\{ A_{ia}, e_j{}^b \right\} = \varepsilon_{ij} \delta_a^b \qquad \left\{ \psi_i, \bar{\psi}_j \right\} = \tfrac{1}{2} \varepsilon_{ij} \, \mathbf{1} \tag{3.7}$$

and the Lagrange multipliers $e_t{}^a$, A_{ta}, $\bar{\psi}_t$ and ψ_t generate the constraints

$$\begin{aligned} \mathcal{H}_a &= \tfrac{1}{2} \varepsilon^{ij} F_{ija} \\ \mathcal{L}_a &= \varepsilon^{ij} D_i e_{ja} - \varepsilon^{ij} \, \bar{\psi}_i \gamma_a \psi_j \\ \mathcal{S} &= 2 \varepsilon^{ij} \, D_i \psi_j \\ \overline{\mathcal{S}} &= 2 \varepsilon^{ij} \, D_i \bar{\psi}_j . \end{aligned} \tag{3.8}$$

Note that their structure is similar to (2.22), but they are much simpler, e.g. only $\mathcal{L}_a$ contains the dreibein. In particular, there are no operator ordering ambiguities and no anomalies will arise in the quantum algebra. In quantum theory it is sufficient to solve $\mathcal{L}_a$, $\mathcal{S}$ and $\overline{\mathcal{S}}$, as $\left\{ \mathcal{S}, \overline{\mathcal{S}} \right\} = -\mathcal{H}_a \gamma^a$.

The complete solution of the corresponding quantum constraints is given in [8]. The operators in the connection representation are

$$\widehat{e_i a} = -\, \mathrm{i}\hbar \varepsilon_{ij} \frac{\delta}{\delta A_{ja}} \qquad \widehat{\bar{\psi}}_i = \tfrac{1}{2} \mathrm{i}\hbar \varepsilon_{ij} \frac{\delta}{\delta \psi_j} . \tag{3.9}$$

The constraints split into a set of multiplication operators $\widehat{\mathcal{H}}_a$ and $\widehat{\mathcal{S}}$, and first order differential operators $\widehat{\mathcal{L}}_a$ and $\widehat{\overline{\mathcal{S}}}$, which generate local Lorentz and chiral supersymmetry transformations ($\delta\psi_i = D_i\epsilon$) on the wavefunctional $\Psi[A_{ia}, \psi_i]$.

The solution to the constraints is given as follows. $\widehat{\mathcal{H}}_a\Psi = 0$ together with $\widehat{\mathcal{S}}\Psi = 0$ imply that Ψ has support only on those fields A_{ia}, ψ_i which have vanishing curvature $F_{ija} = 0$ and vanishing supercurvature $D_{[i}\psi_{j]} = 0$. To simplify the notation we introduce a matrix valued connection field $A_i = \frac{1}{2}A_{ia}\gamma^a$. The covariant derivative of a spinor then reads $D_i\chi = \partial_i\chi + A_i\chi$.

The complete set of curvature-free pairs A_i, ψ_i is parametrized by a $SL(2,\mathbb{R})$ matrix field g and a spinor field ϕ on the covering manifold of the spatial 2-surface, which are subject to certain relations between their values on points mapped to the same point of the 2-surface (see [8]). These conditions are in fact equivalent to the requirement that

$$A_i[g] = g^{-1}\partial_i g \qquad \psi_i[g,\phi] = g^{-1}\partial_i\phi \tag{3.10}$$

are single-valued on the 2-surface. Obviously, $A_i[g]$ and $\psi_i[g,\phi]$ are curvature free.

A complete set of solutions to $\widehat{\mathcal{H}}_a\Psi = 0$ and $\widehat{\mathcal{S}}\Psi = 0$ is given by

$$\Phi_{g,\phi}[A_i, \psi_i] = \prod_x \delta\left(g^{-1}\partial_i g - A_i\right)\prod_x \delta\left(\partial_i\phi - g\psi_i\right). \tag{3.11}$$

This is in fact an overcomplete set, as, e.g. g, ϕ and $g_0 g, g_0\phi + \phi_0$, where g_0 and ϕ_0 are constant, produce the same wavefunctional. Nevertheless, a wavefunction can now be written as

$$\Psi_Y[A_i, \psi_i] = \int \mathrm{D}g\mathrm{D}\phi\, Y[g,\phi]\Phi_{g,\phi}[A_i, \psi_i] \tag{3.12}$$

where the functional integral runs over all fields g and ϕ subject to the restrictions mentioned above, and $Y[g,\phi]$ is an arbitrary functional. The measure we assume to be invariant under multiplication of g with an arbitrary matrix field $g \mapsto gh$ and under addition of a spinor field $\phi \mapsto \phi + \chi$. Thus the expression for Ψ_Y is only formal. But it can be defined properly as follows. Ψ_Y is zero if A_{ma} and ψ_m are not curvature free and takes the value $Y[g,\phi]$ if they are given by (3.10).

The constraints $\widehat{\mathcal{L}}_a$ and $\widehat{\overline{\mathcal{S}}}$ now impose additional conditions on Y: $\widehat{\mathcal{L}}\Psi = 0$ requires that Ψ is invariant under Lorentz transformations. As can be verified, the change of $\Phi_{g,\phi}$ under a Lorentz transformation with parameter $\lambda = \frac{1}{2}\lambda^a\gamma_a$, i.e.

$$\delta A_i = \partial_i\lambda - [\lambda, A_i] \qquad \delta\psi_i = -\lambda\psi_i \tag{3.13}$$

can be compensated by changing g as $\delta g = g\lambda$. Similarly, a supersymmetry transformation generated by $\overline{\mathcal{S}}$ with parameter ϵ is given by

$$\delta\psi_i = D_i\epsilon \tag{3.14}$$

and the variation of $\Phi_{g,\phi}$ can be compensated by $\delta\phi = g\epsilon$. So we have to require that $Y[g,\phi]$ is invariant under Lorentz and supersymmetry transformations

$$\delta g = g\lambda \qquad \delta\phi = g\epsilon \tag{3.15}$$

to obtain the full solution to all constraints.

For a 2-surface with trivial topology these restrictions imply that Y is constant, as every field configuration can be transformed into any other by (3.15). But remember that λ and ϵ are single valued fields on the 2-surface whereas g and ϕ are fields on the covering manifold. For non-trivial topologies there remain finitely many degrees of freedom for Y, called the moduli, which cannot be gauged away by (3.15). For a more detailed discussion we again refer to [8] and references therein.

Thus 2+1 dimensional $N = 2$ supergravity leads to a completely soluble quantum theory which can be obtained by applying the Dirac method to the classical Lagrangian (3.5). The connection representation for the three-dimensional theory, however, is not directly related to that of the four-dimensional theory. In particular, because of the simple structure of the constraints, it is not necessary to multiply the constraints by extra factors of the dreibein to obtain polynomial expressions.

The situation is different when we consider, e.g., the dimensionally reduced version of the four-dimensional $N = 1$ theory, which contains additional matter fields, but whose graviton and gravitino action is the same as that of the $N = 2$ theory discussed here. In [15] we found that for the matter-coupled theory we also have to multiply one of the supersymmetry constraints by an extra factor of $e\gamma^t = -\frac{1}{2}\varepsilon^{ij}\varepsilon_{abc}\gamma^a e_i{}^b e_j{}^c$ to get polynomial constraints, which directly corresponds to the definition of the primed constraints for the four-dimensional theory.

Thus the question arising here is what happens if we start with the constraints given in [15] for the matter-coupled theory and just drop all the matter terms. We then obtain

$$
\begin{aligned}
\mathcal{L}'_a &= \varepsilon^{ij} D_i e_{ja} - \varepsilon^{ij} \bar{\psi}_i \gamma_a \psi_j \\
\mathcal{S}' &= -\varepsilon^{ij}\varepsilon^{kl}\varepsilon_{abc} e_k{}^a e_l{}^b \gamma^c D_i \psi_j \\
\overline{\mathcal{S}}' &= 2\varepsilon^{ij} D_i \bar{\psi}_j .
\end{aligned}
\tag{3.16}
$$

They are just the three-dimensional versions of (2.25) and can be constructed in the same way, i.e. multiplying $\mathcal{S}$ by $e\gamma^t$ and then defining $\mathcal{H}'_a$ as the bracket of $\overline{\mathcal{S}}'$ with $\mathcal{S}'$. *A priori* there is no reason why these constraints could not be regarded as the canonical constraints of $2+1$ supergravity, if (2.25) describes $3+1$ supergravity. They are nothing but their three-dimensional counterparts, and assumed that we did not know that there is a much simpler expression for them, we had to use them to quantize the $2+1$ theory.

The quantized versions of the Lorentz and supersymmetry constraints now become

$$
\begin{aligned}
\widehat{\mathcal{L}'_a} &= \mathrm{i}\hbar D_i \left(\frac{\delta}{\delta A_{ia}}\right) - \tfrac{1}{2}\mathrm{i}\hbar\,(\gamma_a \psi_i)_\alpha \frac{\delta}{\delta \psi_{i\alpha}} \\
\widehat{\mathcal{S}'} &= -\hbar^2 \varepsilon_{abc}\,\gamma^c D_i \psi_j \frac{\delta}{\delta A_{ia}} \frac{\delta}{\delta A_{jb}} \\
\widehat{\overline{\mathcal{S}}'} &= -\mathrm{i}\hbar\, D_i \left(\frac{\delta}{\delta \psi_i}\right) .
\end{aligned}
\tag{3.17}
$$

Their structure is again similar to (2.27) and, in fact, we can also give the Wilson loop solutions to them:

$$
T_\eta = \mathrm{Tr}\ \mathcal{P}\mathrm{exp}\tfrac{1}{2} \oint_\eta \mathrm{d}s\, \dot{\eta}^i(s) A_{ia}(\eta(s))\gamma^a .
\tag{3.18}
$$

It is obtained from the $3+1$ version just by replacing the σ_a matrices by γ_a. Again, $\overline{\mathcal{S}}'$ and $\mathcal{L}'_a$ are solved by T_η because it is Lorentz invariant and does not depend on the gravitino. If the loop is smooth and does not have self-intersections, the remaining constraints are solved because of the same symmetry arguments as in $3+1$ dimensions, thus every smooth loop embedded in the spatial surface yields a state functional.

This is a completely different result from that obtained above, where we found that the state is described by a function Y that depends on finitely many degrees of freedom, the moduli, only. Here the general solution is given as an arbitrary superposition of T_ηs. In particular, the constraints do not imply that the curvature vanishes and thus different loops really produce different wavefunctionals.

At first sight, the problem seems to be that we are dealing with single loops here instead of equivalence classes of loops under diffeomorphisms. We have already mentioned this problem for the four-dimensional theory: it arises because we also multiplied the diffeomorphism constraint by a dreibein factor. The usual way to obtain the diffeomorphism and Wheeler–DeWitt constraint from $\mathcal{H}_a$ would in this case be $\mathcal{D}'_i = e_i{}^a\mathcal{H}_a$ and $\mathcal{H}' = \frac{1}{2}\varepsilon_{abc}\varepsilon^{ij}e_i{}^a e_j{}^b\mathcal{H}^c$.

Then $\mathcal{D}'_i$ in fact generates diffeomorphisms on the 2-surface and requires Ψ to be invariant, and $\mathcal{H}'$ is automatically solved for every solution of $\mathcal{S}'$ and $\overline{\mathcal{S}}'$. Doing this, however, the result is still totally different. Using the loop representation, the wavefunction becomes a function on the generalized knot classes of the 2-surface. But there are still infinitely many, even if the topology is trivial, whereas for the metric, or 'undensitized', connection representation there is only one state Ψ_Y with constant Y.

4. Discussion

In the $2+1$ case described above it is rather obvious that the loop solutions are artificially generated by multiplying the constraints with a factor, which is, in principle, the determinant of the spatial metric. As this factor appears to the right in the quantized constraints and already annihilates the loop functionals, it is not clear what they have to do with the original theory described by (3.8). Here we cannot argue that the polynomialized primed constraints can deal with singular metrics, and thus we get new solutions representing singular metrics: the constraints were polynomial from the beginning, and the unprimed constraints can handle singular metrics, too.

To make this argument more precise, consider the classical theory described by the primed and unprimed constraints. If we allow the dreibein to be singular, there are solutions to the primed constraints which do not solve the unprimed ones. Just take the trivial example $e_i{}^a = 0$, $\psi_i = 0$, but A_{ia} not curvature free. Thus the primed constraints do not describe the classical theory defined by the Lagrangian (3.5) if we allow singular metrics, though the Lagrangian itself does not contain the inverse metric. A Lagrangian which directly leads to the primed constraint can be written in the same way as one usually writes the Einstein–Hilbert Lagrangian in Ashtekar's variables with the time component of the dreibein split into a lapse and shift function. One has to replace the Lagrange multipliers $e_t{}^a$ by $e_t{}^a = n^i e_i{}^a + n\varepsilon^{abc}\varepsilon^{ij}e_{ib}e_{jc}$, and a similar replacement for $\bar{\psi}_t$, but let us drop the fermions now because the following arguments also apply for pure gravity.

Differentiation with respect to n and n^i then directly yields the primed diffeomorphism and Hamiltonian constraints. But observe that the new action is no longer invariant under *spacetime* diffeomorphisms. The transformation of the variables n, n^i and $e_i{}^a$ under the full

$2+1$ dimensional diffeomorphisms involve the inverse of the metric determinant, as n is a density of weight -1.

The argument can be transferred to the four-dimensional theory as well. The Lagrangian in terms of Ashtekar's variables we started from (see (2.18)) was polynomial in $E_M{}^A$, thus it is well defined for singular metrics†. Again, to obtain the primed constraints directly, one has to introduce a lapse and shift function by

$$E_t{}^A = N^m E_m{}^A - \tfrac{1}{6} N \varepsilon^A{}_{BCD} \varepsilon^{mnp} E_m{}^B E_n{}^C E_p{}^D \tag{4.1}$$

replacing the multipliers $E_t{}^A$ by N and N^m. Thus there are two different actions for $3+1$ gravity, one given by (2.18) and the other one by inserting (4.1) and using N, N^m instead of $E_t{}^A$ as the primary field variables. Both actions induce the same equations of motion for non-singular metrics, and both are well defined for (different kinds of) singular metrics.

As in the $2+1$ theory the equations of motion are different for the different actions if the metric becomes singular. It seems that there is *a priori* no reason why one Lagrangian can be better than the other, because we do not know the 'correct' equations of motion. But there is a crucial difference between the two Lagrangians: only one of them is invariant under the full diffeomorphism group of spacetime. This is very easy to see. Let V^M be the generating vector of a diffeomorphism, then the action (2.18) is invariant under

$$\delta E_N{}^A = V^M \partial_M E_N{}^A + \partial_M V^N E_M{}^A \qquad \delta A_{Na} = V^M \partial_M A_{Na} + \partial_M V^N A_{Ma}\,. \tag{4.2}$$

A property of this symmetry is that it remains well defined if the vierbein becomes singular and that it maps non-singular metrics onto non-singular ones, and thus the restriction $E \neq 0$ does not affect the symmetry. But if one now considers the new set of variables $E_m{}^A$, N^m and N, one finds the following transformation law for the lapse function:

$$\delta N = V^M \partial_M N - \partial_m V^m N + \partial_t V^t N - 2\partial_t V^m N^n g_{mn} g^{-1} N^{-1} \tag{4.3}$$

where $g_{mn} = E_m{}^A E_{nA}$ is the spatial metric and g its determinant. Obviously, this is not well defined for singular metrics and there is no natural extension of this symmetry for $g = 0$. Thus, if one requires invariance under all diffeomorphisms *and* allows singular metrics, one has to use the action (2.18). We can conclude that out of the following three properties of classical canonical gravity only two can be realized simultaneously:

- invariance under spacetime diffeomorphism;
- polynomial constraints;
- singular metrics allowed.

This classical argument is in agreement with our conclusion concerning the loop solutions in the quantized theory. We saw that if we also densitize the diffeomorphism constraint, which came out automatically in supergravity, then the loops became solutions to all constraints without considering wave functions that depend on the knot class only. Thus the invariance under spatial diffeomorphisms is destroyed by multiplying the corresponding constraint by extra factors of the dreibein. Now we have seen that in the classical theory the invariance under spacetime diffeomorphisms is lost when introducing the lapse function N of weight -1, i.e. when going over from the unprimed to the primed constraints. Assuming that

† Observe that also the real first order Einstein–Hilbert action is polynomial. But when treating it canonically there arise second class constraints making the Dirac brackets non-polynomial. It is this fact, and not the polynomial action, that simplifies the canonical treatment with Ashtekar's variables.

the arguments above apply to the quantized theory as well, we must conclude that quantum gravity described by the loop states is not invariant under spacetime diffeomorphisms, because it explicitly needs the polynomial constraints and singular metrics. This is rather unsatisfactory, because invariance under spacetime diffeomorphisms is one of the first principles of general relativity: in fact, *it is 'general relativity'*.

So what is the conclusion from this? Let us first note that the problem does not arise with the use of Ashtekar's variables. They can be eliminated by going back to the real action with $E_M{}^A$ and Ω_{MAB} as primary fields. All the arguments concerning the classical theory apply to this action as well, except that there is no reason for introducing a densitized lapse function, because this would not simplify the canonical constraints considerably.

The problem arises when one has to give up one of the three properties listed above. For practical reasons it is most easy to drop general invariance, because in the canonical formulation the *manifest* invariance is lost anyway. But, as already mentioned, this should be the last principle given up when dealing with general relativity. If one insists on 3+1 diffeomorphism invariance and singular metrics, then one has to start with the unprimed constraints, and one is not allowed to multiply them by factors which may become zero. In quantum theory it is even more inappropriate to order these extra factors *to the right*.

On the other hand, if singular metrics are not allowed, then the different actions and different sets of constraints are equivalent. The question arising then is how to implement this restriction in quantum theory. It is sometimes argued that the problem is solved if one considers only such state functionals that are not annihilated by the metric determinant, because they represent states with a non-singular metric. Such states have been found by using special functionals on intersecting loops [16].

However, the primed and unprimed constraints are only equivalent if we restrict the classical phase space to those dreibein fields which have non-vanishing determinant. Translated to the quantum theory this does not simply mean $\widehat{e}(x)\Psi \neq 0$, which, even for the intersecting loop states of [16], is fulfilled at some special points x only. Consider the simple example of a particle moving in a vector space and which is not allowed to be at the origin. The classical phase space is spanned by coordinates $\vec{x}$, $\vec{p}$ with the restriction $\vec{x} \neq 0$. In the $\vec{x}$ representation, which would correspond to the metric representation of gravity, the wavefunction is simply given by a function on the vector space excluding the origin. Thus in the metric representation we could implement the restriction $e \neq 0$ simply by defining the wavefunctional $\Psi[e_{ma}]$ on invertible dreibein fields only. But in the $\vec{p}$ representation the restriction on the wavefunction of the particle would not be $\partial/\partial\vec{p}\,\Psi \neq 0$, which does not even give a vector space, instead we would have to consider wavefunctions with vanishing 'zero modes'. It is by no means clear how to implement the classical restriction $e \neq 0$ in a quantum theory using the connection representation, because e becomes a third order differential operator.

Thus forbidding singular metrics in the connection representation is more than just restricting the solutions to those which are not annihilated by the determinant. The singular metric problem is still present for the intersecting loop states, though it is somehow weaker, because the state is no longer an eigenstate with eigenvalue $e = 0$. We can conclude that the connection representation is defined properly only if the metric is allowed to take singular values. As the loop states require the constraints to be given in the polynomial form, they are solutions to a theory that is not invariant under the full $3 + 1$ dimensional diffeomorphism group, and which therefore should not be called general relativity.

Let us also discuss whether it is possible to use the non-polynomial constraints in the connection representation. We could then allow the dreibein e_{ma} to become singular without loss of general coordinate invariance of the theory. Before considering the non-polynomial constraints let us first change the ordering of the polynomial ones, i.e. let us write the dreibein factors to the left. Note that it is this ordering which leads to a closed algebra of constraints, whereas there is an anomaly in the algebra admitting the loop solutions†. In this ordering, the 'Chern–Simons' state found in [17] solves the constraints (with a cosmological constant).

Now observe that one does not have to make use of the $\tilde{e}_a{}^m$ factors in the Wheeler–DeWitt constraints to show that the Chern–Simons form is a solution. It really 'solves' the non-polynomial constraints, where the quotation marks are used to indicate that the problem is how to define the quantized versions of the non-polynomial constraints properly. These properties of the Chern–Simons solution are easily proved. As it only exists for a non-vanishing cosmological constant, we add a term $\frac{1}{2}\Lambda E$ to L. The diffeomorphism and Hamiltonian constraints then become

$$\mathcal{H}_A = -\,\mathrm{i}E_m{}^B J_{aAB}\left(\varepsilon^{mnp}F_{npa} - \tfrac{1}{3}\Lambda\tilde{e}_a{}^m\right). \tag{4.4}$$

Obviously, these constraints have a simple solution in the connection representation, namely the exponential of the Chern–Simons form, as its derivative yields the field strength.

$$\Psi = \exp\left(\frac{1}{\Lambda}\int \mathrm{d}^3x\, \varepsilon^{mnp}(3A_{ma}\partial_n A_{pa} - \varepsilon^{abc}A_{ma}A_{nb}A_{pc})\right) \tag{4.5}$$

obeys

$$\left(\varepsilon^{mnp}F_{npa} - \tfrac{1}{3}\Lambda\frac{\delta}{\delta A_m{}^a}\right)\Psi = 0. \tag{4.6}$$

We see that Ψ is annihilated by $\widehat{\mathcal{H}}_A$, though this is not yet defined properly, because it is not clear how to represent e_{ma} in the connection representation. But the solution does not depend on how it is defined.

Thus the question arises whether one can make sense out of the non-polynomial expression $\mathcal{H}_A$ in the connection representation. A possible answer is the following. The action of $\widehat{C\mathcal{H}_A}$ on Ψ should give zero for every function $C[e_{ma}]$ that makes $C\mathcal{H}_A$ a polynomial in $\tilde{e}_a{}^m$. Of course, this only makes sense if the extra factors always appear to the left of $\mathcal{H}_A$. This would just lead to the polynomial Wheeler–DeWitt and diffeomorphism constraints with the dreibein factors ordered to the left, as there are only four independent choices for C that make $C\mathcal{H}_A$ a polynomial. So the Chern–Simons functional is a solution of this type, whereas the Wilson loops are not.

Maybe this procedure for treating a non-polynomial constraint is as close as possible to the classical theory which is invariant under the complete $3+1$ diffeomorphism group. Thus there are more reasons to prefer the operator ordering where the dreibein appears to the left than just the fact that only then does the quantum algebra close.

Finally, we should emphasize that, even if we do not use the polynomialized constraints, Ashtekar's variables still simplify the canonical treatment in a way that may have been overlooked, because the polynomial form of the constraints seems to be their simplest form, and only this has been studied so far. But there is still a difference between the constraints $\mathcal{H}_A$ in terms of Ashtekar's variables given in (4.4) and the original Wheeler–DeWitt operator constructed in the metric representation [18, 19], which contains the curvature scalar of the

† See [17] for a discussion of different factor orderings.

spatial metric and thus the inverse metric. However, the only non-polynomial term in $\mathcal{H}_A$ is the dreibein component $e_{ma} = E_m{}^a$, which is not the inverse but a kind of square root of the canonical variable $\tilde{e}_a{}^m$, and a square root of a differential operator is sometimes less a problem then to define its inverse; a well know example is the Dirac operator as square root of the Klein–Gordon operator.

So the question is whether it is possible to define an operator $\widehat{e}_{ma}$ in the connection representation such that

$$\left[\widehat{e}_{ma}, \widehat{e}_{nb}\right] = 0 \qquad \left[\tfrac{1}{2}\varepsilon^{mnp}\varepsilon_{abc}\widehat{e}_{nb}\widehat{e}_{pc}, A_{qd}\right] = \hbar\delta_q^m \eta_{ad}. \tag{4.7}$$

If such a representation exists, $\mathcal{H}_A$ becomes a well defined operator and the Chern–Simons functional becomes a solution to the non-polynomial constraints. As discussed above, this representation would correspond to a classical theory which is invariant under the full spacetime diffeomorphism group, i.e. which could be called 'general relativity'.

References

[1] Dirac P A M 1965 *Lectures on Quantum Mechanics* (New York: Academic)

[2] D'Eath P D 1984 Canonical quantization of supergravity *Phys. Rev.* D **29** 2199

[3] Ashtekar A 1987 New Hamiltonian formulation of general relativity *Phys. Rev.* D 36 1587

[4] Jacobson T 1988 New variables for canonical supergravity *Class. Quantum Grav.* **5** 923

[5] Jacobson T and Smolin L 1988 Non-pertubative quantum geometries *Nucl. Phys.* B **299** 295

[6] Jacobson T and Romano D 1992 Degenerate extensions of general relativity *Class. Quantum Grav.* **9** L119

[7] D'Eath P D 1994 Physical states in $N = 1$ supergravity *Phys. Lett.* **321B** 368

[8] de Wit B, Matschull H-J and Nicolai H 1993 Physical states in $d = 3$, $N = 2$ supergravity *Phys. Lett.* **318B** 115

[9] Page D N 1993 Inconsistency of canonically quantized $N = 1$ supergravity *Preprint* (University of Alberta) Thy–28–93

[10] Deser S and Zumino B 1976 Consistent supergravity *Phys. Lett.* **62B** 335

[11] Ferrara S, Freedman D Z and van Nieuwenhuizen P 1976 Progress towards a theory of supergravity *Phys. Rev.* D **13** 3214

[12] Fukuyama T and Kamimura K 1990 Complex action and quantum gravity *Phys. Rev.* D **41** 1105

[13] Brügmann B and Pullin J 1991 Intersecting N loop solutions of the Hamiltonian constraint of quantum gravity *Nucl. Phys.* B **363** 221

[14] Matschull H-J 1993 Solutions to the Wheeler–DeWitt constraint of canonical gravity coupled to scalar matter fields *Class. Quantum Grav.* **10** L149

[15] Nicolai H and Matschull H-J 1994 Canonical supergravity in three dimensions *Nucl. Phys.* B **411** 609

[16] Brügmann B, Gambini R and Pullin J 1992 Knot invariants as non-degenerate quantum geometries *Phys. Rev. Lett.* **68** 431

[17] Brügmann B, Gambini R and Pullin J 1992 Jones polynomials for intersecting knots as physical states of quantum gravity *Nucl. Phys.* B **385** 587

[18] Wheeler J A 1964 Geometrodynamics and the issue of the final state *Relativity, Groups and Topology* ed C DeWitt and B S DeWitt (New York: Gordon and Breach)

[19] DeWitt B S 1967 Quantum theory of gravity I, II *Phys. Rev.* **160** 1113; **162** 1195

Class. Quantum Grav. **11** (1994) 2411–2415. Printed in the UK

Spin structures on Kleinian manifolds

L J Alty† and A Chamblin‡

Department of Applied Mathematics and Theoretical Physics, University of Cambridge, Cambridge CB3 9EW, UK

Received 25 April 1994, in final form 14 July 1994

Abstract. We derive the topological obstruction to spin–Klein cobordism. This result has implications for signature change in general relativity, and for the $N = 2$ superstring.

PACS numbers: 0240, 0420, 1125

1. Introduction

In this paper we study topological obstructions to the global existence of spin–Klein structure on 4-manifolds. In particular, we wish to determine the topological obstructions to spin–Klein cobordism (in analogy with the recently found obstructions to spin–Lorentz cobordism, see [1–3]).

Let M be any 4-manifold. We will say that a metric on M is of *Kleinian signature* if it has signature $(--++)$, and we will say that M admits *spin structure* if we are able to globally define a fibre bundle over M with structure group Spin(2, 2) (where Spin(2, 2) is the double cover of $SO(2,2)$). It follows [4] that the obstruction to spin structure is the vanishing of the second Stiefel–Whitney class w_2, and we therefore need only understand the obstruction to putting a globally non-singular Klein metric on M.

Let $\{\Sigma_1, \Sigma_2, \ldots, \Sigma_n\}$ be an arbitrary collection of 3-manifolds, then a *cobordism* for $\{\Sigma_1, \Sigma_2, \ldots, \Sigma_n\}$ is a 4-manifold M with $\partial M \cong \Sigma_1 \cup \Sigma_2 \cup \cdots \cup \Sigma_n$ (where $\cup$ denotes disjoint union). We are interested in cobordisms that possess more structure. In particular, we shall say that M is a *spin–Klein cobordism* for $\{\Sigma_1, \Sigma_2, \ldots, \Sigma_n\}$ if and only if M is a cobordism for $\{\Sigma_1, \Sigma_2, \ldots, \Sigma_n\}$, and M admits both spin structure and a globally defined non-singular metric of Kleinian signature.

One motivation for this work is the recent interest in spacetimes in which 'signature change' is allowed [5, 6]. These signature change studies have centred upon cosmological solutions to the classical Einstein field equations in which the metric changes signature from $(++++)$ to $(-+++)$. It could be argued that this is too limiting, and more general signature transitions, for example those from $(-+++)$ to $(--++)$ should be considered [7]. Our cobordism result is relevant to those cases in which the spacetime metric changes to a Kleinian signature.

Our result may also prove to be of importance to the $N = 2$ superstring. The Weyl anomaly for this theory cancels provided that the string propagates in a four-dimensional target space. If the worldsheet is chosen to be of Lorentzian signature then the target space must have Kleinian signature [8].

† E-mail address: lja13@amtp.cam.ac.uk
‡ E-mail address: hac1002@amtp.cam.ac.uk

2. Klein metric homotopy

In order to understand the topology of Klein metrics, we first need to understand the obstruction to the existence of global Kleinian structure. Indeed, we have from [9]

Lemma 1. (Steenrod). Let M be a smooth 4-manifold without boundary. Then M admits a globally defined (non-singular) Klein metric if and only if there exists a globally defined (non-singular) field of 2-planes.

A field of 2-planes may be defined in the following way. Let $G_{2,4} \cong S^2 \times S^2$ denote the set of 2-planes in $\mathbb{R}^4$, then a field of 2-planes is just a section of the fibre bundle which has fibre $G_{2,4}$. If the section is everywhere non-vanishing then the field of 2-planes is non-singular, otherwise the field of 2-planes is singular. These definitions provide a natural generalization of a (non-singular/singular) vector field, which is a (non-vanishing/vanishing) section of a sphere bundle. We will use the expressions 'field of 2-planes' and 'plane field' interchangeably.

Each singularity of a singular plane field has an index associated with it. Suppose that x is a point in our manifold M at which some plane field P becomes singular, and let $S^3(x)$ denote a little sphere about x. Then the index indicates the homotopy-type of the map (defined by P) from $S^3(x)$ to $G_{2,4}$. However, such homotopy classes are in one-to-one correspondence with elements of $\pi_3(G_{2,4}) \simeq \mathbb{Z} \oplus \mathbb{Z}$. Thus the index of P at x is classified by a pair of integers.

Let M be an oriented compact 4-manifold. Let H denote the free Abelian group $H^2(M, \mathbb{Z})$/torsion subgroup, and let S denote the intersection pairing on H defined by the cup-product (i.e. S defines a map from $H \otimes H$ to $\mathbb{Z}$ by taking the cup-product of elements in H and evaluating them on the fundamental orientation class of M). Define the co-set $W \subseteq H/2H$ by $w \in W$ if $S(w, x) = S(x, x) \bmod 2$ for all $x \in H$. Finally, let Ω denote the set of integers $\{S(w, w) | w \in W\}$. We now recall an important theorem of Hirzebruch and Hopf [10] (also see [11]).

Theorem 1. (Hirzebruch and Hopf). Let M be an oriented compact 4-manifold without boundary. Then M has a field of 2-planes with finite singularities. The total index of such a field is given by a pair of integers (a, b). The following integers and only these, occur as the index for some plane field on M:

$$a = \tfrac{1}{4}(\alpha - 3\sigma - 2X) \qquad b = \tfrac{1}{4}(\beta - 3\sigma + 2X)$$

where $\alpha, \beta \in \Omega$, $X = X(M)$ denotes the Euler number of M, and $\sigma = \sigma(M)$ denotes the Hirzebruch signature of M.

Now suppose we are given a cobordism M for some collection of closed, orientable 3-manifolds $\{\Sigma_1, \Sigma_2, \ldots, \Sigma_n\}$. Then we can always form the double of M, denoted by $2M$, in the usual way [12]. The double has no boundary, and its Euler number is given by $X(2M) = 2X(M)$. It also satisfies $\sigma(2M) = 0$ (since the two 'halves' of the double will have opposite orientations). Furthermore, we know from [13] that

$$\alpha, \beta = \sigma(2M) \bmod 8\,.$$

Combining these results with theorem 1 we see that the total index of any plane field on $2M$ has the following form:

$$\text{total index} = (2n - X(M), 2m + X(M))$$

where $m, n \in \mathbb{Z}$.

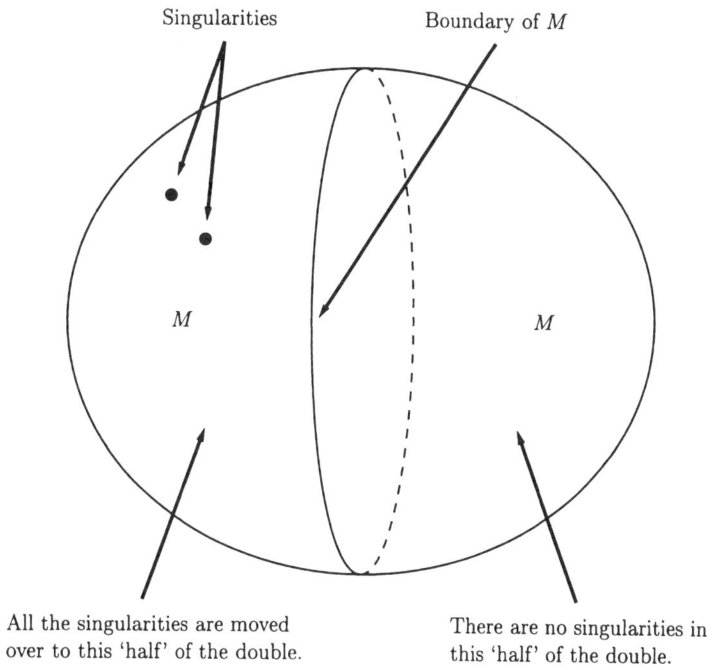

Figure 1. The plane field singularities will generically lie in both 'halves' of the double of M. By continuously deforming the plane field, all singularities can be moved into one 'half' of the double.

If the singularities of the plane field lie in both 'halves' of $2M$, then push all of the singularities over to one of the 'halves' as shown in figure 1. Thus, at least one of the 'halves' of $2M$ is free of singularities, and by taking this 'half', we have constructed a non-singular plane field on M. The degree of the map from ∂M to $G_{2,4}$ (defined by the plane field) must be, by construction,

$$(2n - \chi(M), 2m + \chi(M)) \qquad \text{for} \quad m, n \in \mathbb{Z}.$$

Recalling that a non-singular plane field defines a Klein metric g_K up to homotopy, we shall (in analogy with the Lorentz kink [3]) call this degree the *Klein kink* of the metric with respect to ∂M and denote it by kink$(\partial M; g_K)$. Thus we have shown

Lemma 2. Let M be any orientable manifold with $\partial M \cong \Sigma_1 \cup \Sigma_2 \cup \cdots \cup \Sigma_n \neq \emptyset$. Then there are always globally defined non-singular plane fields on M, and every non-singular plane field on M must have Klein kink on ∂M equal to

$$(2n - \chi(M), 2m + \chi(M)) \qquad \text{for some} \quad m, n \in \mathbb{Z}.$$

We now combine lemma 2 with the spin obstruction to obtain our main result.

3. Obstruction to spin–Klein cobordism

To begin, recall [1], the result that for manifolds M with boundary we have the formula

$$(u(\partial M) + \chi(M)) = \hat{w}_2 \bmod 2$$

where the mod 2 Kervaire semi-characteristic $u(\partial M)$ is defined as

$$u(\partial M) = \dim_{\mathbb{Z}_2} (H_0(\partial M; \mathbb{Z}_2) \oplus H_1(\partial M; \mathbb{Z}_2)) \bmod 2$$

and $\hat{w}_2$ is defined in terms of the second Stiefel–Whitney class w_2 as

$$\hat{w}_2 = \begin{cases} 0 & \text{iff} \quad w_2[c] = 0 \quad \text{for all 2-cycles} \quad c \in H_2(M) \\ 1 & \text{iff} \quad w_2[c] = 1 \quad \text{for some 2-cycle} \quad c \in H_2(M)\,. \end{cases}$$

Now, suppose we are given a collection of 3-manifolds $\{\Sigma_1, \Sigma_2, \ldots, \Sigma_n\}$ and a field of 2-planes (i.e. a Klein metric $\bar{g}_K$) on a collar neighbourhood of the Σ_i. Then the field of 2-planes will define a map from $\Sigma_1 \cup \Sigma_2 \cup \cdots \cup \Sigma_n$ to $G_{2,4}$ and we will denote the degree of this map by $\text{kink}(\Sigma_1 \cup \Sigma_2 \cup \cdots \cup \Sigma_n; \bar{g}_K)$. A manifold with boundary M is a spin–Klein cobordism for the Σ_i if M admits a spin structure and a globally defined Klein metric g_K such that $\text{kink}(\partial M; g_K) = \text{kink}(\Sigma_1 \cup \Sigma_2 \cup \cdots \cup \Sigma_n; \bar{g}_K)$. We then have

Theorem 2. Let $\{\Sigma_1, \Sigma_2, \ldots, \Sigma_n\}$ be some collection of closed orientable 3-manifolds with Klein metric $\bar{g}_K$ defined on a collar neighbourhood of $\Sigma_1 \cup \Sigma_2 \cup \cdots \cup \Sigma_n$. Let $\text{kink}(\Sigma_1 \cup \Sigma_2 \cup \cdots \cup \Sigma_n; \bar{g}_K) = (k_1, k_2)$, then there exists a spin–Klein cobordism M for the Σ_i if and only if

$$(u(\partial M) + k_i) = 0 \bmod 2$$

where k_i is either of k_1 or k_2.

Proof. Suppose such a spin–Klein cobordism M exists. Then $w_2 = 0$, and so $u(\partial M) = \chi(M) \bmod 2$. Furthermore, by lemma 2 we know that

$$\text{kink}(\partial M; g_K) = (2n - \chi, 2m + \chi)\,.$$

We see that the parity of $2n - \chi$ or $2m - \chi$ is determined solely by χ (i.e. $k_i = \chi \bmod 2$), and hence $(u(\partial M) + k_i) = 0 \bmod 2$.

Conversely, suppose $(u(\partial M) + k_i) = 0 \bmod 2$. Take any globally defined Klein metric g_K on M then we have

$$k_1 = k_2 = \chi \bmod 2\,.$$

Hence $(u(\partial M) + \chi) = 0 \bmod 2$ and so $w_2 = 0$. Thus M is spin–Klein. □

4. Examples and conclusions

Some examples of our obstruction to spin–Klein cobordism are as follows.

Example 1. Let S^2 be the two-sphere and D^2 a closed two-disk, then we may put a Klein metric g_K on the product manifold $M \cong S^2 \times D^2$ by simply taking the product metric induced by a signature $(++)$ metric on S^2 and a signature $(--)$ metric on D^2. The boundary of M is then given by $\partial M \cong S^1 \times S^2$, and the induced metric on ∂M is non-singular with signature $(-++)$. Since the Klein kink of g_K on ∂M could also be calculated using intersection theory (by counting the number of degenerate points), it follows that $\text{kink}(\partial M; g_K) = (0, 0)$. Hence, since $u(S^1 \times S^2) = 0$, we have that $S^2 \times D^2$ is a spin–Klein cobordism for $S^1 \times S^2$ with zero Klein kink.

Example 2. Take $\mathbb{R}^4$ with ordinary flat metric g_K of signature $(--++)$, and let $M \cong D^4$ be some closed 4-ball in $\mathbb{R}^4$. Then M is itself a Kleinian manifold with boundary $\partial M \cong S^3$. In this case, we can clearly identify the plane field P (corresponding to g_K) with a pair of non-vanishing vectors $\{v_1, v_2\}$ which span P. In other words, the section P of $G_{2,4}$ reduces to a section of the Stiefel manifold $V_{2,4}$. It follows that

$$\text{kink}(\partial M; g_K) = (\text{kink}(\partial M; v_1), \text{kink}(\partial M; v_2))$$

where $\text{kink}(\partial M; v_1) = \text{kink}(\partial M; v_2) = 1$ denote the kink of the respective vector field on ∂M. Hence, since $u(S^3) = 1$, we see that D^4 is an example of a spin–Klein cobordism for S^3 with $\text{kink}(S^3; g_K) = (1, 1)$.

Example 3. As was noted in the introduction, our result may be applied to spacetimes that change signature from $(-+++)$ to $(--++)$. Consider a spacetime $M \cong M_L \cup M_K$ with $\partial M_L = \partial M_K = \Sigma \neq \emptyset$, such that the metric g_L on M_L is of signature $(-+++)$ and the metric g_K on M_K is of signature $(--++)$. We require that the two metrics agree on Σ, and hence the induced metric on Σ is of signature $(-++)$. Now, restricting our attention to (M_K, g_K), it follows that $\text{kink}(\partial M_K; g_K) = (0, 0)$ and hence, from our theorem, M_K has spin structure if and only if $u(\partial M_K) = 0$. Thus, since $u(S^3) = 1$, we see that it is not possible for 'bubbles' of Kleinian signature to form across an S^3 surface whilst maintaining spin structure.

Finally, we note that there is probably a generalization of this work to non-orientable manifolds in the form of a topological obstruction to *pin*-Klein cobordism (cf [4]). In analogy with the Lorentzian case, this will presumably require us to pass to non-oriented plane fields (i.e. sections of the bundle of non-oriented planes).

Acknowledgments

The authors thank Gary Gibbons for many helpful discussions. AC is supported by NSF Graduate Fellowship no RCD–9255644.

References

[1] Chamblin A 1994 Some applications of differential topology in general relativity *J. Geom. Phys.* **13** 357–77
[2] Gibbons G W and Hawking S W 1992 Selection rules for topology change *Commun. Math. Phys.* **148** 345–52
[3] Gibbons G W and Hawking S W 1992 Kinks and topology change *Phys. Rev. Lett.* **69** 1719–21
[4] Chamblin A On the obstructions to non-Cliffordian pin structures *Commun. Math. Phys.* **164** 65–87
[5] Ellis G, Sumeruk A, Coule D and Hellaby C 1992 Change of signature in classical relativity *Class. Quantum Grav.* **9** 1535–54
[6] Hayward S A 1992 Signature change in general relativity *Class. Quantum Grav.* **9** 1851–62
[7] Alty L J 1994 *Class. Quantum Grav.* **11** 2523–36
[8] Barrett J, Gibbons G W, Perry M J, Pope C N and Ruback P 1994 Kleinian geometry and the $N = 2$ superstring *Int. J. Mod. Phys.* A **9** 1457–93
[9] Steenrod N 1951 *Topology of Fibre Bundles (Princeton Mathematical Series)* (Princeton, NJ: Princeton University Press)
[10] Hirzebruch F and Hopf H 1958 Felder von flächenelementen in 4-dimensionalen mannigfaltigkeiten *Math. Ann.* **136** 156–72
[11] Thomas E 1969 Vector fields on manifolds *Bull. Am. Math. Soc.* **75** 643–68
[12] Spivak M 1979 *Comprehensive Introduction to Differential Geometry* vol 5 2nd edn (Berkeley, CA: Publish or Perish)
[13] Hirzebruch F and Mayer K H 1968 $O(n)$—mannigfaltigkeiten, exotische sphären and singularitäten *Lecture Notes in Mathematics 57* (Berlin: Springer)

Class. Quantum Grav. **11** (1994) 2417–2434. Printed in the UK

New loop representations for 2 + 1 gravity

Abhay Ashtekar and R Loll†

Center for Gravitational Physics and Geometry, Physics Department, Pennsylvania State University, University Park, PA 16802, USA

Received 25 May 1994, in final form 29 July 1994

Abstract. Since the gauge group underlying (2 + 1)-dimensional general relativity is non-compact, certain difficulties arise in the passage from the connection to the loop representations. It is shown that these problems can be handled by appropriately choosing the measure that features in the definition of the loop transform. Thus, 'old-fashioned' loop representations—based on ordinary loops—do exist. In the case when the spatial topology is that of a 2-torus, these can be constructed explicitly; *all* quantum states can be represented as functions of (homotopy classes of) loops and the scalar product and the action of the basic observables can be given directly in terms of loops.

PACS numbers: 0460K

1. Introduction

Theories of connections are playing an increasingly important role in the current description of all fundamental interactions in nature. Such theories are of interest from a mathematical viewpoint as well: many of the recent advances in the understanding of the topology of low-dimensional manifolds have come from theories of connections, in particular from the analysis of Yang–Mills instantons and expectation values of Wilson loop functionals in Chern–Simons theories.

In these theories, the configuration space generally arises as a space $\mathcal{A}$ of connection 1-forms $A(x)$ on a Cauchy surface Σ, taking values in the Lie algebra of a gauge group G. The corresponding phase space $T^*\mathcal{A}$ is then naturally parametrized by canonically conjugate pairs of fields $(A(x), E(x))$, where E is a vector density of weight one on Σ, taking values in the dual of the Lie algebra of G, which can be thought of as a generalized 'electric' field conjugate to the gauge potential A. Gauge invariance is enforced by a Gauss constraint $\mathrm{Div}_A E(x) = 0$. As a consequence, the physical configuration space is $\mathcal{A}/\mathcal{G}$, the quotient of $\mathcal{A}$ by the group $\mathcal{G}$ of local gauge transformations. The physical observables are the gauge-invariant functions on phase space.

Gambini and Trias [1] were the first to point out that a convenient set of such observables can be associated with loops, i.e. closed curves α in Σ as follows. Choose a representation of dimension N of the gauge group G, and set

$$\begin{aligned}(T^0(\alpha))[A] &= \frac{1}{N}\mathrm{Tr}\Big(\mathrm{P}\exp\oint_\alpha A\Big)\\ (T^1(\alpha, s))^a\ [A, E] &= \mathrm{Tr}\Big(E^a(\alpha(s))\mathrm{P}\exp\oint_\alpha A\Big)\end{aligned} \tag{1.1}$$

† Present address: Sezione INFN di Firenze, Largo E Fermi 2, Arcetri, I-50125 Firenze, Italy.

0264-9381/94/102417+18$19.50

where P denotes path ordering along the loop α. $T^0(\alpha)$ is labelled by the loop α and represents a (gauge-invariant) configuration variable since it only depends on the connection A. $(T^1(\alpha, s))^a$ is labelled by a loop α and a point $\alpha(s)$ on α, and is a vector density at the point $\alpha(s)$. Being linear in the electric field, it represents a (gauge-invariant) momentum observable. These configuration and momentum variables are closed under the Poisson bracket and constitute a complete† set of functions on the phase space [2, 3] (in the sense that their gradients suffice to span the tangent space of the phase space almost everywhere). Therefore, in an algebraic approach to quantization [3], they can be chosen as the 'elementary variables' which define the 'basic' operators of the quantum theory. The task of quantization is then reduced to that of finding appropriate representations of the commutator algebra of these operators which mirror the Poisson bracket algebra of $T^0(\alpha)$ and $(T^1(\alpha, s))^a$.

The obvious way to represent states is by suitable functionals $\Psi[A]_{\mathcal{G}}$ of gauge equivalence classes of connections. This is the configuration or the connection representation. The operators $\hat{T}^0$ act by multiplication and the $\hat{T}^1$ by (Lie-) derivation. There is, however, another possibility: states can also be represented as suitable functions of closed loops. This is suggested by the possibility of making a 'Fourier-type' transform from the connection representation to a loop representation via

$$\psi(\alpha) := \int_{\mathcal{A}/\mathcal{G}} \mathrm{d}\mu\ ([A]_{\mathcal{G}}) T^0_A(\alpha)[A]_{\mathcal{G}} \Psi[A]_{\mathcal{G}} \tag{1.2}$$

where μ is a measure on $\mathcal{A}/\mathcal{G}$. This *loop transform* was first introduced in the context of Yang–Mills theories by Gambini and Trias [1] and later, but independently, in the context of general relativity by Rovelli and Smolin [5]. In both cases, however, it was a heuristic technique because the measure μ was not specified and the required integration theory did not exist. Nonetheless, it has played a powerful role as a heuristic device, especially in the context of general relativity.

Recent mathematical developments have made such considerations rigorous in the case when the gauge group is compact (for an overview see, for example, [4]). These results can be summarized as follows. In the connection representation, states are complex-valued functions on an appropriate completion $\overline{\mathcal{A}/\mathcal{G}}$ of $\mathcal{A}/\mathcal{G}$. This domain space of quantum wavefunctions, $\overline{\mathcal{A}/\mathcal{G}}$, carries a natural diffeomorphism-invariant measure μ_o which can be used to rigorously define the loop transform (1.2). Operators such as the $\hat{T}^0$ and the $\hat{T}^1$ can then be taken to the loop side and used in various constructions. Thus, in the case when the gauge group is compact, loop representations exist rigorously‡.

Loop representations have several appealing features. For example, irrespective of the choice of the gauge group and the precise physics contained in the theory, quantum states arise as suitably regular functions on the loop space. That is, the domain space of quantum states in various physical theories is the same. The regularity conditions on loop states of course change from one theory to another. Nonetheless, since the domain space is 'unified',

† Actually, these variables are *over*complete. This occurs because the variables $T(\alpha)$ for different α are not all independent; partly this is an expression of the fact that there are 'many more' loops α than points x in Σ. In the algebraic quantization method discussed below, the relations among these variables have to be imposed in an appropriate fashion on the quantum algebra.

‡ Note that each choice of measure defines a connection and an equivalent loop representation. (In the connection representation, the measure defines the inner product.) However, different choices of measures μ on $\overline{\mathcal{A}/\mathcal{G}}$ give rise to different representations. If the systems under consideration have an infinite number of degrees of freedom, the resulting representations are not generally expected to be unitarily equivalent. It is the underlying physics that must determine the appropriate measure and hence the appropriate representation.

mathematical techniques can be shared between the various theories. For diffeomorphism-invariant theories, one has the further advantage that the action of the diffeomorphism group is coded more easily in the loop space. In topological field theories where the connections under consideration are all flat, one encounters an interesting interplay between the quantum theory and the first homotopy group of the manifold Σ. Finally, if Σ is three-dimensional, one has an avenue to explore knot invariants via theories of connections.

It is therefore natural to ask if the loop representations can also be developed rigorously for physical theories—such as general relativity in three and four dimensions—in which the gauge group G is non-compact.

A number of difficulties arise immediately. First, if the gauge group is non-compact, the techniques [4] used to develop integration theory over $\overline{\mathcal{A}/\mathcal{G}}$ fail at a rather early stage. The problems here are not insurmountable. However, a number of new ideas are needed and, in general, the theory is likely to be considerably more complicated unless, *as is the case in four-dimensional general relativity*, the gauge group is the complexification of a compact Lie group. It is therefore natural to first restrict oneself to a context where these difficulties do not arise. One such setting is provided by three-dimensional general relativity. Here, the connections of interest turn out to be flat and one can replace $\overline{\mathcal{A}/\mathcal{G}}$ by the moduli space of flat connections. Since these spaces are finite-dimensional, one does not have to develop the integration theory; the problems mentioned above do not arise. The moduli space has several 'sectors'. On the sector where the traces of holonomies T^0_α are all bounded, the transform can be defined and the loop representation can be constructed in a straightforward fashion [6]. However, it turns out that this sector does not correspond to 'geometrodynamics'. On the sector which does, a new difficulty arises: although the integration theory is straightforward, the traces of holonomies T^0_α fail to be square-integrable with respect to 'natural' measures, making the loop transform analogous to (1.2) ill-defined. Thus, on these physically interesting sectors, a new strategy is needed.

This problem was first pointed out by Marolf [7] who also suggested a way of tackling it in the special case where the manifold Σ is a torus, T^2. The purpose of this paper is to suggest an alternative solution, which consists of suitably modifying the measure that appears in the transform. This solution is conceptually simpler in the sense that with the new measures various problems are avoided right from the beginning. There is also a technical simplification. While in the Marolf approach, one first restricts oneself to a suitable dense subspace of the connection Hilbert space, defines the transform and then extends it to the full Hilbert space, with the modified measures, the transform exists on the full Hilbert space from the outset. More importantly, in our approach *the final result is a genuine loop representation* in the sense that all states in the Hilbert space are represented as functions of loops. By contrast, in the Marolf approach the limiting states, which are not contained in the initial dense subspace, cannot, in general, be represented as functions of loops. This result had given rise to some concern about the utility and the role of loops in the representation of quantum states in the case when the gauge group is non-compact. Our analysis clarifies this issue and shows that 'old-fashioned' loop representations, without the need for any smearing, do exist even in the geometrodynamical sector of three-dimensional gravity. Our solution does, however, have a drawback: our expressions for the $\hat{T}^1$-operators in the loop representation are more complicated. In spite of these differences, the final theory we obtain is unitarily equivalent to Marolf's for $\Sigma = T^2$. Therefore, the choice between the two strategies is primarily a matter of taste and convenience.

The outline of this paper is as follows. In section 2 we recall the basic structure of $2+1$ gravity on spacetimes $M = \Sigma^g \times \mathbb{R}$, with Σ^g a compact Riemann surface of genus g. Section 3 discusses quantization in the connection representation and presents the general strategy for modifying the measure to make the loop transform well defined. This strategy always leads to a 'regular' loop representation. In the case when the sector of the moduli space of flat connections under consideration is compact and the traces of holonomies T^0_α are bounded functions, the modification of the measure is unnecessary. However, if one chose to follow this route, the resulting loop representation would be unitarily equivalent to that of [6]. In the non-compact case, on the other hand, the strategy appears to be essential to obtain a genuine loop representation. In section 4, the procedure is carried out in detail for the case when Σ is a 2-torus. In particular, we present a family of new measures which make the loop transform well defined and obtain the modified expressions of the $\hat{T}^0$- and $\hat{T}^1$-operators as well as the explicit form of the scalar product in the loop representation for this family. Section 5 contains our conclusions. In the appendix we present some partial results for the genus-2 case.

2. Preliminaries

In this section, we will collect those results from the classical Hamiltonian formulation of the $(2+1)$-dimensional (Lorentzian) general relativity that will be needed in the main part of the paper in sections 3 and 4. The discussion will also serve to fix our notation. Note that this is not meant to be an exhaustive summary; further details may be found, for example, in [3, 8–10].

Since we are interested primarily in the canonical quantization of the spatially compact case, we will restrict ourselves to three-dimensional spacetimes M of the form $M = \Sigma^g \times \mathbb{R}$, where Σ^g is a two-dimensional compact Riemann surface of genus g. In the connection dynamics version, the Hamiltonian formulation of the theory leads to two sets of first-class constraints [8, 6, 3], one linear in momenta and the other independent of momenta. Consequently, the Dirac and the reduced phase space quantization methods lead to equivalent quantum theories. For convenience of presentation, we will use the reduced phase space method here. Because of the simplicity of the constraints, the reduced phase space is a cotangent bundle over a reduced configuration space, which, in turn, is just the moduli space $\mathcal{A}^{\mathrm{F}}/\mathcal{G}$ of flat $PSU(1,1)$-connections on Σ^g ($PSU(1,1) = SU(1,1)/\mathbb{Z}_2$). We will first recall relevant facts about $\mathcal{A}^{\mathrm{F}}/\mathcal{G}$ and then go on to discuss the structure of the reduced phase space $T^\star(\mathcal{A}^{\mathrm{F}}/\mathcal{G})$.

A flat $PSU(1,1)$-connection A on Σ^g is determined by the values of the $2g$ holonomies U_i, $i = 1\ldots 2g$, around representatives α_i of the $2g$ homotopy generators $\{\alpha_i\}$, $i = 1\ldots 2g$, on Σ^g. Explicitly, we have

$$U_i = \mathrm{P}\exp \oint_{\alpha_i} A\,. \tag{2.1}$$

Without loss of generality we will assume these representatives to be based at a fixed point $p \in \Sigma^g$ and evaluate the holonomies U_i at p. Because of the fundamental relation between the generators of the homotopy group, these holonomies are subject to the condition

$$U_1 U_2 U_1^{-1} U_2^{-1} U_3 U_4 U_3^{-1} U_4^{-1} \ldots U_{2g-1} U_{2g} U_{2g-1}^{-1} U_{2g}^{-1} = \mathbb{1}\,. \tag{2.2}$$

Our task is to construct the moduli space of flat connections, i.e. to determine the structure of the orbit space $\mathcal{A}^{\mathrm{F}}/\mathcal{G}$. For this, we note that $\mathcal{G}$ now acts on the holonomies $(U_1, \ldots, U_{2g})$ at the base point p by the adjoint action of $U(p) \in PSU(1,1)$ according to

$$(U_1, U_2, \ldots, U_{2g}) \to U(p)(U_1, U_2, \ldots, U_{2g})U(p)^{-1}\,. \tag{2.3}$$

The moduli space is therefore obtained by incorporating (2.2) and taking the quotient with respect to (2.3).

A key point for us is that the moduli space contains a finite number of components, often referred to as 'sectors'. (For more precise statements, see, for example, [9].) This comes about because the isotropy group I of a holonomy U_i, i.e. the subgroup of $SU(1,1)$ leaving fixed a particular U_i under the adjoint action, is not universal but depends on (certain properties of) U_i. We can distinguish the following cases†:

(i) Rotations about a time-like axis; $I = SO(2)$, e.g.

$$I = \left\{ \begin{pmatrix} e^{i\omega} & 0 \\ 0 & e^{-i\omega} \end{pmatrix}, \omega \in [0, 2\pi] \right\} . \tag{2.4}$$

A general time-like vector can be obtained by conjugation of U with an arbitrary group element, in which case the isotropy group I likewise changes by conjugation. The same remark applies to the cases below.

(ii) Boosts about a space-like axis; $I = \mathbb{R} \times \mathbb{Z}_2$, e.g.

$$I = \left\{ \begin{pmatrix} \epsilon \cosh \omega & \sinh \omega \\ \sinh \omega & \epsilon \cosh \omega \end{pmatrix}, \ \omega \in \mathbb{R}, \epsilon \in \mathbb{Z}_2 \right\} \tag{2.5}$$

with the group law $(\epsilon_1, \omega_1)(\epsilon_2, \omega_2) = (\epsilon_1\epsilon_2, \epsilon_1\omega_1 + \epsilon_2\omega_2)$;

(iii) Null rotations; $I = \mathbb{R} \times \mathbb{Z}_2$, e.g.

$$I = \left\{ \begin{pmatrix} \epsilon + i\omega & \omega \\ \omega & \epsilon - i\omega \end{pmatrix}, \ \omega \in \mathbb{R}, \epsilon \in \mathbb{Z}_2 \right\} \tag{2.6}$$

with the group law $(\epsilon_1, \omega_1)(\epsilon_2, \omega_2) = (\epsilon_1\epsilon_2, \epsilon_1\omega_2 + \epsilon_2\omega_1)$.

In the sector that corresponds to geometrodynamics, all the U_i are boosts [10], corresponding to case (ii) above‡. It turns out that for $g > 1$ this sector is identical with the well known Teichmüller space $\mathcal{T}(\Sigma^g)$ associated with the Riemann surface Σ^g. This is a finite-dimensional space diffeomorphic to $\mathbb{R}^{6g-6}$. For $g = 1$, this sector is diffeomorphic to $\mathbb{R}^2$. Thus, the geometrodynamic sector of the reduced configuration space—on which we will focus from now on—is precisely the Teichmüller space $\mathcal{T}(\Sigma^g)$.

Let us now turn to the reduced phase space $T^*\mathcal{T}(\Sigma^g)$. We will give an explicit description of the moduli space of flat connections in terms of gauge-invariant loop variables.

We begin by recalling [6, 11] that there is an (over)complete set of Dirac observables on the phase space $T^*(\mathcal{A}^F/\mathcal{G})$, given by the loop variables

$$T^0(\alpha)[A] = \tfrac{1}{2}\mathrm{Tr}\, U_\alpha \qquad T^1(\alpha)[A, E] = \oint_\alpha \mathrm{d}\alpha^a \, \eta_{ab} \mathrm{Tr}(E^b U_\alpha) \tag{2.7}$$

where the canonical pairs $([A], [E])$ coordinatize the cotangent bundle $T^*\mathcal{A}^F$ of the space $\mathcal{A}^F$ of flat connections, $U_\alpha = \mathrm{P}\exp \oint_\alpha A$ is the holonomy around the loop α evaluated at the base point, and where η_{ab} is the totally anti-symmetric Levi-Civita density on Σ^g. (Note that we have integrated out $(T^1(\alpha, s))^a$ in (1.1) over the loop α to obtain a momentum

† This classification is equivalent to that of [9] which is based on the Lie algebra elements generating $U_i \in SU(1, 1)$.

‡ For brevity, here and in the rest of this section, we have ignored certain subtleties that arise in the $g = 1$ case [9], and have summarized the situation for the generic, $g \geqslant 2$ case.

observable $T^1(\alpha)[A, E]$; the vector density index a and the dependence on the marked point $\alpha(s)$ are lost in the integration.) Since all connections under consideration are flat, the variables (2.7) depend only on the homotopy of the loop α, and we may substitute α by its corresponding homotopy group element.

The loop variables T^I, $I = 0, 1$, form a closed Poisson algebra with respect to both the canonical symplectic structure on $T^*\mathcal{A}^{\mathrm{F}}$, and to the induced symplectic structure on the reduced, physical phase space $T^*(\mathcal{A}^{\mathrm{F}}/\mathcal{G})$. The algebra is given by

$$\begin{aligned} \{T^0(\alpha), T^0(\beta)\} &= 0 \\ \{T^0(\alpha), T^1(\beta)\} &= -\tfrac{1}{2}\sum_i \Delta_i(\alpha,\beta)\left(T^0(\alpha\circ_i\beta) - T^0(\alpha\circ_i\beta^{-1})\right) \\ \{T^1(\alpha), T^1(\beta)\} &= -\tfrac{1}{2}\sum_i \Delta_i(\alpha,\beta)\left(T^1(\alpha\circ_i\beta) - T^1(\alpha\circ_i\beta^{-1})\right). \end{aligned} \tag{2.8}$$

The sums in (2.8) are over all points i of intersection of the loops α and β, with $\Delta_i(\alpha, \beta) = 1$ $(= -1)$ if the two tangent vectors $(\dot\alpha, \dot\beta)$ form a right- (left-)handed dyad at i and zero if the tangent vectors are parallel. The algebra (2.8) is independent of the representatives chosen in the homotopy classes $\{\alpha\}$ and $\{\beta\}$, and the representatives can be chosen to originate and intersect at a fixed base point $p \in \Sigma$. For this reason we will, from now on, identify the loop composition $\circ_i$ with the group multiplication $\circ$ in $\pi_1(\Sigma^g)$.

Because of the identities that hold among the traces of 2×2-matrices, the T^I are not all independent. They are subject to the following algebraic relations:

$$\begin{aligned} T^0(\alpha)T^0(\beta) &= \tfrac{1}{2}\left(T^0(\alpha\circ\beta) + T^0(\alpha\circ\beta^{-1})\right) \\ T^0(\alpha)T^1(\beta) + T^0(\beta)T^1(\alpha) &= \tfrac{1}{2}\left(T^1(\alpha\circ\beta) + T^1(\alpha\circ\beta^{-1})\right). \end{aligned} \tag{2.9}$$

For a general gauge group, relations of this type are also known as Mandelstam constraints. Since the reduced configuration space, the Teichmüller space $\mathcal{T}$, is topologically trivial, one can attempt to find a global chart on it using the T^0-functions. One way of achieving this is to eliminate the redundancy inherent in the Mandelstam constraints (2.9) and to re-express and solve the condition (2.2) on holonomies as conditions on the *traces* of the holonomies of the $2g$ homotopy generators. Such sets of independent and complete Wilson loop invariants do indeed exist for any genus and are the subject of a forthcoming publication [11].

3. Quantization

This section is divided into two parts. In the first, we recall quantization in the connection representation. This discussion will facilitate the introduction of the loop transform and also serve to bring out some subtleties. In the second, we first point out the difficulties associated with the loop transform and then sketch our proposal for overcoming them. In the next section, this strategy is carried out in detail for the genus-1 case.

3.1. Connection representation

By a quantization of 2 + 1-gravity we shall mean a representation of the algebra (2.8) as the commutator algebra of self-adjoint operators $\hat{T}^I$ on some Hilbert space, such that the conditions (2.9) hold (the products of operators on the left side being replaced by their anti-commutators).

In the connection representation, we can proceed as one normally does when dealing with quantum mechanics of systems whose configuration space is a manifold. We can take the states to be densities $\tilde{\Psi}[A]_{\mathcal{G}}$ of weight one-half on the reduced configuration space $\mathcal{T}$, and let the scalar product be the obvious one:

$$\langle\tilde{\Psi}|\tilde{\Phi}\rangle := \int_{\mathcal{T}} \overline{\tilde{\Psi}}[A]_{\mathcal{G}} \; \tilde{\Phi}[A]_{\mathcal{G}} \tag{3.1}$$

where a choice of a volume element on $\mathcal{T}$ is not necessary because the integrand is a density of weight one. Since the T^0 are configuration variables, they are represented by multiplication operators. Similarly, being momentum variables, the T^1 are represented by Lie derivatives,

$$(\hat{T}^0_\alpha \circ \tilde{\Psi})[A]_{\mathcal{G}} = T^0_\alpha[A]_{\mathcal{G}}\tilde{\Psi}[A]_{\mathcal{G}} \qquad (\hat{T}^1_\alpha \circ \tilde{\Psi})[A]_{\mathcal{G}} = \frac{\hbar}{\mathrm{i}}\mathcal{L}_{X(\alpha)}\tilde{\Psi}[A]_{\mathcal{G}} \tag{3.2}$$

where $X(\alpha)$ is the vector field on the reduced configuration space $\mathcal{T}$ which defines the classical momentum variable $T^1(\alpha)$. (Recall that T^1_α, being linear in momentum, is a contraction of $[E]$ with a vector field on $\mathcal{T}$. Thus, $X(\alpha)$ is the projection to $\mathcal{T}$ of the vector field δA on $\mathcal{A}^{\mathrm{F}}$, defined by $\delta A = \oint_\alpha \mathrm{d}\alpha^a\, \eta_{ab}\, \delta^2(x, \alpha(s))\, \mathrm{Tr}\, U_\alpha(s)(\delta/\delta A_b(x))$.) It is straightforward to verify that with this definition the $\hat{T}^I_\alpha$ satisfy the canonical commutation relations that arise from (2.8) and the algebraic conditions coming from (2.9).

In this description, the states naturally arose as scalar densities. Given a volume element, one can 'de-densitize' them and represent them as the more familiar wavefunctions. This can be achieved using *any* volume element $\mathrm{d}V$ on $\mathcal{T}$. Furthermore, for any $\mathrm{d}V$, the action of the operators $\hat{T}^I$ can be translated in a canonical fashion to $L^2(\mathcal{T}, \mathrm{d}V)$.

It turns out that the Teichmüller space $\mathcal{T}$ admits a natural volume element. To see this, we first note that there is a natural symplectic structure Ω on the space of connections $\mathcal{A}$ given by

$$\Omega(\delta A, \delta A') = \int_\Sigma \mathrm{d}^2x\; \eta^{ab}(\delta A)^i_a(\delta A')_{bi} \tag{3.3}$$

for any two tangent vectors δA, $\delta A'$ to $\mathcal{A}$. This form Ω can be pulled back to the space $\mathcal{A}^{\mathrm{F}}$ of flat connections. Being gauge-invariant, the pull-back, in turn, projects down to Teichmüller space, where it coincides with the well known Weil–Petersson symplectic form ω†. This gives rise to a natural volume element $\mathrm{d}V$ on $\mathcal{T}$, namely, the $(3g-3)$-fold exterior product $\mathrm{d}V_o = \omega \wedge \ldots \wedge \omega$. It is then natural to represent states as complex-valued *functions* $\Psi[A]_{\mathcal{G}}$ and define the inner product as

$$\langle\Psi|\Phi\rangle := \int_{\mathcal{T}} \mathrm{d}V_o \overline{\Psi}[A]_{\mathcal{G}}\Phi[A]_{\mathcal{G}}\,. \tag{3.4}$$

On this Hilbert space, the $\hat{T}^0_\alpha$ can be represented, as before, by multiplication operators. They are densely defined and symmetric. Normally, the definition of the momentum operators $\hat{T}^1(\alpha)$ would require a modification: if the Lie derivative of the volume element

† A global coordinate chart on Teichmüller space which is convenient for our purposes is given by the Fenchel–Nielsen coordinates (see, for example, [14]). For $g > 1$, they are a set of length and angle coordinates $[l_i, \tau_i]$, $i = 1 \ldots 3g - 3$, associated with a pants decomposition of the Riemann surface along a set of $3g - 3$ minimal geodesics. Moreover, they are canonical coordinates for the symplectic form ω, i.e. $\omega = \sum_{i=1}^{3g-3} \mathrm{d}l_i \wedge \mathrm{d}\tau_i$.

with respect to the vector fields $X(\alpha)$ does not vanish, we would have to add a multiple of the divergence of the vector field to ensure that the resulting operator is symmetric. However, it turns out [15] that the vector fields $X(\alpha)$ are, in fact, the Hamiltonian vector fields on $\mathcal{A}^{\mathrm{F}}$ for the symplectic structure ω, where the Hamiltonians are simply the functions T^0_α on $\mathcal{T}$. Hence, in particular, the Lie derivative of the Liouville volume element $\mathrm{d}V_{\mathrm{o}}$ with respect to any vector field $X(\alpha)$ vanishes. Therefore, we can continue to represent $\hat{T}^1(\alpha)$ simply by the Lie derivative. Thus, the representation of the basic operators is the same as in (3.2), although the states are now wavefunctions $\Psi[A]_{\mathcal{G}}$ on $\mathcal{T}$ rather than half-densities $\tilde{\Psi}[A]_{\mathcal{G}}$. This formulation of the connection representation will constitute the starting point for the discussion of the loop transform in the next subsection.

3.2. The loop transform

We are now ready to construct the loop representation. The key idea is to define this representation through a loop transform of the type (1.2). In the present case, using the fact that the reduced configuration space can be identified with the Teichmüller space $\mathcal{T}$, we can simplify the transform to

$$\psi(\alpha) = \int_{\mathcal{T}} \mathrm{d}V\, T^0(\alpha)[l,\tau]\Psi[l,\tau] \tag{3.5}$$

where $\mathrm{d}V$ is a volume element on $\mathcal{T}$ and we have used the Fenchel–Nielsen coordinates [14] to parametrize $\mathcal{T}$ globally. Thus, just as in the general context of section 1 the transform needed a measure on $\overline{\mathcal{A}/\mathcal{G}}$, the transform now requires the introduction of a volume element on $\mathcal{T}$. Could we not have avoided the introduction of this *ad hoc* structure? After all, the transform has the form of an inner product of $T^0(\alpha)$ with a wavefunction of connections Ψ and the connection representation *could* be constructed intrinsically (i.e. without any additional structure such as the volume element) if the states were represented by densities $\tilde{\Psi}$ of weight one-half on $\mathcal{T}$. Unfortunately, even if we replaced the $\Psi[A]$ in (3.5) by $\tilde{\Psi}[A]$, *because the integral kernel of the transform, $T^0(\alpha)$, is a function* rather than a density of weight one-half, we would still need an additional structure (say, a fiducial density of weight one-half) to make the integral well defined. Thus, while the connection representation itself does not require the choice of a volume element, the passage to the loop representation does†.

We saw at the end of subsection 3.1 that there *is* a natural volume element $\mathrm{d}V_{\mathrm{o}}$ on $\mathcal{T}$ which arises from the Weil–Petersson symplectic form. Therefore, a simple solution to the problem would be to just choose this $\mathrm{d}V_{\mathrm{o}}$ for the required volume element in (3.5). This strategy would work if we were interested in the 'time-like' sector of 2 + 1 gravity [6]. However, as noted in section 2, in this paper we are interested in the 'space-like' sector which corresponds to geometrodynamics. It is this choice that led us to take the Teichmüller space $\mathcal{T}$ as the reduced configuration space. Now, $\mathcal{T}$ is non-compact and, in general, the $T^0(\alpha)$, being unbounded, fail to be square-integrable on $(\mathcal{T}, \mathrm{d}V_{\mathrm{o}})$. Hence, for a general quantum state Ψ in the connection representation—which belongs to $L^2(\mathcal{T}, \mathrm{d}V_{\mathrm{o}})$—the integral in the transform would not be well defined. In the general setting considered here, this is the problem that was first noted by Marolf [7] in the explicit context of the torus topology for Σ.

† One might imagine defining the transform intrinsically by using densities of weight one (rather than one-half) as connection states. However, one would still need a volume element to decide which of these densities of weight one are normalizable, i.e. qualify to feature in the transform in the first place.

A way out would be to first restrict the transform to a dense subspace D of $L^2(\mathcal{T}, \mathrm{d}V_0)$—such as the one spanned by the smooth wavefunctions $\Psi[l, \tau]$ of compact support—on which the transform *is* well defined, obtain the loop states and then take the Cauchy completion of this space. While this procedure appears to be simple at first sight, a detailed examination [7] shows that there are two key problems. First, for the loop representation to exist, the dense space D has to satisfy *three* conditions: (i) the integral on the right of (3.5) must be well defined for all Ψ in D; (ii) on D, the transform should be faithful; and (iii) D should remain invariant under the action of the $\hat{T}^I$-operators. Although one does expect such dense subspaces D to exist, already in the torus case it is quite a non-trivial problem to find them. The second and conceptually more important problem is that when one takes the Cauchy completion of the image of D, one finds that it admits states which cannot, in a natural way, be represented as functions of loops. Consequently, the sense in which such a representation can be considered a 'loop representation' becomes rather obscure.

Our proposal, therefore, is to try a new strategy. The key idea is to exploit the freedom in the choice of the volume element $\mathrm{d}V$. Since we are regarding $\mathcal{T}$ as a manifold, volume elements correspond to $(6g-6)$-forms on $\mathcal{T}$. Hence, any two are related by a (suitably regular) function. Thus, we can set $\mathrm{d}V = m[l, \tau]\mathrm{d}V_0$ for some non-negative, smooth function $m[l, \tau]$. Following the terminology common in physics, we will refer to $m[l, \tau]$ as a 'measure'. The idea therefore is to choose an appropriately damped measure to make the loop transform well defined.

What conditions does $m[l, \tau]$ have to satisfy? First, as already noted, it should ensure a sufficient damping so that the loop transform is well defined. More precisely, we will require that $m[l, \tau]$ be such that the traces of holonomies $T^0(\alpha)$ are in $L^2(\mathcal{T}, \mathrm{d}V = m\,\mathrm{d}V_0)$. Then, if we *define* the connection representation using $\mathrm{d}V$—which we are free to do—we will be led to a well defined loop transform. However, we also need the action of the operators $\hat{T}^I(\alpha)$ on loop states to be well defined and manageable. For a general measure $m[l, \tau]$, the action of the $\hat{T}^I(\alpha)$ reduces to

$$\begin{aligned}(\hat{T}^0(\alpha)\Psi)[l, \tau] &= T^0(\alpha)[l, \tau]\Psi[l, \tau] \\ (\hat{T}^1(\alpha)\Psi)[l, \tau] &= -\mathrm{i}\hbar\big(\mathcal{L}_{X(\alpha)} + \tfrac{1}{2}\mathcal{L}_{X(\alpha)}\ln m\big)\Psi[l, \tau]\end{aligned} \tag{3.6}$$

where, as before, $\mathcal{L}_{X(\alpha)}$ is the Lie derivative along the vector field $X(\alpha)$ corresponding to the loop momentum observable $T^1(\alpha)$, and where the second term on the right-hand side of the relation for $\hat{T}^1$ compensates for the fact that the Lie derivative of the volume element $\mathrm{d}V = m\,\mathrm{d}V_0$ with respect to $X(\alpha)$ may not vanish.

These operator actions can be translated into the loop representation via the transform (3.5). In order that the action of the resulting operators be manageable—so that in the final picture the loop representation can exist in its own right—it is necessary that the term $\hat{T}^I(\beta)T^0(\alpha)$ in

$$\begin{aligned}(\hat{T}^I(\beta)\psi)(\alpha) &:= \int_{\mathcal{T}} \mathrm{d}V\, m[l, \tau]\, T^0(\alpha)\big(\hat{T}^I(\beta)\Psi[l, \tau]\big) \\ &= \int_{\mathcal{T}} \mathrm{d}V\, m[l, \tau]\,\big(\hat{T}^I(\beta)T^0(\alpha)\big)\Psi[l, \tau]\end{aligned} \tag{3.7}$$

be expressible as some linear combination $\sum a_j T^0(\alpha_j)$ (where α_j denotes any homotopy group element). For $\hat{T}^I = \hat{T}^0$ this condition is automatically fulfilled, thanks to the algebraic relation (2.9). For $I = 1$, the term under consideration takes the form

$$\big(\mathcal{L}_{X_\beta} + \tfrac{1}{2}\mathcal{L}_{X_\beta}\ln m\big)T^0(\alpha) = \big\{T^1(\beta), T^0(\alpha)\big\} + \tfrac{1}{2}\big\{T^1(\beta), \ln m\big\}T^0(\alpha) \tag{3.8}$$

where the curly brackets denote the Poisson brackets on phase space. The first term on the right-hand side is already in the required form, again due to the algebraic relations (2.9). Next, it follows from (2.8) and (2.9) that the second term, $\{T^1(\beta), \ln m\}$, would have the desired form if it were expressible as a linear combination of the T^0. Thus, the action of $\hat{T}^1$ would be manageable on loop states if the measure $m[l, \tau]$ were of the form

$$m = \exp\left(\sum_i b_i\, T^0(\alpha_i)\right) \tag{3.9}$$

for some fixed real constants b_i and some fixed homotopy generators α_i. (Note that we could also have chosen to use for $m[l, \tau]$ the exponential of a product of $T^0(\alpha_i)$ since the product can always be re-expressed as a sum, using (2.9).)

Our strategy is therefore to use a measure $m[l, \tau]$ of the form (3.9) (both in the definition of the connection representation and) in the definition of the loop transform. The key question is whether one can choose a finite number of b_i and α_i such that the measure damps sufficiently fast for the transform (3.5) to be well defined for *any* element α of the homotopy group. At a heuristic level, it would seem that the freedom in the choice of α_i is so large that it should be easy to meet this damping condition. However, because we do not have sufficient control over the behaviour of traces of holonomies $T^0(\alpha)$ on the Teichmüller spaces of higher genus, we have been able to demonstrate explicitly the existence of the measures of the required type only in the $g = 1$ case. In the general genus case, if b_i and α_i can be chosen to ensure the existence of the integral in (3.5) for all α, the existence of a loop representation with the required properties *is* ensured. In particular, all normalizable states in such a representation would arise as functions of loops; generalized loops would not be necessary.

4. The torus case

Let us briefly review the explicit structure of the reduced phase space $T^*\mathcal{T}$ in the case when the two-manifold Σ is a torus. (For further details, see [3, 7].) We are interested in the sector where the holonomies U_i, $i = 1, 2$ of both homotopy generators (a_1, a_2) are rotations about space-like axes in the three-dimensional Minkowski space. The fundamental relation (2.2) implies that U_1 and U_2 commute and are therefore rotations about the same axis. Without loss of generality we may choose this axis to lie along the vector $(0, 1, 0)$. The reduced configuration space is therefore the two-dimensional space $\mathcal{T}$ of flat connections on a torus T^2 and can be parametrized by $\boldsymbol{a} \in \mathbb{R}^2$, with opposite signs identified ($\boldsymbol{a} \sim -\boldsymbol{a}$), i.e. $\mathcal{T} = \mathbb{R}^2/\mathbb{Z}_2$. The corresponding reduced phase space is its cotangent bundle, parametrized globally by the canonical variable pair $(\boldsymbol{a}, \boldsymbol{p})$.

The loop variables $T^i(\alpha)$ depend on only the homotopy class $\{\alpha\}$ of the loop α, which for $\Sigma = T^2$ can be labelled by two integers $\boldsymbol{n}$, characterizing the decomposition $\{\alpha\} = n_1\{\alpha_1\} + n_2\{\alpha_2\}$. The variables T^0 and their associated momentum variables T^1 form an overcomplete set of observables on phase space. Their explicit form is

$$T^0(\boldsymbol{k})[\boldsymbol{a}] = \cosh(\boldsymbol{k}\cdot\boldsymbol{a}) \qquad T^1(\boldsymbol{k})[\boldsymbol{a}, \boldsymbol{p}] = \sinh(\boldsymbol{k}\cdot\boldsymbol{a})\, \boldsymbol{k}\times\boldsymbol{p} \tag{4.1}$$

where $\boldsymbol{k}\times\boldsymbol{p} = k_1 p_2 - k_2 p_1$. Their Poisson algebra can be written down explicitly

$$\begin{aligned}
&\{T^0(\boldsymbol{m}), T^0(\boldsymbol{n})\} = 0 \\
&\{T^0(\boldsymbol{m}), T^1(\boldsymbol{n})\} = -\tfrac{1}{2}\,(\boldsymbol{m}\times\boldsymbol{n})\left(T^0(\boldsymbol{m}+\boldsymbol{n}) - T^0(\boldsymbol{m}-\boldsymbol{n})\right) \\
&\{T^1(\boldsymbol{m}), T^1(\boldsymbol{n})\} = -\tfrac{1}{2}\,(\boldsymbol{m}\times\boldsymbol{n})\left(T^1(\boldsymbol{m}+\boldsymbol{n}) - T^1(\boldsymbol{m}-\boldsymbol{n})\right).
\end{aligned} \tag{4.2}$$

The overcompleteness of these variables is due to the Mandelstam constraints (2.9), which now simplify to

$$
\begin{aligned}
&T^0(m)T^0(n) = \tfrac{1}{2}\big(T^0(m+n) + T^0(m-n)\big) \\
&T^0(m)T^1(n) + T^0(n)T^1(m) = \tfrac{1}{2}\big(T^1(m+n) + T^1(m-n)\big)\,.
\end{aligned}
\tag{4.3}
$$

Note, as an aside, that in the 'time-like' sector where the moduli space of connections is compact, the relevant loop observables are obtained from (4.1) by substituting the hyperbolic functions with the corresponding trigonometric functions, and the relation (4.2) and (4.3) remain the same.

For the two generators of the homotopy group $\pi_1(T^2)$, $\alpha_1 = (1, 0)$ and $\alpha_2 = (0, 1)$, one finds

$$
\begin{aligned}
&T^0(\alpha_1)[a] = \cosh a_1 \qquad T^0(\alpha_2)[a] = \cosh a_2 \\
&\tfrac{1}{2}\big(T^0(\alpha_1 \circ \alpha_2) - T^0\big(\alpha_1 \circ \alpha_2^{-1}\big)\big)[a] = \sinh a_1 \sinh a_2\,.
\end{aligned}
\tag{4.4}
$$

It follows that $\frac{1}{2}\big(T^0(\alpha_1 \circ \alpha_2) - T^0\big(\alpha_1 \circ \alpha_2^{-1}\big)\big)$ together with one of $T^0(\alpha_1)$, $T^0(\alpha_2)$ parametrizes $\mathcal{T}$ globally and therefore constitutes a good choice of independent loop variables in terms of which all other $T^0(\alpha)$ can be expressed. At first sight, the easiest choice for an independent set may seem to take $T^0(\alpha_1)$ and $T^0(\alpha_2)$. However, they do not form a good global chart (a similar statement holds for the genus-2 case, see the appendix).

In the coordinates a_i on $\mathcal{T}$, the natural Liouville volume element is simply $\mathrm{d}V_\mathrm{o} = \mathrm{d}a_1\, \mathrm{d}a_2$. Our objective is to choose an appropriate measure $m(a_i) = \exp(-M(a_i))$ such that the transform and the resulting loop representation are well defined. To follow the strategy outlined in subsection 3.2, let us begin by writing the analogues of (3.5)–(3.9) explicitly. First, we have

$$
\psi(n) = \langle T^0(n), \Psi \rangle = \int_0^\infty \mathrm{d}a_1 \int_0^\infty \mathrm{d}a_2\, \mathrm{e}^{-M(a)}\, T^0(n)\, \Psi(a)
\tag{4.5}
$$

using the scalar product notation. (Note that since the $T^0(n)$ are all real, complex conjugation is unnecessary in the integral.) According to our reasoning in section 3, the function M on $\mathcal{T}$ should have the form $M = \sum b_i T^0(\alpha_i)$ (cf (3.9)), where the b_i and the α_i are such that holonomies around arbitrary loops are square-integrable with respect to $\mathrm{d}V = (\exp -M)\mathrm{d}V_\mathrm{o}$. In the case of the torus, an obvious choice is

$$
M = c\big(T^0(q_1) + T^0(q_2)\big)
\tag{4.6}
$$

where $c \in \mathbb{R}$, $c > 0$, and q_1 and q_2 are linearly independent homotopy classes. It turns out that for *any* positive c and any q_1, q_2, (4.6) leads to a loop representation with all the desired properties discussed in section 3. More precisely, we have the following.

(i) It is straightforward to verify that every $T^0(k)$ of (4.1) belongs to $L^2(\mathcal{T}, \mathrm{d}V)$. Choosing this Hilbert space as the space of states in the connection representation, it follows that the transform, being simply the inner product, is a well defined, continuous map from the connection states to loop states. In particular, while in the connection representation the states Ψ are equivalence classes of functions on $\mathcal{T}$ (where two are equivalent if they differ by a set of measure zero), their images $\psi(\alpha)$ in the loop representation are genuine functions of homotopy classes of loops.

(ii) The transform is *faithful*. To see this, note first that for any choice of a measure M, $L^2(\mathcal{T}, \mathrm{d}V)$ provides an irreducible representation of the algebra of operators $\hat{T}^I$, defined by

$$\hat{T}^0(\boldsymbol{n}) = \cosh(\boldsymbol{n} \cdot \boldsymbol{a}) \qquad \hat{T}^1(\boldsymbol{n}) = -\mathrm{i}\hbar \sinh(\boldsymbol{n} \cdot \boldsymbol{a})\, \boldsymbol{n} \times \left(\frac{\partial}{\partial \boldsymbol{a}} + \tfrac{1}{2}\boldsymbol{\nabla} M\right). \tag{4.7}$$

In particular, this is true for our choice (4.6). Now, suppose the transform has a kernel $\mathcal{K}$. Then, $\mathcal{K}$ is a closed subspace of $L^2(\mathcal{T}, \mathrm{d}V)$ and, since (4.6) is of the type (3.9), it follows that $\mathcal{K}$ remains invariant under the action of the $\hat{T}^I$. Hence, $\mathcal{K}$ must be either the zero subspace or the full Hilbert space. It cannot be the full Hilbert space because, in particular, the elements $T^0(\alpha)$ of $L^2(\mathcal{T}, \mathrm{d}V)$ cannot lie in the kernel for any α since its norm is positive definite. Hence $\mathcal{K}$ must contain only the zero vector.

(iii) It follows from our discussion of the connection representation that for any positive M, the representation (4.7) of the $\hat{T}^I$-algebra on $L^2(\mathcal{T}, \mathrm{d}V = (\exp -M)\mathrm{d}V_\mathrm{o})$ is unitarily equivalent to the representation on $L^2(\mathcal{T}, \mathrm{d}V_\mathrm{o})$. Hence, in particular, it follows that the loop representations of the $\hat{T}^I$-algebra obtained by using different $c \geqslant 0$, and independent $\boldsymbol{q}_1$, $\boldsymbol{q}_2$ are also unitarily equivalent. In particular then, these 'genuine' loop representations are unitarily equivalent to the one constructed by Marolf [7].

The transform enables us to endow the loop states with an inner product and to represent the observables $\hat{T}^I$ directly on the loop states. The action (4.7) of the $\hat{T}^I(\alpha)$-operators translates to the loop representation yielding

$$\begin{aligned}
\big(\hat{T}^0(\boldsymbol{k})\psi\big)(\boldsymbol{n}) &= \tfrac{1}{2}\big(\psi(\boldsymbol{n}+\boldsymbol{k}) + \psi(\boldsymbol{n}-\boldsymbol{k})\big) \\
\big(\hat{T}^1(\boldsymbol{k})\psi\big)(\boldsymbol{n}) &= -\tfrac{1}{2}\mathrm{i}\hbar(\boldsymbol{k}\times\boldsymbol{n})\big(\psi(\boldsymbol{k}+\boldsymbol{n}) - \psi(\boldsymbol{k}-\boldsymbol{n})\big) + \tfrac{1}{4}\mathrm{i}c\hbar \sum_{i=1,2}(\boldsymbol{k}\times\boldsymbol{q}_i) \\
&\quad \times \big(\psi(\boldsymbol{k}+\boldsymbol{n}+\boldsymbol{q}_i) - \psi(\boldsymbol{k}+\boldsymbol{n}-\boldsymbol{q}_i) + \psi(\boldsymbol{k}-\boldsymbol{n}+\boldsymbol{q}_i) - \psi(\boldsymbol{k}-\boldsymbol{n}-\boldsymbol{q}_i)\big).
\end{aligned} \tag{4.8}$$

Using these expressions, one may check explicitly that the commutation relations that result from (2.8) and the algebraic conditions that arise from (2.9) are satisfied by the $\hat{T}^I$ in this loop representation.

Finally, let us exhibit the inner products between loop states. For simplicity, let us consider the subset of loop states $\{T^0_A(\boldsymbol{k}),\ \boldsymbol{k} \in \mathbb{R}^2/\mathbb{Z}_2\} \subset L^2(\mathcal{T}, m\,\mathrm{d}V_\mathrm{o})$, and, in the measure, set $\boldsymbol{q}_1 = (1,0)$ and $\boldsymbol{q}_2 = (0,1)$. The general expression for the scalar product is then

$$\begin{aligned}
\langle T^0(\boldsymbol{k}), T^0(\boldsymbol{n})\rangle &= 2\int_0^\infty \mathrm{d}a_1 \int_0^\infty \mathrm{d}a_2\, \mathrm{e}^{-c(\cosh a_1 + \cosh a_2)}\, T^0(\boldsymbol{k})T^0(\boldsymbol{n}) \\
&= F(k_1+n_1)F(k_2+n_2) + F(k_1-n_1)F(k_2-n_2) \\
&\quad + G(k_1+n_1)G(k_2+n_2) + G(k_1-n_1)G(k_2-n_2)
\end{aligned} \tag{4.9}$$

where

$$F(0) = K_0(c) \qquad F(1) = K_1(c)$$

$$\begin{aligned}
F(n) &= K_0(c)\left(2^{n-1} + n\sum_{k=1}^{n/2}(-1)^k \frac{1}{k}\binom{n-k-1}{k-1} 2^{n-2k-1}\right) \\
&\quad + \sum_{m=1}^{n/2}(2m-1)!!\frac{1}{c^m}K_m(c)\Bigg(2^{n-1}\binom{n/2}{n/2-m} \\
&\quad + n\sum_{k=1}^{(n/2)-m}(-1)^k\frac{1}{k}\binom{n-k-1}{k-1}2^{n-2k-1}\binom{n/2-k}{n/2-k-m}\Bigg) \qquad n \text{ even}
\end{aligned}$$

$$F(n) = K_1(c)\left(2^{n-1} + n\sum_{k=1}^{(n-1)/2}(-1)^k\frac{1}{k}\binom{n-k-1}{k-1}2^{n-2k-1}\right) \tag{4.10}$$

$$+\sum_{m=1}^{(n-1)/2}(2m-1)!!\frac{1}{c^m}K_{m+1}(c)\left(2^{n-1}\binom{(n-1)/2}{(n-1)/2-m}\right.$$

$$\left.+n\sum_{k=1}^{(n-1)/2-m}(-1)^k\frac{1}{k}\binom{n-k-1}{k-1}2^{n-2k-1}\binom{(n-1)/2-k}{(n-1)/2-k-m}\right) \qquad n \text{ odd}$$

$$G(0) = 0$$

$$G(n) = \mathrm{e}^{-c}\sum_{m=1}^{n}\frac{1}{c^m}\sum_{k=0}^{E((n-m)/2)}(-1)^k\binom{n-k-1}{k}2^{n-2k-1}\frac{(n-2k-1)!}{(n-2k-m)!}\,.$$

Here, the functions K_n are the modified Bessel functions of integer order [16]. It is important to notice that, unlike in the case of the 'time-like sector' [6], the scalar product is *not* proportional to $\delta_{k,n}$. Note also that, as expected, the above expressions diverge as $c \to 0$ (corresponding to the trivial measure $M = 0$), due to the divergence of the Bessel functions and of the terms proportional to c^{-m} in (4.14). Let us give a few special cases of (4.13) for illustration:

$$\begin{aligned}
\langle T^0(0,0), T^0(0,0)\rangle &= 2K_0(c)^2 \\
\langle T^0(1,0), T^0(1,0)\rangle &= \langle T^0(0,1), T^0(0,1)\rangle = 2K_0(c)\left(\frac{1}{c}K_1(c) + K_0(c)\right) \\
\langle T^0(1,0), T^0(0,0)\rangle &= \langle T^0(0,1), T^0(0,0)\rangle = 2K_0(c)K_1(c) \\
\langle T^0(1,0), T^0(0,1)\rangle &= 2K_1(c)^2 \\
&\vdots
\end{aligned} \tag{4.11}$$

Finally we note that, although the wavefunctions $T^0(k,n)$ do not form an orthonormal set, since the weight function $\mathrm{e}^{-M(a)}$ is everywhere positive on $\mathcal{T}$, there exists a sequence of polynomials $P_m(a_1)Q_n(a_2)$, where

$$P_m(a_1) = \sum_{k=1}^{m} P_{mk}T^0(k,0) \qquad Q_m(a_2) = \sum_{k=1}^{m} Q_{mk}T^0(0,k) \tag{4.12}$$

with constants P_{mk}, Q_{mk}, such that

$$\langle P_kQ_l, P_mQ_n\rangle = \delta_{km}\delta_{ln}\,. \tag{4.13}$$

Such a basis would still have a discrete labelling, but the direct geometric interpretation of the wavefunctions in terms of loops on T^2 would be lost.

One may object to the introduction of a non-trivial measure factor $\exp(-M)$ on the grounds that the trivial measure $\mathrm{d}a_1 \wedge \mathrm{d}a_2$ is distinguished by its modular invariance, i.e. invariance under the action of the modular group, whose generators act on the connection variables (a_1, a_2) according to

$$(a_1, a_2) \to (a_2, -a_1) \qquad (a_1, a_2) \to (a_1, a_1 + a_2)\,. \tag{4.14}$$

Modular invariance (i.e. invariance under large diffeomorphisms), however, is not a physical requirement of the 2 + 1-theory, and, in general, its imposition leads to orbifold singularities in the reduced configuration space [17]. Although our modified measures are not modular invariant, their corresponding quantum representations still allow for a unitary implementation of the modular group. The action of the generators on wavefunctions is given by

$$\begin{aligned}\Psi(a_1,a_2) &\to \mathrm{e}^{(M(a_1,a_2)-M(a_2,-a_1))/2}\,\Psi(a_2,-a_1)\\ \Psi(a_1,a_2) &\to \mathrm{e}^{(M(a_1,a_2)-M(a_1,a_1+a_2))/2}\,\Psi(a_1,a_1+a_2)\,.\end{aligned}\tag{4.15}$$

For our particular choice $M = c(T^0(1,0) + T^0(0,1))$ one obtains

$$\Psi(a_1,a_2) \to \Psi(a_2,-a_1) \qquad \Psi(a_1,a_2) \to \mathrm{e}^{(c/2)(T^0(0,1)-T^0(1,1))}\,\Psi(a_1,a_1+a_2)\,.\tag{4.16}$$

This action takes on a simple form in the loop representation only if one makes a change of basis (to one in which the operators $\hat{T}^0(\boldsymbol{k})$ are diagonal).

5. Conclusion

Let us begin with a summary of the main results. While the connection representation can be constructed without a volume element or a measure on $\mathcal{T}$, the loop transform does require this additional structure. The Teichmüller space $\mathcal{T}$ is equipped with a natural, Liouville volume element $\mathrm{d}V_o$. Unfortunately, because the traces of holonomies grow unboundedly in the coordinates which are canonical for the Liouville form, the integral in the loop transform defined using $\mathrm{d}V_o$ is, in general, ill defined. Our strategy was to exploit the freedom available in the choice of the volume element to introduce a damping factor $\exp(-M)$ ($M \geqslant 0$) and define the transform using $\mathrm{d}V = (\exp(-M))\mathrm{d}V_o$ instead. The transform is then well defined if M is chosen so that all the traces of holonomies T^0_α belong to $L^2(\mathcal{T}, \mathrm{d}V)$. The requirement that the observables $\hat{T}^I$ have manageable expressions in the loop representation further restricts the damping factor M: it has to be of the type $\sum b_i T(\alpha_i^o)$ for some real numbers b_i and homotopy generators $\{\alpha_i^o\}$. The key question then is whether the two requirements on the damping factor can be met simultaneously. In the genus $g = 1$ case, we saw that it was quite straightforward to achieve this. In the more general case, the issue remains open although the available freedom in the choice of constants and homotopy generators seems large enough to meet these conditions.

The solution we propose here does have an inelegant feature: the expressions of the operators $\hat{T}^1(\alpha)$ in the loop representation now involve not only the homotopy generator α labelling the operator but also the fiducial loops α_i^o we fixed to define the measure. Could we have avoided this by modifying the strategy slightly? For example, in the definition of the transform, $\psi(\alpha) = \int(\exp -M)\mathrm{d}V_o\, T^0(\alpha)\Psi$, could we not have constructed the damping factor M from the homotopy generator α itself, without introducing any fiducial generators α_i^o? This is a tempting strategy since it avoids all references to fiducial loops. However, it does not work, essentially because the transform no longer has the form of an inner product of Ψ with $T^0(\alpha)$. More specifically, in the resulting loop representation, it is not possible to express the action of even the $\hat{T}^0$ operators in a manageable way!

A second strategy [5] would be to avoid the loop transform altogether and introduce the loop representation *ab initio*. Thus, one may begin with the quantum algebra of the $\hat{T}^I$-operators and attempt to find a representation directly on a vector space of suitable functions $\psi(\alpha)$ of (homotopy classes of) loops. Unfortunately, any ansatz which avoids the introduction of the transform and reference to the connection representation faces two important problems. First, in such representations, it seems difficult to incorporate the numerous identities and inequalities satisfied by the loop variables [13]. Second, it is difficult to simply guess the class of 'suitable' functions $\psi(\alpha)$ one has to begin with. In the case of 2 + 1 gravity on a torus, for example, it would *a priori* seem natural to begin with functions $\psi(n) = \sum c_i \delta_{n,n_i}$, obtained by taking linear combinations of characteristic functions of homotopy classes. When the quantization program [3] is completed, however, one finds that the spectrum of all the $\hat{T}^0(n)$-operators is bounded between $(-1, 1)$ while classically, on the geometrodynamic sector, the classical $T^0(\alpha)(n)$ take values precisely in the complement of this interval! That is, harmless assumptions on the initial 'regularity' conditions end up having unexpected, physically important and often undesired consequences. In the example just described, the quantum theory can be constructed but it corresponds to the 'time-like sector' of the moduli space of flat connections which has no geometrodynamic analogue. More importantly, the 'correct' regularity conditions that will finally lead one to the desired sector may be quite involved and difficult to guess because one's favourite loop states—such as the characteristic functions of homotopy classes—may not belong to the physical Hilbert space. This is the case both for the representations presented in the last section and the ones found by Marolf [7]. All these problems are avoided if one constructs the loop representation via the loop transform.

2 + 1 gravity is a 'toy model' for the physical 3 + 1 theory. What lessons can one learn from its analysis? First, we found that, once appropriate care is taken in defining measures, 'old fashioned' loop representations *do* exist for the 2 + 1 theory. Results of [7] had been used by some to question the existence and utility of the loop representation in cases when the gauge group is non-compact. Our analysis removes these objections. It does point out, however, that even in absence of infinite-dimensional, field-theoretic problems the task of choosing physically interesting sectors can be a rather delicate issue in loop representations. In the connection representation of the 2 + 1 theory, restricting oneself to the geometrodynamic sector was straightforward: we simply restricted the wavefunctions to have support in this sector. In the loop representation, by contrast, the restriction is implemented by imposing different 'regularity conditions' on loop states which, in turn, lead to quite different inner products. Without recourse to the transform, it would have been hopeless to unravel this subtle intertwining between the physics of the representation and the mathematical regularity conditions. The second lesson therefore is that it would be 'safer' to construct the loop representation via the transform also in the 3 + 1 theory. The relation between the mathematical assumptions and their physical implications would then be more transparent and the numerous identities and inequalities [13] between traces of holonomies automatically incorporated in the loop representation. This, in turn, suggests that the construction of suitable measures on $\mathcal{A}/\mathcal{G}$ would be a central problem in the 3 + 1 theory as well. This is the third lesson. The details of the required strategies in 3 + 1 dimensions will, however, be quite different from those that have been successful in the present analysis. The 3 + 1-problem is both more difficult and easier. It is more difficult because the space $\mathcal{A}/\mathcal{G}$ is now *infinite*-dimensional. However, it is also easier because the gauge group $\mathbb{C}SU(2)$ of the 3 + 1 theory is the complexification of a *compact* group, and states in the connection representation are *holomorphic* functions of connections. It has

recently been shown [21] that the integration theory on $\mathcal{A}/\mathcal{G}$ developed for compact groups [4] admits a natural generalization to this case. This approach leads to measures of a quite different sort than the ones introduced in this paper and, in particular, does not refer to any fiducial loops.

Acknowledgments

We would like to thank Jorma Louko and Don Marolf for discussions and Steve Carlip for pointing out reference [14]. This work was supported in part by the NSF grant PHY93-96246, the Eberly research fund of Penn State University, and a DFG fellowship to RL.

Appendix

In this appendix we present some details on the genus-2 case. In particular, we will demonstrate that the problems encountered in the genus-1 case persist and therefore a measure with an appropriate damping factor is needed again.

The reduced configuration space for the genus-2 case is the six-dimensional Teichmüller space and may be parametrized globally by the Fenchel–Nielsen coordinates [14]. They are a set of length and angle coordinates of a pants decomposition of the genus-g surface. The surface is cut along a set of $3g$ simple geodesic loops and, at the ith cut, $l_i \in \mathbb{R}^+$ measures the intrinsic length of the border curve and $\tau_i \in \mathbb{R}$ the relative twisting angle of the opposite sides of the cut. As was shown by Wolpert [18], the Weil–Petersson symplectic form ω in these coordinates is simply given by $\omega = \sum_{i=1}^{3g-3} \mathrm{d}l_i \wedge \mathrm{d}\tau_i$.

To find a suitable damping factor, we need estimates on how fast the traces of holonomies diverge on $\mathcal{T}$. Thus, we have to express the loop holonomies as functions of the $[l_i, \tau_i]$. For $g \geqslant 2$ this task has been carried out by Okai [19].

We will adopt the notation of [19] and denote the six Fenchel–Nielsen parameters by $(l_{-\infty}, l_0, l_\infty, \tau_{-\infty}, \tau_0, \tau_\infty)$. Next one has to choose a set of six loops on Σ such that the traces of their holonomies are independent, i.e. parametrize the Teichmüller space locally. There are clearly many different ways of doing this. The simplest choice we have found is illustrated in figure A1.

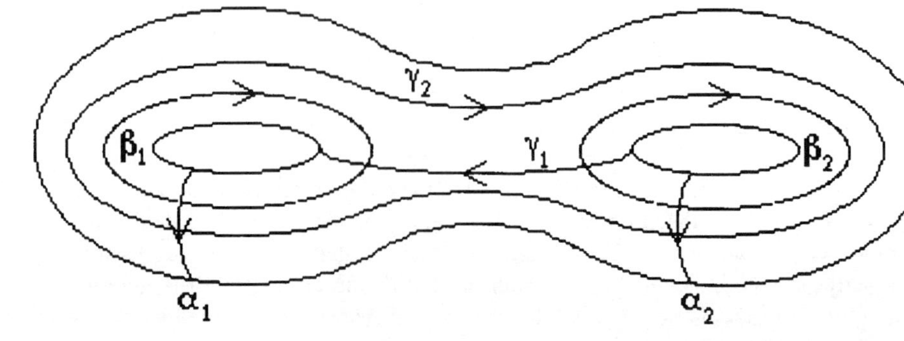

Figure A1.

The loops $(\alpha_1, \alpha_2, \beta_1, \beta_2)$ are the usual homotopy generators, and in addition we have the two loops (γ_1, γ_2). The normalized traced holonomies of these loops are the following functions of (l_i, τ_i):

$$\begin{aligned}
&T^0(\alpha_1) = \cosh\frac{l_{-\infty}}{2} \qquad T^0(\alpha_2) = \cosh\frac{l_\infty}{2} \qquad T^0(\gamma_1) = \cosh\frac{l_0}{2}\\
&T^0(\beta_1) = \sinh\frac{\tau_{-\infty}}{2}\sinh\frac{\tau_0}{2} + \cosh s_\infty \cosh\frac{\tau_{-\infty}}{2}\cosh\frac{\tau_0}{2}\\
&T^0(\beta_2) = \sinh\frac{\tau_\infty}{2}\sinh\frac{\tau_0}{2} + \cosh s_{-\infty} \cosh\frac{\tau_\infty}{2}\cosh\frac{\tau_0}{2}\\
&T^0(\gamma_2) = \sinh\frac{\tau_{-\infty}}{2}\sinh\frac{\tau_\infty}{2} + \cosh s_0 \cosh\frac{\tau_{-\infty}}{2}\cosh\frac{\tau_\infty}{2}
\end{aligned} \tag{A.1}$$

where the length parameters $s_{-\infty}$, s_0 and s_∞ are functions of the l_i alone,

$$\begin{aligned}
\cosh s_{-\infty} &= \frac{\cosh(l_\infty/2)\cosh(l_0/2) + \cosh(l_{-\infty}/2)}{\sinh((l_\infty)/2)\sinh(l_0/2)}\\
\cosh s_0 &= \frac{\cosh(l_\infty/2)\cosh(l_{-\infty}/2) + \cosh(l_0/2)}{\sinh(l_\infty/2)\sinh((l_{-\infty})/2)}\\
\cosh s_\infty &= \frac{\cosh((l_{-\infty})/2)\cosh(l_0/2) + \cosh(l_\infty/2)}{\sinh((l_{-\infty})/2)\sinh(l_0/2)}\,.
\end{aligned} \tag{A.2}$$

Note that the particular T^0 chosen in (A.1), unlike the Fenchel–Nielsen variables, are not good global coordinates on Teichmüller space, since a simultaneous sign change of the τ_i leads to the same values for the independent loop variables. It is obvious from (A.1) and (A.2) that a problem similar to that encountered in the genus-1 case arises here too (and, in fact, for any higher genus), since none of the loop variables in (A.1) are square-integrable with respect to the measure $\Pi_i\, \mathrm{d}l_i\, \mathrm{d}\tau_i$. Moreover, there is now an additional problem for small l, namely, another divergence in the loop transform coming from terms like $\cosh s_\infty$. This problem also occurs in some calculations in string theory (see, for example, [20]), and may be dealt with by introducing a cut-off for small lengths.

Unfortunately, it appears difficult to extract from (A.1) and (A.2) estimates for the asymptotic growth of traces of holonomies around arbitrary elements of the homotopy group. This happens because—although possible, in principle—it is in practice difficult to express arbitrary $T^0(\alpha)$ as functions of the independent set. Thus, while it is tempting to choose the damping factor simply as

$$\exp\left(-\sum_{i=1}^{2}\left[T^0(\alpha_i) + T^0(\beta_i) + T^0(\gamma_i)\right]\right) \tag{A.3}$$

there is no guarantee that this damping will suffice to make the loop transform well defined for any α in the homotopy group.

Even if one restricts the quantum wavefunctions to sums of tensor products of the six basic loop functions $T^0(\alpha_1), \ldots, T^0(\gamma_2)$, the situation is still non-trivial. Since some of those functions themselves have a complicated, coupled dependence on the Fenchel–Nielsen parameters, it is not immediately clear whether (A.3) gives a sufficient damping in all of the asymptotic regions. One could probably get more control over the measure by re-expressing it in terms of the independent variables T^0, but we have not found a way of rewriting the

Jacobian of this transformation in terms of the independent set $\{T^0\}$. The algebraic problems encountered here are not insurmountable, however, a detailed case-by-case analysis appears to be necessary to obtain a complete control on the asymptotic behaviour of the traces of holonomies around arbitrary loops. Only then will one be able to construct explicit measures that make the loop transforms for higher-genus surfaces well defined.

References

[1] Gambini R, and Trias A 1986 Gauge dynamics in the C-representation *Nucl. Phys.* B **278** 436–48

[2] Goldberg J, Lewandowski J and Stornaiolo C 1992 Degeneracy in loop variables *Commun. Math. Phys.* **148** 377

Ashtekar A and Lewandowski J 1993 Completeness of Wilson loop functionals on the moduli space of $SL(2, \mathbb{C})$ and $SU(1, 1)$ connections *Class. Quantum Grav.* L69–74

[3] Ashtekar A 1991 *Lectures on Non-Perturbative Canonical Gravity* (Singapore: World Scientific)

[4] Ashtekar A, Marolf D and Mourão J Integration on the space of connections modulo gauge transformations *Proc. Lanczos Centenary Conf.* ed D Brown *et al* (SIAM)

[5] Rovelli C and Smolin L 1990 Loop space representation of quantum general relativity *Nucl. Phys.* B **331** 80–152

[6] Ashtekar A, Husain V, Rovelli C, Samuel J and Smolin L 1989 2 + 1 gravity as a toy model for the 3 + 1 theory *Class. Quantum Grav.* **6** L185–93

Smolin L Loop representation for quantum gravity in 2 + 1 dimensions *Knots, Topology and Quantum Field Theory (Proc. 12th Johns Hopkins Workshop)*

Ashtekar A Lessons from 2 + 1 dimensional quantum gravity *Strings 90* ed R Arnowitt *et al* (Singapore: World Scientific)

[7] Marolf D M 1993 Loop representations for 2 + 1 gravity on a torus *Class. Quantum Grav.* **10** 2625–47

[8] Witten 1989 E 2 + 1 dimensional gravity as an exactly soluble system *Nucl. Phys.* B **311** 46–78

[9] Carlip S 1990 Observables, gauge invariance and time in (2 + 1)-dimensional quantum gravity *Phys. Rev.* D **42** 2647–54

Louko J and Marolf D M 1994 The solution space of 2 + 1-gravity on $\mathbb{R} \times T^2$ in Witten's connection formulation *Class. Quantum Grav.* 11 311–30

[10] Moncrief V 1989 Reduction of the Einstein equations in 2 + 1 dimensions to a Hamiltonian system over Teichmüller space *J. Math. Phys.* **30** 2907–14

Mess G Lorentz spacetimes of constant curvature *Preprint* Institut des Hautes Études Scientifiques IHES/M/90/28

[11] Loll R Independent loop invariants for 2 + 1 gravity *CGPG preprint* in preparation

[12] Giles R 1981 Reconstruction of gauge potentials from Wilson loops *Phys. Rev.* D **24** 2160–8

[13] Loll R 1993 Loop variable inequalities in gravity and gauge theory *Class. Quantum Grav.* **10** 1471–6

[14] Abikoff W 1980 The real analytic theory of Teichmüller space *Lecture Notes in Mathematics* **820** (Berlin: Springer)

[15] Ashtekar A and Romano J 1989 Chern–Simons and Palatini actions and 2 + 1 gravity *Phys. Lett.* **229B** 56–60

[16] Gradsteyn I S and Ryzhik I M 1980 *Table of Integrals, Series, and Products* corrected and enlarged edition (San Diego: Academic)

[17] Moncrief V 1993 Recent advances in ADM reduction *Directions in General Relativity* vol 1 ed B L Hu *et al* (Cambridge: Cambridge University Press) pp 231–43

[18] Wolpert S 1982 On the Weil–Petersson geometry of the moduli space of curves *Ann. Math.* **115** 501–28

[19] Okai T 1992 An explicit description of the Teichmüller space as holonomy representations and its applications *Hiroshima Math. J.* **22** 259–71

[20] D'Hoker E and Phong D H 1988 The geometry of string perturbation theory *Rev. Mod. Phys.* **60** 917–1065

[21] Ashtekar A, Lewandowski J, Marolf D, Mourão J and Thiemann T Coherent state transform on the space of connections *CGPG preprint*

Class. Quantum Grav. **11** (1994) 2435–2442. Printed in the UK

Scattering of fermions off dilaton black holes

Bjørn Jensen†§ and Ulf Lindström‡¶

† Institute of Physics, University of Oslo, PO Box 1048, N-0316 Blindern, Oslo 3, Norway

‡ ITP, University of Stockholm, Box 6730, S-113 85, Stockholm, Sweden

Received 7 April 1994, in final form 4 August 1994

Abstract. We discuss how various properties of dilaton black holes depend on the dilaton coupling constant a. In particular we investigate the a-dependence of certain mass parameters both outside and in the extremal limit and discuss their relation to thermodynamical quantities. To further illuminate the role of the coupling constant a we look at a massless point particle in a dilaton black hole geometry as well as the scattering of (neutral) fermions. In this latter case we find that the scattering potential vanishes for the zero angular momentum mode which seems to indicate a catastrophic deradiation when $a > 1$.

PACS numbers: 0365S, 0420C, 0460, 0462

Dilaton black holes are studied because they arise as solutions to the low-energy string effective field theory [1], and because they shed new light on the old question of black holes as elementary particles [2].

The dilaton black holes are solutions to field equations that describe a scalar (dilaton) field and an electromagnetic field coupled to Einstein gravity. It has proven useful to let the dilaton coupling depend on a non-negative parameter a. The value $a = 0$ corresponds to the usual Einstein theory whereas the coupling $a = 1$ is the value dictated by string theory. Furthermore, the value $a = \sqrt{3}$ seems to be special. For this value of a rotating solutions [3] as well as a string black hole [4] have been found. It also arises in dimensional reduction of $d = 5$ supergravity studied in the context of supersymmetric solitons [5]. In general many features of the solution (but not all) depend on the value of a. The present work corroborates this picture.

The main problem treated in this paper is the scattering of spinning particles off of dilaton black holes. We derive the corresponding potential and discuss its behaviour as a function of a. In particular we find that it generally diverges at the horizon in the extremal limit for $a > 1$. For $a = \sqrt{3}$ there is an additional pole, interestingly enough, although the relation to the above-mentioned rotating solution, string solution or soliton solutions is unclear.

This paper also contains a derivation and discussion of certain mass parameters: the ADM mass, the Tolman mass and the mass of the black hole. Although we relate the usual definition of the temperature and entropy of the black hole to its ADM mass, we use the vanishing of the black hole mass as the main indication for the vanishing of the black hole in the extremal limit. This is because it has been argued that the thermodynamical description breaks down in that limit [2, 6].

§ E-mail: BJensen@boson.uio.no.

¶ E-mail: UL@vana.physto.se and ulfl@boson.uio.no.

The action we consider is

$$S = \int \mathrm{d}^4x\sqrt{-g}(-R + 2(\nabla\Psi)^2 + \mathrm{e}^{-2a\Psi}F^2) \tag{1}$$

where R is the Ricci curvature scalar, Ψ the dilaton field, F^2 the square of the electromagnetic field strength and a a non-negative real parameter. The corresponding field equations are

$$\nabla_\mu(\mathrm{e}^{-2a\Psi}F^{\mu\nu}) = 0 \tag{2}$$

$$\nabla^2\Psi + \tfrac{1}{2}a\mathrm{e}^{-2a\Psi}F^2 = 0 \tag{3}$$

$$R_{\mu\nu} = 2\nabla_\mu\Psi\nabla_\nu\Psi + 2\mathrm{e}^{-2a\Psi}F_{\mu\,\rho}F_\nu{}^\rho - \tfrac{1}{2}g_{\mu\nu}\mathrm{e}^{-2a\Psi}F^2. \tag{4}$$

The line element for an electrically charged black hole solution is [7, 8]

$$\mathrm{d}s^2 = \Delta\sigma^{-2}\mathrm{d}t^2 - \sigma^2(\Delta^{-1}\mathrm{d}r^2 + r^2\mathrm{d}\Omega_2^2) \tag{5}$$

with Δ and σ given by

$$\Delta = \left(1 - \frac{r_+}{r}\right)\left(1 - \frac{r_-}{r}\right) \qquad \sigma^2 = \left(1 - \frac{r_-}{r}\right)^{2a^2/(1+a^2)}. \tag{6}$$

Here $\mathrm{d}\Omega_2^2$ is the metric in the unit 2-sphere S^2, r_+ and r_- are implicitly given by

$$r_+r_- = Q^2(1+a^2) \qquad (2M - r_+)\left(\frac{1+a^2}{1-a^2}\right) = r_-. \tag{7}$$

M represents the metric mass and Q is the electric charge on the hole. The electric field is given by

$$\underline{F} = \frac{Q}{r^2}\underline{dt} \wedge \underline{dr} \tag{8}$$

and the dilaton by

$$\mathrm{e}^{2a\Psi} = \sigma^2. \tag{9}$$

The scalar charge Σ is defined as $\Psi \sim \Psi_0 - \Sigma/r$ for large r which gives

$$r_- = \frac{(1+a^2)\Sigma}{a} \Rightarrow r_+ = a\frac{Q^2}{\Sigma}. \tag{10}$$

The metric mass can thus be written

$$M = \frac{a^2Q^2 + (1-a^2)\Sigma^2}{2a\Sigma}. \tag{11}$$

Some aspects of these solutions can be better understood when one analytically continues the Schwarzschild coordinates. To this end we define an advanced null coordinate v by $v = t + r^*$ where

$$\frac{\mathrm{d}r^*}{\mathrm{d}r} = \frac{\sigma^2}{\Delta}. \tag{12}$$

In terms of this coordinate we easily obtain the advanced Eddington–Finkelstein form of the metric equation (5)

$$ds^2 = \frac{\Delta}{\sigma^2}dv^2 - 2dvdr - r^2\sigma^2 d\Omega_2^2. \tag{13}$$

It is clear that this metric in general is free from singularities when $a \leqslant 1$. Off extremality ($r_- \neq r_+$) the analytic continuation is singular at $r = r_-$ when $a > 1$. However this singularity generally disappears in the extremal limit, i.e. in this limit the metric is non-singular all the way down to the singularity at $r = 0$.

We finally give the Kruskal form of the metric. Defining the (U, V) coordinates in the usual way we get

$$ds^2 = -32M^3 \frac{d}{dr} e^{-r^*/2M}(dU^2 - dV^2) - r^2\sigma^2 d\Omega_2^2. \tag{14}$$

From this expression one can easily convince oneself that the (U, V) part of the metrics of the $a = 0$ and $a = 1$ solutions are in fact identical (when $2M$ is replaced by r_+ in the $a = 1$ solution). Hence, when $a = 1$ the sole effect of the singularity at $r = r_-$ (which coincides with the event horizon in the extremal limit), is the vanishing of the angular part of the metric. The structure of the $a \neq 1$ extensions is much more complicated and will not be considered here.

Now we want to discuss the energy content of the solutions and to this end we calculate certain mass parameters. Let ξ^a be a time translation Killing vector field which is timelike near infinity such that $\xi^a\xi_a = 1$ and which has vanishing norm $\xi^a\xi_a = 0$ at the horizon. Let S denote the region outside the event horizon. The boundary ∂S of S is taken to be the event horizon ∂B of the black hole and a 2-surface ∂S_∞ at infinity. Due to the asymptotic flatness of the dilaton black hole solutions it follows that the ADM mass of the black hole configuration is given by [9]

$$M^\infty = \frac{1}{8\pi}\int_S (2T^a{}_b - T\delta^a{}_b)\xi^b d\Sigma_a + \frac{1}{4\pi}\int_{\partial B} \xi^{a;b} d\Sigma_{ab} \equiv M^M + M^H. \tag{15}$$

The last integral is just the surface gravity κ multiplied with the surface area A^H of the event horizon. M^M and M^H are naturally interpreted as the gravitational mass of the matter distribution outside the event horizon and the gravitational mass of the black hole, respectively. In the following we will set $\boldsymbol{\xi} = \partial_t$ since this vector always vanishes on the event horizon independent of the value of a. It then follows that the first integral M^M is the usual Tolman mass of the matter outside the event horizon [10].

From the action equation (1) we derive the energy–momentum tensor $T_{\mu\nu}$ for the dilaton and the electromagnetic field

$$T^t{}_t = \frac{\sigma^2}{\Delta}\left(\frac{\Delta Q^2}{\sigma^4 r^4} + \frac{\Sigma^2\Delta^2}{\sigma^4 r^2(r - r_-)^2}\right) \tag{16}$$

$$T^r{}_r = -\frac{\Delta}{\sigma^2}\left(\frac{\Sigma^2}{r^2(r - r_-)^2} - \frac{Q^2}{\Delta r^4}\right) \tag{17}$$

$$T^\Omega{}_\Omega = -\frac{Q^2}{\sigma^2 r^4} + \frac{\Delta\Sigma^2}{\sigma^2 r^2(r - r_-)^2}. \tag{18}$$

The Tolman mass density is then given by

$$\mathcal{M}^M = T^t{}_t - T^r{}_r - 2T^\Omega{}_\Omega = \frac{2Q^2}{\sigma^2 r^4}. \tag{19}$$

This expression reduces to the usual one in the Reissner–Nordström geometry when a is set to zero. From the above expression it follows that $\mathcal{M}^M$ generally diverges on $r = r_-$ when $a > 0$ thus giving rise to a curvature singularity there. However, the total energy outside the event horizon is always positive and finite. It is rather surprising that it is only the electric field that contributes to the gravitational energy. The dilaton contribution cancels due to the large pressures in the angular directions.

It is interesting to relate the temperature and the entropy of the black hole to its ADM mass. Let $\boldsymbol{N}$ denote a second Killing vector field (normalized to unity $N^a N_a = -1$) orthogonal to the event horizon. The surface gravity κ can then be written $\kappa^2 = N_b \xi^a \nabla_a \xi^b$ [9]. The area of the event horizon is $A^H = 4\pi r_+^2 \sigma^2 (r = r_+)$. This implies that the contribution to the total gravitational mass of the system from behind the event horizon is

$$M^H = 2\pi r_+ \left(1 - \frac{r_-}{r_+}\right). \tag{20}$$

This quantity always vanishes when $r_- = r_+$. In this sense the black hole disappears completely in the extremal limit. The red-shifted temperature T^H measured by a distant observer is formally given by

$$T^H = \frac{\hbar}{2\pi k_B}|\kappa| = \frac{\hbar}{4\pi k_B r_+}\left(1 - \frac{r_-}{r_+}\right)^{(1-a^2)/(1+a^2)} = \frac{\hbar}{4\pi k_B r_+}\left(\frac{M^H}{2\pi r_+}\right)^{(1-a^2)/(1+a^2)}. \tag{21}$$

From this it is clear that T^H vanishes when $a < 1$, it is finite when $a = 1$ and infinite whenever $a > 1$ in the extremal limit. By simply using the event horizon area of the black hole as a measure of the entropy content S^H of the black hole we get

$$S^H = \tfrac{1}{4}A^H = \pi\left[r_+\left(\frac{M^H}{2\pi r_+}\right)^{a^2/(1+a^2)}\right]^2 = \frac{\hbar}{8\pi k_B}\frac{M^H}{T^H}. \tag{22}$$

It follows that the entropy vanishes in the extremal state independent of the value of a. It has been argued that the thermodynamic interpretation breaks down near the extremal limit [2, 6]. Nevertheless we feel that the above relations give an indication of the behaviour of the thermodynamical entities in this limit, but we empasize the vanishing of M^H.

The qualitative thermodynamical properties of the dilaton holes in the extremal limit depend crucially on the value of the coupling constant a. It is interesting to investigate whether this is reflected in the dynamics of particles propagating in these geometries. Since the case of a real massless scalar field is treated in [2] we shall consider a classical photon and the propagation of neutrinos.

Consider a photon propagating in a general dilaton black hole background with fixed θ coordinate. Define $E = p_t$ and $L = p_\phi$. Then $g_{\mu\nu}p^\mu p^\nu = 0$ implies

$$\dot{r}^2 = E^2 - \frac{L^2}{r^2}\left(1 - \frac{r_+}{r}\right)\left(1 - \frac{r_-}{r}\right)^{(1-3a^2)/(1+a^2)} \equiv E^2 - V^2. \tag{23}$$

From this it follows that

$$\ddot{r} = -\frac{1}{2}\frac{\mathrm{d}V^2}{\mathrm{d}r} = \frac{L^2}{r^3}\left(1-\frac{r_-}{r}\right)^{(1-3a^2)/(1+a^2)}\left[1-\frac{3r_+}{2r}-\frac{r_-}{2r}\left(\frac{1-3a^2}{1+a^2}\right)\left(\frac{r-r_+}{r-r_-}\right)\right]. \tag{24}$$

In the extremal limit the analysis of this expression is particulary simple. Due to the exponent appearing in this expression it is natural to concentrate the attention to three a-parameter intervals: I, $0 \leqslant a < 1/\sqrt{3}$; II, $a = 1/\sqrt{3}$ and III, $a > 1/\sqrt{3}$. The second derivative $\ddot{r}$ may vanish outside r_+ when a takes values in the parameter intervals I and II while turning points only exist outside r_+ in region III if we have the additional restriction $a < 1$. When $a = 1$ the turning point coincides with r_+ and when $a > 1$ the turning point is behind r_+. Hence, in the extremal limit closed circular photon orbits can only exist outside $r = r_+$ when $a < 1$.

Another interesting question is whether $\ddot{r}$ may diverge. In I and II this function is finite for all non-zero values of r. When $1/\sqrt{3} < a < 1$ $\ddot{r}$ diverges on r_+. With $a = 1$ the radial acceleration is finite on r_+ while for $a > 1$ it diverges again for this value of the radial coordinate. The significance of the $a = 1/\sqrt{3}$ solution is somewhat unclear. It does not seem to have any remarkable thermodynamical properties in the extremal limit. However, it is worth noting that only $a = 1$ and not $a = 1/\sqrt{3}$ appears as a special value for the scalar field dynamics in the extremal limit [2].

Let us now turn to the scattering of particles with spin. In flat space fermions can be described by two, in general coupled, equations of two 2-component spinors P^A and $\overline{Q}_{\dot{B}}$. Following [11] we write $P^0 \equiv F_1$, $P^1 \equiv F_2$, $\overline{Q}^{\dot{1}} \equiv G_1$ and $\overline{Q}^{\dot{0}} \equiv -G_2$. In the Newman–Penrose formalism the equations governing the dynamics of massless fermions can be written as

$$(D+\epsilon-\rho)F_1 + (\delta^* + \pi - \alpha)F_2 = 0 \tag{25}$$

$$(\Delta+\mu-\gamma)F_2 + (\delta+\beta-\tau)F_1 = 0 \tag{26}$$

$$(D+\epsilon^*-\rho^*)G_2 - (\delta+\pi^*-\alpha^*)G_1 = 0 \tag{27}$$

$$(\Delta+\mu^*-\gamma^*)G_1 - (\delta^*+\beta^*-\tau^*)G_2 = 0. \tag{28}$$

Using the standard null-frame notation $(\boldsymbol{l}, \boldsymbol{n}, \boldsymbol{m}, \boldsymbol{m}^*)$ it follows that $D = \boldsymbol{l}$, $\Delta = \boldsymbol{n}$, $\delta = \boldsymbol{m}$ and $\delta^* = \boldsymbol{m}^*$. It is also assumed that $l^a n_a = -m^a m^*_a = -1$ and $l^a m_a = n^a m_a = 0$. We will let $\boldsymbol{l}$ represent the 4-velocity of an outwards radially moving massless and electrically uncharged particle with vanishing orbital angular momentum ($L = 0$)

$$l^a = \frac{\sigma^2}{\Delta}\delta^a_0 + \delta^a_1. \tag{29}$$

Further, n^a is identified with the 4-velocity of the corresponding inwards moving particle

$$n^a = \tfrac{1}{2}\delta^a_0 - \frac{\Delta}{2\sigma^2}\delta^a_1. \tag{30}$$

We also define the complex vector m^a ;

$$m^a = \frac{1}{\sqrt{2}r\sigma}\left(\delta^a_2 + \frac{\mathrm{i}}{\sin\theta}\delta^a_3\right). \tag{31}$$

The non-zero spin coefficients are then given by

$$\rho = \frac{-1}{r\sigma}\partial_r(r\sigma) \qquad \mu = \frac{-\Delta}{2r\sigma^3}\partial_r(r\sigma) \tag{32}$$

$$\gamma = \tfrac{1}{4}\partial_r\left(\frac{\Delta}{\sigma^2}\right) \qquad \alpha = -\beta = \frac{-\cot\theta}{2\sqrt{2}r\sigma}. \tag{33}$$

Let $f_1 = r\sigma F_1$, $f_2 = F_2$, $g_1 = G_1$ and $g_2 = r\sigma G_2$. Also define

$$\hat{L} = \partial_\theta + \frac{\mathrm{i}}{\sin\theta}\partial_\phi + \tfrac{1}{2}\cot\theta \tag{34}$$

$$\hat{V} = \mathrm{i}\omega\frac{\sigma^2}{\Delta} + \partial_r \tag{35}$$

$$\hat{W} = \tfrac{1}{2}\mathrm{i}\omega - \frac{\Delta}{2\sigma^2}\partial_r - \frac{\Delta}{2r\sigma^3}\partial_r(r\sigma) - \tfrac{1}{4}\partial_r\left(\frac{\Delta}{\sigma^2}\right). \tag{36}$$

We concentrate our attention on the F-equations since the G-equations are just the complex conjugate of these. Then equation (25) becomes

$$\hat{V}f_1 + 2^{-1/2}\hat{L}^\dagger f_2 = 0 \tag{37}$$

and equation (26) can similarly be written

$$r^2\sigma^2\hat{W}f_2 + 2^{-1/2}\hat{L}f_1 = 0. \tag{38}$$

By utilizing that

$$\hat{W} = -\frac{\sqrt{\Delta}}{2r\sigma^2}\hat{V}^\dagger(r\sqrt{\Delta}) \tag{39}$$

equation (38) can be brought to the more useful form

$$r\sqrt{\Delta}\hat{V}^\dagger(r\sqrt{\Delta}f_2) - 2^{1/2}\hat{L}f_1 = 0. \tag{40}$$

Define

$$f_1 = R_{1/2}(r)S_{1/2}(\theta)\mathrm{e}^{\mathrm{i}\omega t}\mathrm{e}^{\mathrm{i}m\phi} \qquad g_1 = R_{-1/2}(r)S_{1/2}(\theta)\mathrm{e}^{\mathrm{i}\omega t}\mathrm{e}^{\mathrm{i}m\phi} \tag{41}$$

$$f_2 = R_{-1/2}(r)S_{-1/2}(\theta)\mathrm{e}^{\mathrm{i}\omega t}\mathrm{e}^{\mathrm{i}m\phi} \qquad g_2 = R_{1/2}(r)S_{-1/2}(\theta)\mathrm{e}^{\mathrm{i}\omega t}\mathrm{e}^{\mathrm{i}m\phi} \tag{42}$$

where m takes on positive and negative integer values. Then the two equations above separate into

$$\hat{V}R_{1/2} - \lambda R_{-1/2} = 0 \tag{43}$$

$$r^2\sigma^2\hat{W}R_{-1/2} + \lambda R_{1/2} = 0 \tag{44}$$

and

$$\hat{L}^\dagger S_{-1/2} + \sqrt{2}\lambda S_{1/2} = 0 \tag{45}$$

$$\hat{L}S_{+1/2} - \sqrt{2}\lambda S_{-1/2} = 0 \tag{46}$$

where λ is the separation constant. Define $P_{-1/2} = \sqrt{2}r\sqrt{\Delta}R_{-1/2}$ and $P_{1/2} = R_{1/2}$ and let $\sqrt{2}\lambda \to \lambda$. The radial equations then become

$$r\sqrt{\Delta}\hat{V}P_{1/2} = \lambda P_{-1/2} \tag{47}$$

$$r\sqrt{\Delta}\hat{V}^{\dagger}P_{-1/2} = \lambda P_{1/2}. \tag{48}$$

Further define two functions $Z_{\pm}$ by $Z_{\pm} = P_{1/2} \pm P_{-1/2}$. Then a one-dimensional wave equation in r^* follows

$$\left(\frac{\mathrm{d}^2}{\mathrm{d}r^{*2}} + \omega^2\right) Z_{\pm} = V_{\pm} Z_{\pm} \tag{49}$$

where

$$V_{\pm} = \lambda^2 \frac{\Delta}{r^2\sigma^4} \pm \lambda \frac{\mathrm{d}}{\mathrm{d}r^*}\left(\frac{\sqrt{\Delta}}{r\sigma^2}\right). \tag{50}$$

Solving this equation for $Z_{\pm}$ we can easily obtain the solutions for $P_{\pm 1/2}$. When $a = 0$ and $Q = 0$ this expression reduces to the one found in the Schwarzschild geometry. According to standard scattering theory V_- and V_+ will give rise to the same reflection and transmission coefficients [11]. It is easily seen that $V_{\pm}$ vanishes on $r = r_+$ in the extremal limit when $a < 1$ and is finite outside. When $a = 1$ it is finite on $r = r_+$. When $a > 1$ the potential is singular at the horizon in the extremal limit. This is exactly what happens in the scalar field case. In addition to this singularity structure the potential above exhibits an additional pole at $r = r_+$ for $a = \sqrt{3}$. This pole did not appear neither in the point particle case nor in the dynamics of the real scalar field. It is interesting to note that the value $a = \sqrt{3}$ arises in Kaluza-Klein dilaton black hole solutions [3–5] and string solutions [4].

The equations for the $S_{\pm 1/2}$ functions are independent of a. It follows that these functions are identical to the corresponding functions in the Schwarzschild geometry, i.e. spherical harmonics $Y_{lm}(\theta, \phi)$. The regularity of these functions on the unit sphere implies that $\lambda = l(l+1)$. Hence, for the lowest angular momentum mode $l = 0$ it follows that $V_{\pm}$ vanishes. This feature is also present in the massless point particle case but it is absent in the dynamics of a real massless scalar field. An immediate consequence of the vanishing of $V_{\pm}$ is that the outgoing Hawking radiation for $a > 1$ holes in the extremal state apparently gives rise to a divergent change of the gravitational mass M^H relative to infinity

$$\left.\frac{\mathrm{d}M^H}{\mathrm{d}t}\right|_{l=0} = \int \frac{\Gamma(l=0,\omega)\omega d\omega}{\mathrm{e}^{\omega/T^H} + 1} \to \infty. \tag{51}$$

Since $V_{\pm} = 0$ this implies that the transmission coefficient $\Gamma(l = 0, \omega)$ equals unity in the extremal limit and seems to indicate a catastrophic deradiation of these configurations. We interpret the result to mean that we are outside the domain of validity of the semiclassical approximation and that we are no longer free to ignore the back reaction on the geometry. However, it would seem that we could retreat sufficiently far away from extremality for the thermodynamical results to be valid and still obtain a very large rate for the outgoing Hawking radiation. Naively this would seem to eventually result in a negative gravitational mass of the black hole when M^H is sufficiently small†.

† Note that although the potential (50) vanishes also for a Reissner–Nordström black hole there is no catastrophic mass loss in that case. This is because the temperature of those holes vanishes in the extremal limit.

It is interesting to note that the zero angular momentum mode of the massless uncharged fermion destroys the picture of extreme black holes as elementary particles which was contemplated in [2]. This is due to the fact that the vanishing of the effective potential implies that such a fermion will be absorbed by the extremal structure with unit probability in sharp contrast with the usual view of elementary particles as structure-less entities. It is also worth noting that it is essential for the fermion to be uncharged since the infall of a electrically charged fermion leads to the existence of a strong electric field 'tail' behind the particle which will hinder the particle being absorbed by the black hole [12].

Acknowledgments

BJ thanks the University of Stockholm for hospitality during the time in which parts of this work were carried out. UL acknowledges partial support from the NFR under grant no 4038-300 and from NorfA under grant no 93.35.088/00.

References

[1] Witten E 1991 *Phys. Rev.* D **44** 314–24
[2] Holzhey C F and Wilczek F 1992 *Nucl. Phys.* B **380** 447–77
[3] Frolov V P, Zelnikov A I and Bleyer U 1987 *Ann. Phys., Lpz.* **44** 371–7
[4] Duff M J, Khuri R, Minasian R and Rahmfeld J 1993 New black hole, string and membrane solutions of the four-dimensional heterotic string *Texas A& M Preprint* CTP/TAMU-49/93 and *Cern Preprint* CERN-TH.7088/93
[5] Gibbons G W, Kastor D, London, L A J and Townsend P K 1993 Super symmetric self-gravitating solitons *Cambridge Preprint* DAMTP R/93/27 and *Massachusetts Preprint* UMMEP-371
[6] Preskill J, Schwarz P, Shapere A, Trivedi S and Wilczek F 1991 *Mod. Phys. Lett.* **6A** 2353–61
[7] Gibbons G W and Maeda K 1988 *Nucl. Phys.* B **298** 741–75
[8] Garfinkle D and Horowitz G T 1991 *Phys. Rev.* D **43** 3140–3
[9] Bardeen J M, Carter B and Hawking S W 1973 *Commun. Math. Phys.* **31** 161–70
[10] Tolman R C 1930 *Phys. Rev.* **35** 875–95
[11] Chandrasekhar S 1992 *The Mathematical Theory of Black Holes* (Oxford: Oxford University Press,)
[12] Alford M G and Strominger A 1992 *Phys. Rev. Lett.* **69** 563–6

Class. Quantum Grav. **11** (1994) 2443–2453. Printed in the UK

A version of quantum measure in Regge calculus in three dimensions

V Khatsymovsky†‡
Institute of Theoretical Physics, Box 803, S-751 08 Uppsala, Sweden

Received 3 November 1993, in final form 22 February 1994

Abstract. We propose a version of quantum measure in 3D Regge calculus based on principles of analyticity, canonical quantization and equivalence of different coordinates. This measure is specified by the following requirement. Suppose a Regge manifold is made continuous in some direction by tending corresponding dimensions of simplices to zero. Then the measure of interest is required to tend to that corresponding to the canonical quantization with continuous coordinate playing the role of time. (If a continuous coordinate has spacelike signature, the same is required to hold for analytical continuation of the measure of interest obtained by multiplying link-lengths along the considered direction by $\sqrt{-1}$.) By equivalence of different coordinates this property of measure is assumed to hold for any coordinate being made continuous.

We construct the above measure in three dimensions using the tetrad-connection form of Regge calculus. It is remarkable that in the 3D case such a measure does really exist and, moreover, is defined by the considered requirement practically uniquely (in fact, a family of similar measures is obtained). Averaging with the help of the constructed measure gives finite expectation values for links.

PACS numbers: 0460

1. Introduction

To quantize continuum field theory the canonical quantization prescription is usually used. However, the latter does not work in discrete theory, such as Regge calculus formulation of general relativity (suggested by Regge in 1961 [1]; see recent review [2]) because of the lack of a regular continuous coordinate playing the role of time. Although quantum theory can still be formulated with the help of the functional integral in this case, the absence of a strict framework leaves the quantum measure practically arbitrary. In fact, various measures are used in applications (e.g., in phase analysis of simplicial quantum gravity [3]).

To obtain more restrictions on the measure, one should specify the properties that the quantum model should possess. For example, in [4] the partition function of 3D Regge calculus is taken as a discrete sum of the products of $6j$-symbols corresponding to angular momenta j_i for separate tetrahedrons with link-lengths $j_i + \frac{1}{2}$ (for further developments of this approach see, e.g., [5–8]). The basis for such a choice is that the exponential of 3D Regge action arises in the asymptotic form of $6j$-symbols at large j_i. Besides that, the statistical weight in this sum can be chosen so that the partition function be invariant under the refinement of any tetrahedron into four smaller tetrahedrons. Thus, quantum measure is specified here by topological requirements. Therefore it is quite natural for this model to

† Present address: Budker Institute of Nuclear Physics, Novosibirsk, Russia.
‡ Present e-mail address: khatsym@inp.nsk.su

0264-9381/94/102443+11$19.50

have correspondence [8] with the continuum 3D gravity [9] which possesses only topological degrees of freedom.

Here we would like to propose another choice of quantum measure in 3D Regge calculus based on other principles: canonical quantization, coordinate equivalence and analyticity. Thereby we restrict the measure to having a quite definite limiting form (fixed by canonical quantization); besides, any coordinate can be chosen as a time coordinate (due to coordinate equivalence) perhaps with inversion of signature (allowed by the suggested analyticity) if required. This model possesses some non-trivial microscopic (at the scale of Planck length) structure: link-lengths have finite expectation values, although the theory remains trivial in the sense that there is no interaction between neighbouring links† Note for comparison that in the Ponzano–Regge model vacuum averages of link-lengths should be infinite. This may be viewed, for example, from formulae of [8] where detailed derivation of the partition function is given: corresponding sums are saturated by infinitely large values of the link-lengths.

The whole construction proceeds in two steps. For the first step we write out the canonical quantization measure of the 3D Regge calculus; for the second step this measure is considered as a limiting form of the measure of interest, and thus the latter is derived. The first step is considered in section 2 where the Hamiltonian form is presented and in section 3 where the canonical quantization measure is written out. It is shown that Hamiltonization of the tetrad-connection Regge calculus leads, upon changing notations, to the system of constraints suggested for $2+1$ gravity by Waelbroeck from symmetry considerations [10]. The second step is considered in section 4 where the full discrete measure is recovered. In section 5 consequences of the measure obtained for the link averages are discussed. As for the global degrees of freedom of the model, it is more appropriate to study these in the continuous time formalism where we just have a precise definition of what is the degree of freedom. But that formalism was shown in [10] to possess the set of observables the same as that in the continuum Witten's model [9].

2. Hamiltonian form

Continuous time Regge calculus has been studied in a number of works [11–16]. For our purposes it is more convenient to use the tetrad-connection form of Regge calculus according to which the 3D Regge action

$$S(l) = \sum l\varphi(l) \tag{1}$$

is presented in the form [17]

$$S(l, \Omega) = \sum l \arcsin \frac{l * R(\Omega)}{2l} \qquad l * R \stackrel{\text{def}}{=} l^a R^{bc} \epsilon_{abc}. \tag{2}$$

Equation (1) is the sum over links of link-length weighted angle defects $\varphi(l)$ which are functions of link-lengths l. In (2) these angle defects are expressed as functions of

† Here one might point out the apparent contradiction between this separate quantization of links and triangle inequalities. The fact of the matter is that we work in the triad-connection representation of Regge calculus which provides the closure of links into triangles only as a consequence of equations of motion for the connections, i.e. on a classical level.

new independent variables Ω—orthogonal ($SO(3)$ in the Euclidean and $SO(2,1)$ in the Minkowski case) matrices defined on triangles. Here defined on links

$$R(\Omega) = \prod \Omega^{\pm 1} \tag{3}$$

are curvature matrices—path-ordered products of Ω's on the triangles sharing a given link. It is thought of each tetrahedron that there is a local frame associated to it. The vector $\boldsymbol{l}$ corresponding to each link is defined in the frame of some one of the tetrahedrons sharing this link.

A formal expression for the measure will suffer from divergences, and it is a non-trivial problem to find a region in the complex plane of field variables where the measure is well-defined and can be continued to the physical region. Let us start from Euclidean theory. The measure of interest can be defined as an analytical continuation from the region obtained from the Euclidean one by rescaling all the link vectors by $\sqrt{-1}$:

$$\boldsymbol{l} \Rightarrow -\mathrm{i}\boldsymbol{l}. \tag{4}$$

This trick, while leaving all three coordinates equivalent, allows one to solve several problems at once. First, after such transformation each coordinate is suitable for using it as formal time in canonical analysis, as if it had timelike signature. Second, we obtain factor i multiplying the action in the path integral exponential

$$\exp(\mathrm{i}S)$$

which makes all the intermediate integrals conditionally convergent. Third, the rotational group remains $SO(3)$; therefore integrations over connections are finite. At the same time upon rescaling the line element the resulting theory can still be viewed as Euclidean.

Let $A, B, C, \ldots$ denote vertices in a 3D Regge manifold, then (AB), (ABC), $(ABCD)$ will denote link, triangle and tetrahedron, respectively, containing vertices enumerated in round brackets. The latter will denote unordered sequences of vertices. More detailed notation for the vector of link (AB) will be $\boldsymbol{l}_{(AB)}$, for connection on a triangle (ABC)—$\Omega_{(ABC)}$. To point out the tetrahedron $(ABCD)$ in the frame of which the link vector is defined we shall write

$$\boldsymbol{l}_{(AB)(CD)}.$$

Note that for each (AB) there is only one $(ABCD)$ for which $\boldsymbol{l}_{(AB)(CD)}$ enters the action.

The form (2) can be reduced easily to the canonical one. For such a reduction being a non-singular one, some special type of arrangement of vertices should issue when passing to the continuous time limit. Namely, each tetrahedron should contain an infinitesimal link (i.e. that tending to zero in the continuous time limit). Divide the manifold into 2D leaves of vertices in such a way that each tetrahedron would lay between the two neighbouring leaves. Leaves are labelled by a parameter t playing the role of time such that its difference between the neighbouring leaves $\mathrm{d}t \to 0$ in the continuous time limit. In the leaf at a moment t denote vertices $i, j, k, \ldots$. Then for each vertex i there should exist both its image i^+ and pre-image i^- in subsequent and preceding leaves, respectively, such that there are links (ii^+) and (ii^-) of the length $O(\mathrm{d}t)$; also links (ik), (i^+k^+), (i^-k^-) exist or do not simultaneously. In other words, space between any two neighbouring leaves is

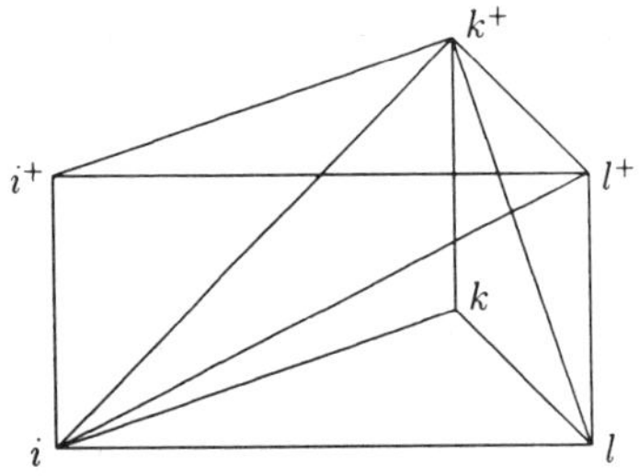

Figure 1. Infinitesimal 3-prism.

divided into infinitesimal prisms which we shall denote just as their triangle bases (ikl), see figure 1.

The behaviour of quantities $\boldsymbol{l}$, Ω at $\mathrm{d}t \to 0$ depends on the type of simplices on which these are defined. We shall conventionally refer to the simplices completely contained in one leaf as we do to the spacelike ones; those which contain vertices of two (neighbouring) leaves will be called timelike if their measure (length, area) tends to zero as $\mathrm{d}t \to 0$ and diagonal otherwise. Connections on spacelike and diagonal triangles are discrete analogues of the continuum connections for the parallel vector transport at a distance $O(\mathrm{d}t)$ and can be naturally assumed to be infinitesimal. For the same reasons local frames should be attributed to infinitesimal prisms (ikl) rather than to the separate tetrahedrons. Thus on spacelike simplices we have the quantities

$$\boldsymbol{l}_{(ik)l} \qquad \Omega_{(ikl)} \stackrel{\mathrm{def}}{=} 1 + f_{(ikl)}\,\mathrm{d}t \qquad (f^{ab}_{(ikl)} = -f^{ba}_{(ikl)}) \tag{5}$$

($\boldsymbol{l}_{(ik)l}$ is the vector of link (ik) defined in the frame of prism (ikl)) and similar ones on diagonal simplices,

$$\begin{aligned} \boldsymbol{l}_{(ik^+)l} &\stackrel{\mathrm{def}}{=} \boldsymbol{l}_{(ik)l} + \boldsymbol{N}_{kil}\,\mathrm{d}t \\ \Omega_{(ik^+l)} &\stackrel{\mathrm{def}}{=} 1 + f_{(ik^+l)}\,\mathrm{d}t \\ \Omega_{(ik^+l^+)} &\stackrel{\mathrm{def}}{=} 1 + f_{(ik^+l^+)}\,\mathrm{d}t \end{aligned} \tag{6}$$

whereas, on the contrary, timelike objects carry infinitesimal $\boldsymbol{l}$ and finite Ω:

$$\boldsymbol{l}_{(ii^+)(kl)} \stackrel{\mathrm{def}}{=} \boldsymbol{N}_{i(kl)}\,\mathrm{d}t \qquad \Omega_{(ii^+k)} \stackrel{\mathrm{def}}{=} \Omega_{ik}. \tag{7}$$

The resulting set of variables $\boldsymbol{l}$, Ω, $\boldsymbol{N}$, f seems to be rather large for the canonical quantization for it contains superfluous variables. Let us exclude $\boldsymbol{N}_{ikl}$ and one of the connections Ω_{ik}, Ω_{ki} for each link (ik). This is easy to do, because N^a_{ikl} enters resulting at the $\mathrm{d}t \to 0$ Lagrangian linearly as the factor at the finite part of curvature on the diagonal link,

$$\epsilon_{abc}[(\Omega_{ik}\Omega^{\delta_1}_{ki})^{\delta_2}]^{bc}.$$

Here δ_1, $\delta_2 = \pm 1$, $\Omega^{-1} \equiv \bar{\Omega}$. Thus Ω_{ik} and Ω_{ki} are, up to possible inversion, the same and one of them will be denoted as $\Omega_{(ik)}$, some link variable. Then $\boldsymbol{N}_{ikl}$ drops out.

The remaining set $\boldsymbol{l}$, Ω, $\boldsymbol{N}$, f is still too large, but f_{ABC} enters the Lagrangian being summed over the interior of the infinitesimal prism:

$$h_{(ikl)} \stackrel{\mathrm{def}}{=} \pm f_{(ikl)} \pm f_{(ik^+l)} \pm f_{(ik^+l^+)}. \tag{8}$$

The $h_{(ikl)}$ is the true analogue of the connection ω_0 of the continuum theory.

Now, of the set $\boldsymbol{l}$, Ω, $\boldsymbol{N}$, h, it would be appropriate to consider $\boldsymbol{l}_{(ik)}$, $\Omega_{(ik)}$ as conjugate variables and $\boldsymbol{N}_i$, $h_{(ikl)}$ as Lagrange multipliers. Note that we have denoted $\boldsymbol{N}_{i(kl)}$ simply as $\boldsymbol{N}_i$, because for any i there is only one (ikl) for which $\boldsymbol{N}_{i(kl)}$ enters the Lagrangian. The only trouble is that it enters not as a Lagrange multiplier, but in the form

$$N_i \arcsin \frac{\boldsymbol{N}_i * R_i(\Omega)}{2N_i}. \tag{9}$$

However, equations of motion for $\boldsymbol{N}_i$ have the same solution, $R_i = 1$, as if we had simply deleted here the symbol 'arcsin'†.

This gives the Lagrangian in the form

$$L = L_{\dot\Omega} + L_h + L_N \tag{10}$$

$$L_{\dot\Omega} = \tfrac{1}{2}\sum_{(ik)} \boldsymbol{l}_{(ik)} * (\bar\Omega_{(ik)}\dot\Omega_{(ik)}) \tag{11}$$

$$L_h = \tfrac{1}{2}\sum_{(ikl)}\left[\sum_{\text{cycle perm } ikl} \varepsilon_{(ik)l}\Omega_{(ik)}^{\delta_{(ik)l}}\boldsymbol{l}_{(ik)}\right] * \boldsymbol{h}_{(ikl)} \stackrel{\text{def}}{=} C(h) \qquad \delta \stackrel{\text{def}}{=} (1+\varepsilon)/2 \tag{12}$$

$$L_N = \tfrac{1}{2}\sum_i \boldsymbol{N}_i * \boldsymbol{R}_i \stackrel{\text{def}}{=} \boldsymbol{R}(\boldsymbol{N}) \qquad R_i = \Omega_{(ik_n)}^{\varepsilon_{ik_n}}\cdots\Omega_{(ik_1)}^{\varepsilon_{ik_1}} \qquad \varepsilon_{ik_j} = -\varepsilon_{(ik_j)k_{j-1}} = \varepsilon_{(ik_j)k_{j+1}}. \tag{13}$$

The form of the kinetic term $L_{\dot\Omega}$ uniform for each link is obtained by making, if necessary, the change of variables $\Omega \to \bar\Omega$ and/or $\boldsymbol{l} \to \Omega^{\pm 1}\boldsymbol{l}$. Here $\varepsilon_{(ik)l} = \pm 1$ is a sign function which specifies for each link vector $\boldsymbol{l}_{(ik)}$ the choice of one of two prisms (ikl) in the frame of which it is defined. We may observe that upon changing notations the Lagrangian obtained describes the system suggested for 2+1 gravity by Waelbroeck [10] from symmetry considerations. It is not surprising since 3D Regge gravity due to its local triviality should possess these symmetries. Now we have seen how Waelbroeck's system arises from Regge calculus. This will also be important when making the backward steps to recover the full discrete measure from the canonical continuous time one.

3. Canonical quantization measure

How should we write down the canonical quantization anzatz for the Lagrangian (10) in the functional integral form? Note, that there exists an analogue of Poisson brackets for the specific form of the kinetic term $L_{\dot\Omega}$,

$$\{f, g\} = \frac{\partial f}{\partial \boldsymbol{l}} \times \frac{\partial g}{\partial \boldsymbol{l}} \cdot \boldsymbol{l} - \frac{\partial f}{\partial \boldsymbol{l}} * \left(\bar\Omega\frac{\partial g}{\partial \Omega}\right) + \frac{\partial g}{\partial \boldsymbol{l}} * \left(\bar\Omega\frac{\partial f}{\partial \Omega}\right) \tag{14}$$

† This implies the convention that another solution, $R_i = -1$, corresponding to the angle defect $\varphi = \pi$, should be ignored.

(summation over links is implied) such that

$$\frac{\partial f}{\partial t} = \{f, H\} \qquad H = -L_h - L_N. \tag{15}$$

The $\boldsymbol{C}$, $\boldsymbol{R}$ act via these brackets as rotations and translations, respectively. These are class I constraints and form a closed algebra w.r.t. these brackets. Therefore the formal functional integral measure for this system follows immediately†:

$$\mathrm{d}\mu = \exp\left(\mathrm{i}\int L_{\dot\Omega}\,\mathrm{d}t\right)\delta(\boldsymbol{C})\delta(\boldsymbol{R})\mathrm{D}\boldsymbol{l}\mathcal{D}\Omega, \qquad \mathrm{D}\boldsymbol{l} \stackrel{\mathrm{def}}{=} \prod_{(ik)} \mathrm{d}^3\boldsymbol{l}_{(ik)} \qquad \mathcal{D}\Omega \stackrel{\mathrm{def}}{=} \prod_{(ik)} \mathcal{D}\Omega_{(ik)} \tag{16}$$

where Lebesgue $\mathrm{d}^3\boldsymbol{l}$ and Haar $\mathcal{D}\Omega$ measures arise as the only ones invariant w.r.t. the symmetries generated by $\boldsymbol{C}$, $\boldsymbol{R}$. Raise $\boldsymbol{C}$, $\boldsymbol{R}$ from δ-functions to exponent with the help of Lagrange multipliers $\boldsymbol{N}$, h and rewrite $\mathrm{d}\mu$ as

$$\mathrm{d}\mu = \exp\left(\mathrm{i}\int L\,\mathrm{d}t\right)\mathrm{D}\boldsymbol{l}\mathrm{D}\boldsymbol{N}\mathcal{D}\Omega\mathrm{D}h. \tag{17}$$

The only thing which we have yet not done needed to reduce the measure to the form standard for systems subject to class I constraints is to fix the gauge and divide it by the volume of the symmetry group generated by constraints. Because of the compactness of $SO(3)$ such a procedure for the rotational subgroup is unnecessary. As for the volume of the translational subgroup, it is infinite, and in some cases separating it out can help in making the functional integral convergent. To break the translational symmetry let us impose partial gauge condition $\boldsymbol{f} = 0$ where

$$\boldsymbol{f} = \{\boldsymbol{l}_{(ik)} - \boldsymbol{a}_{(ik)} | (ik) \in F\}. \tag{18}$$

Here $\boldsymbol{a}_{(ik)}$ are constant vectors, F is a set of links arranged in non-intersecting and non-self-intersecting broken lines passing through all the vertices, see figure 2.

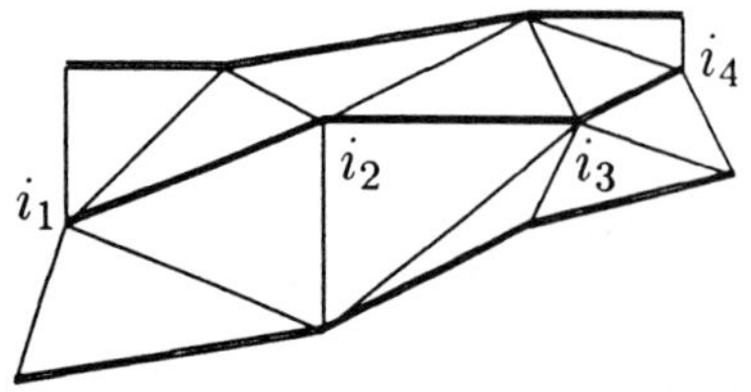

Figure 2. Gauge fixing: links of the set F (solid lines) have fixed vectors.

† Of course, the same can be obtained in a more rigorous way introducing, as in [10], the variables $P_{ab} = l^c \Omega^f_a \epsilon_{cfb}$ and Ω^{ab} for each link, canonically conjugated in the usual sense. There will be δ-functions in the functional integral taking into account class II constraints to which P, Ω are subject, $\delta^6(\bar\Omega\Omega - 1)\delta^6(\bar\Omega P + \bar P \Omega)$. The invariant measure $\mathrm{D}\boldsymbol{l}\mathcal{D}\Omega$ just arises on integrating out these δ's. The $\{f, g\}$ turn out to be Dirac brackets w.r.t. this system of class II constraints.

By the standard rule of the Faddeev–Popov ansatz, after dividing by the translational group volume we get, apart from $\delta(\boldsymbol{f})$, the Jacobian factor $\det \|\{\boldsymbol{R}, \boldsymbol{f}\}\|$, for which we have

$$\det \|\{\boldsymbol{R}, \boldsymbol{f}\}\|^{-1} = \int \delta(\{\boldsymbol{R}(\boldsymbol{N}), \boldsymbol{f}\}) \mathrm{D}\boldsymbol{N}. \tag{19}$$

But along any broken line we have the sequence of gauge conditions $\boldsymbol{f}_j = \boldsymbol{l}_{i_j i_{j+1}} - \boldsymbol{a}_{i_j i_{j+1}}$ (where $\{i_j | j = \ldots, 1, 2, 3, \ldots\}$ is a sequence of vertices, see figure 2) and product of δ-functions

$$\delta(\{\boldsymbol{R}(\boldsymbol{N}), \boldsymbol{f}_j\}) = \delta^3(\varepsilon_{i_j i_{j+1}} \Gamma_{i_j i_{j+1}} \boldsymbol{N}_{i_j} + \varepsilon_{i_{j+1} i_j} \Gamma_{i_{j+1} i_j} \boldsymbol{N}_{i_{j+1}}) \tag{20}$$

where Γ_{ik} is some $SO(3)$ matrix (product of Ωs) required to express $\boldsymbol{N}_i$ in the frame where $\boldsymbol{l}_{(ik)}$ is defined. The $\boldsymbol{N}_{i_j}$ are consecutively integrated out in (19) under appropriate boundary conditions along each broken line giving unity for the Jacobian factor of interest. As for $\delta(\boldsymbol{f})$ in the Faddeev–Popov ansatz, this is eliminated by integrations over $\mathrm{d}^3\boldsymbol{l}_{(ik)}$, $(ik) \in F$.

As a result, the whole effect of our gauge fixing amounts to omitting integration in the expression for the canonical measure $\mathrm{d}\mu$ over links in the 2D leaf constituting the set of broken lines F. It is natural to call the latter the *field* of links in analogy with the continuum vector field which picks out a definite direction at each vertex. From the viewpoint of coordinate equivalence it is natural to expect that integration over timelike links can be equally omitted. Indeed, due to conservation in time the constraint $\boldsymbol{R}$ may be imposed only as an initial condition; there $\mathrm{D}\boldsymbol{N}$ integration giving $\delta(\boldsymbol{R})$ at finite time is not required and can be omitted.

Thus, in addition to the form (17) of the canonical measure we can work also with any other form where integration over some field of links, spacelike or timelike, is omitted.

4. Recovering the full discrete measure

Now we can restore the full discrete measure by means of identity transformation of the functional integral by going back and inserting into $\mathrm{d}\mu$ the variables earlier eliminated by use of equations of motion and/or by neglecting infinitesimal contributions to action.

First recall that the contribution of timelike links enters the exponential of the canonical measure, not as term (9) in the original discrete action, but with the symbol 'arcsin' omitted. By coordinate equivalence, when passing to the full discrete measure we should write down the contribution of all other links to the exponential also without 'arcsin'. (Eventually, such a possibility is connected with local triviality of classical 3D gravity.)

This allows us to introduce integration over diagonal links $\mathrm{D}\boldsymbol{l}_{(ik^+)}$ and over additional connection for each link so that instead of $\mathcal{D}\Omega_{(ik)}$ we shall have $\mathcal{D}\Omega_{ik}\mathcal{D}\Omega_{ki}$. This is an identity transformation if simultaneously we add the terms $\frac{1}{2}\boldsymbol{l}_{(ik^+)} * R_{(ik^+)}$ in the exponential. Indeed, integrating over $D\boldsymbol{l}_{(ik^+)}$ results in the δ-function

$$\delta^3(R_{(ik^+)} - \bar{R}_{(ik^+)}).$$

Integration of this δ over one of $\mathcal{D}\Omega_{ik}$, $\mathcal{D}\Omega_{ki}$ due to invariance of the Haar measure is equivalent to integration over $\mathcal{D}R_{(ik^+)}$ since $R_{(ik^+)}$ has the form $\Gamma_1 \Omega^{\pm 1} \Gamma_2$ where Ω is Ω_{ik} or Ω_{ki} and $\Gamma_1, \Gamma_2 \in SO(3)$. Therefore this integration gives a constant.

However, the number of such δ-functions is too large if one integrates over *all* the links meeting at any given vertex A: the product of these δ-functions

$$\prod_{B} \delta^3(R_{(AB)} - \bar{R}_{(AB)})$$

is singular, because arguments of δs are related by Bianchi identities [1]. Therefore integration over at least one vector of links meeting at any vertex should be excluded. Simple consideration shows that integration should be excluded for the set of links $\mathcal{F}$ forming non-intersecting and non-self-intersecting broken lines passing through each vertex of the now full discrete Regge manifold. We can speak of $\mathcal{F}$ as a field of links having in view some one-dimensional set which picks out a definite vector at each vertex.

The above considered ways of gauge fixing in the canonical measure $\mathrm{d}\mu$ just correspond to the different choices (w.r.t. the set $\mathcal{F}$) of the continuous coordinate playing the role of time. Namely, if we recover the full discrete measure by the above way from $\mathrm{d}\mu$ with $\mathrm{D}N$ omitted, the set $\mathcal{F}$ turns out to be timelike, while omitting integrations over spacelike set F gives $\mathcal{F} = F$. Even if we issue from the most symmetrical form for the canonical measure (17) with all the link integrations available, in order to avoid singularities we should not insert integrations over some field of diagonal links.

The integration element $\mathrm{d}^3\boldsymbol{N}_i$ should be converted, upon rescaling it by a power of $\mathrm{d}t$, into $\mathrm{d}^3\boldsymbol{l}_{(ii^+)}$. This is the only possible way of recovering the finite element $\boldsymbol{l}_{(ii^+)}$ while keeping the equivalence of different links.

It remains only to recover finite connections on the spacelike and diagonal triangles. These are now infinitesimal and enter the theory being summed over the groups of spacelike and diagonal triangles into $h_{(ikl)}$, see (8). Here we note that due to $SO(3)$-invariance we can rotate the frame in each tetrahedron so that in each such group of connections all but one connection would become trivial; say, non-trivial is $\Omega_{(ikl)}$. This is a kind of gauge choosing already in full discrete theory. Correspondingly, in the continuous time in each sum for $h_{(ikl)}$ we have only one term, $f_{(ikl)}$. The $\mathrm{d}^3 f_{(ikl)}$ is converted, upon rescaling it by a power of $\mathrm{d}t$, into $\mathcal{D}\Omega_{(ikl)}$. It is the only way of recovering $\Omega_{(ikl)}$ if one sticks to the equivalence of different triangles. Due to the gauge freedom the result of integrating over $\mathcal{D}\Omega_{(ikl)}$ will not depend on the diagonal triangle connections, and including integrations over them, $\mathcal{D}\Omega_{(ikl^+)}\mathcal{D}\Omega_{(ik^+l^+)}$, into the measure results simply in some constant normalization factor. It is important here that in our method of giving sense to the ill-defined theory the rotational group is compact.

Thus we can write out the resulting form of the full discrete measure:

$$\mathrm{d}\mathcal{M}_{\mathcal{F}} = \exp(\mathrm{i}\tilde{S}) \prod_{(AB)\in\mathcal{F}} \mathrm{d}^3\boldsymbol{l}_{(AB)} \prod_{(ABC)} \mathcal{D}\Omega_{(ABC)} \tag{21}$$

$$\tilde{S} \stackrel{\mathrm{def}}{=} \frac{1}{2} \sum_{(AB)} l^a_{(AB)} \epsilon_{abc} R^{bc}_{(AB)}. \tag{22}$$

We observe that it is impossible to get the full discrete measure completely symmetrical w.r.t. the different coordinates since the set $\mathcal{F}$ picks out some coordinate and thus violates this symmetry. However, there is an arbitrariness in the choice of $\mathcal{F}$, so we can speak of a kind of spontaneous symmetry breaking as a result of this choice.

5. Discussion. Link averages

Let us try to find some quantum averages using the obtained measure $\mathrm{d}\mathcal{M}_{\mathcal{F}}$. For a non-degenerate manifold the number of triangles is not less than the number of links (this follows from simple combinatorial consideration taking into account that each link is shared by no less than three triangles). So it is tempting to try to change variables so that some Ωs were replaced by an equal number of curvatures R. Since curvatures on links meeting at a vertex are related by the Bianchi identity, we can use curvatures as independent variables only on links not belonging to some set $\mathcal{F}'$. This set is constructed completely analogously to $\mathcal{F}$ and has the same structure (for example, one can take $\mathcal{F}' = \mathcal{F}$). Now consider $R_{(AB)}$, $(AB) \notin \mathcal{F}'$ as independent variables instead of the same number of appropriately chosen Ωs. The dependence of given R on given Ω (if any) has the form $R = \Gamma_1 \Omega^{\pm 1} \Gamma_2$, $\Gamma_1, \Gamma_2 \in SO(3)$, and, by the properties of invariant measure, $\mathcal{D}R = \mathcal{D}\Omega$. Subsequently substituting Rs instead of Ωs we find the Jacobian of this change to be unity.

Now, integration over $\mathcal{D}\Omega$ in $\mathrm{d}\mathcal{M}_{\mathcal{F}}$ is substituted by that over $\mathcal{D}R'\mathcal{D}\Omega'$ where the prime means connections of the reduced set and curvatures on all the links not belonging to $\mathcal{F}'$. But the measure still seems not to simplify considerably: apart from simple terms $l * R'$ in the exponential there are also $l_{(AB)} * R_{(AB)}$ at $(AB) \in \mathcal{F}'$ in which $R_{(AB)}$ is the bulky product of some R', Ω'. Recall, however, that integration over links of the set $\mathcal{F}$ is absent, therefore these should be fixed by hand. Making use of this circumstance to avoid unnecessary complications let us put $\mathcal{F}' = \mathcal{F}$ and rescale the links of this set to zero. This corresponds to the most detailed possible description of the quantum Regge manifold when the coordinate which defines the distance along the field of links $\mathcal{F}$ is becoming continuous and we have at our disposal sections of the manifold at arbitrary values of this coordinate.

Now we see that for such a description the measure $\mathrm{d}\mathcal{M}_{\mathcal{F}}$ factorizes into the product of independent ones,

$$\mathrm{d}\mathcal{M}_l = \exp(\tfrac{1}{2}\mathrm{i}l_{(AB)} * R_{(AB)})\,\mathrm{d}^3 l_{(AB)}\,\mathcal{D}R_{(AB)} \qquad (AB) \notin \mathcal{F} \tag{23}$$

referred to separate links. There is also integration over $\mathcal{D}\Omega'$, but Ω's enter only through $R_{(AB)}$, $(AB) \in \mathcal{F}'$, and upon setting $l_{(AB)} = 0$ this integration is trivial.

To find the Euclidean quantum average of the invariant l^{2k} one should apply the measure $\mathrm{d}\mathcal{M}_l$ to $(-\mathrm{i}l)^{2k} = (-l^2)^k$ (recall that thus far we have worked in the region obtained from the Euclidean one by substituting $l \to -\mathrm{i}l$). Due to $SO(3)$-invariance of $\mathrm{d}\mathcal{M}_l$ this is sufficient for averaging any link function $g(l)$. Integral over $\mathrm{d}^3 l$ gives derivatives of the δ-function

$$\delta^3\left(\frac{R-\bar{R}}{2}\right) = \delta^3\left(\frac{\sin\phi}{\phi}\phi\right)$$

where ϕ parametrizes $R : R \stackrel{\mathrm{def}}{=} \exp(\epsilon^{abc}\phi_c)$. Then the δ-function is integrated out over

$$\mathcal{D}R = \frac{\sin^2(\phi/2)}{4\pi^2\phi^2}\,\mathrm{d}^3\phi$$

in the vicinity of $\phi = 0$ (another root at $\phi = \pi$ should be omitted according to our convention, see the footnote on p 2447). The (normalized by $\langle 1 \rangle = 1$) result reads

$$\langle l^{2k} \rangle_E = (-1)^k \langle l^{2k} \rangle = \frac{4^{-k}\Gamma(2k+2)^2}{\Gamma(k+2)\Gamma(k+1)} \tag{24}$$

for the Euclidean quantum average. It is finite at $k > -1$ and non-zero. This allows us to recover the quantum average for any function of link:

$$\langle g(l)\rangle_E = \int \frac{\mathrm{d}o_l}{4\pi} \int_0^\infty g(l)\nu(l)\,\mathrm{d}l \qquad \nu(l) = \frac{2l}{\pi}\int_0^\pi \exp\left(-\frac{l}{\sin\varphi}\right)\mathrm{d}\varphi. \tag{25}$$

This is the result of the author's unpublished work [18]. Thus, quantum average $\langle\cdot\rangle$ in our region obtained by analytical continuation $\boldsymbol{l} \to -i\boldsymbol{l}$ where expressions for measure are well defined proves to be non-positive. Remarkably, however, is that on returning back to the Euclidean region we get a positively defined measure.

Up to now, the word 'time' was used to denote simply a continuous coordinate or that which should be made such. If one continues formulae (24), (25) to the Minkowski region, $\langle l^2\rangle$ remains positive, although now l^2 is an indefinite quadratic form of components l^a of signature $(+, +, -)$. Let us now use the word 'time' in its strict sense implying that timelike interval is negative. In this sense fluctuating links are always spacelike ones. Therefore timelike links may be only non-fluctuating ones, i.e. those from the set $\mathcal{F}$. Thus, time does not fluctuate, which is quite natural. It is amusing that this fact arises here as a consequence of quantum Regge calculus. In addition, time cannot be more than one-dimensional, since set $\mathcal{F}$ picks out only one direction at each point. In turn, occurrence of $\mathcal{F}$ and its one-dimensionality can be traced to Bianchi identities. Thus, one-dimensionality of time and its non-quantizability in 3D spacetime follow from the Bianchi identities!

Note, Regge calculus is a limiting case of the smooth Riemannian geometry, and vice versa, smooth Riemannian geometry is a limiting case of the Regge one when all the link-lengths tend to zero. So we can speak of the two to be *a priori* equally possible formulations of general relativity. Then our result on the link expectation values allows us to adopt the following point of view on the origin of divergences in the continuum gravity: continuum formulation is inadequate, for were it not so, link-length expectation values would be zero. Thus we can say that gravity is becoming discrete *dynamically*.

Acknowledgments

I am grateful to Professors A Niemi, S Yngve and the personnel of the Institute of Theoretical Physics at Uppsala University for warm hospitality and support during the work on this paper.

References

[1] Regge T 1961 *Nuovo Cimento* **19** 558
[2] Williams R M and Tuckey P A 1992 *Class. Quantum Grav.* **9** 1409
[3] Hamber H W 1993 *Nucl. Phys.* B **400** 347
[4] Ponzano G and Regge T 1968 *Spectroscopic and Group Theoretical Methods in Physics* ed F Block (Amsterdam: North-Holland)
[5] Hasslacher B and Perry M J 1981 *Phys. Lett.* **103B** 21
[6] Turaev V G and Viro O Y 1990 *Topology* **31** 865; 1992 State sum invariants of 3-manifolds and quantum $6j$-symbols *Preprint*
[7] Archer F G and Williams R M 1991 *Phys. Lett.* **273B** 438
[8] Ooguri H 1992 *Nucl. Phys.* B **382** 276
[9] Witten E 1988 *Nucl. Phys.* B **311** 46
[10] Waelbroeck H 1990 *Class. Quantum Grav.* **7** 751
[11] Collins P A and Williams R M 1973 *Phys. Rev.* D **7** 965; 1974 *Phys. Rev.* D **10** 3537

[12] Williams R M 1986 *Class. Quantum Grav.* **3** 853
[13] Friedman J L and Jack I J 1986 *J. Math. Phys.* **27** 2973
[14] Piran T and Williams R M 1986 *Phys. Rev.* **33** 1622
[15] Porter J 1987 *Class. Quantum Grav.* **4** 375, 391, 651
[16] Brewin L 1987 *Class. Quantum Grav.* **4** 899; 1988 *Class. Quantum Grav.* **5** 839
[17] Khatsymovsky V 1989 *Class. Quantum Grav.* **6** L249
[18] Khatsymovsky V 1991 3D Regge gravity as an exactly soluble system *Novosibirsk Preprint* INP 91-60

Class. Quantum Grav. **11** (1994) 2455–2481. Printed in the UK

Conformally invariant cosmology based on Riemann–Cartan spacetime*

Yuri V Shtanov†‡ and Sergei A Yushchenko†
† Bogolyubov Institute for Theoretical Physics, Kiev, 252143, Ukraine
‡ Department of Physics, Brown University, Providence, RI 02912, USA

Received 3 November 1993, in final form 18 July 1994

Abstract. A conformally invariant field-theoretical model including gravity based on the Riemann–Cartan spacetime U_4 is considered. The cosmological scenario that follows from this model is discussed and a standard quantum gravitational formalism in the Arnowitt–Deser–Misner form is developed. The general formalism is then illustrated using the Bianchi-IX minisuperspace cosmological model. Wavefunctions of the universe in the de Sitter minisuperspace model with Vilenkin and Hartle–Hawking boundary conditions are considered and corresponding probability distributions for the scalar field values are calculated.

PACS numbers: 0460, 9880

1. Introduction

When unifying gravity with other interactions, concepts of symmetries undoubtedly play an important role. Among various symmetries that have been put forward at different times, the symmetry with respect to local conformal transformations remains of special significance and keeps attracting the attention of many researchers. First, because theories with such a symmetry typically exhibit better quantum behaviour, and the issue of renormalizability might lead to conformal invariance. Second, because such kind of symmetry might help to resolve the cosmological constant problem (the usual Λ-term breaks the conformal symmetry of the action and therefore is excluded simply by a demand of such a symmetry).

The original idea of local conformal invariance belongs to Weyl [1] who introduced a compensating gauge vector field for the transformations under consideration. The corresponding geometric structure is usually called Weylian. It was afterwards developed in various directions [2]. A conformally invariant theory of gravity based on the metric structure alone was considered in [3].

Invariance with respect to local conformal transformations can be introduced into a theory of gravity in different ways [4]. One of the rather simple ways of building such a theory consistent with observations is to base it on Riemann–Cartan spacetime. Riemann–Cartan geometry naturally arises within the framework of the Poincaré gauge theory of gravity and is of great interest by itself, regardless of the conformal invariance issue. The effective action and renormalization group in Riemann–Cartan spacetime have been extensively studied in [5].

Some work has already been done in studying conformally invariant theories on a Riemann–Cartan geometric background. Conformally invariant dynamics of spinor fields

* Work supported in part by the Department of Energy under contract DE-FG02-91ER40688—Task A.

in Riemann–Cartan space were studied in [6]. In [7] a simple example of the gravitational part of such a theory has been considered. It incorporated a scalar field into the term $\varphi^2 R$ in the Lagrangian. The aim of the present paper is to generalize these proposals and to consider a complete theory which possesses Weyl invariance. In doing this we take into account that the vector trace Q_μ of the torsion tensor transforms under the action of the local conformal group in a similar way to the Weyl vector. Making use of this property we construct a generic field-theoretical model, based on the Riemann–Cartan geometric structure, which contains a scalar field multiplet and which is invariant with respect to local conformal transformations. The key difference between the theory considered here and Weyl's theory is that in the latter a special vector field is introduced as a compensating gauge field for the local conformal transformations whereas in our theory this role is played by the torsion trace vector. We therefore avoid having to introduce new specific entities to make our theory conformally invariant.

The role of the scalar fields in the theory considered here, on the one hand, is the same as in standard particle physics models. They give masses to gauge vector bosons and to fermions through the Higg's symmetry breaking mechanism, making it possible to preserve the (ordinary) gauge invariance of the theory. On the other hand, they allow one to use the term of type $\varphi^2 R$ in the Lagrangian and thus to generate the gravitational coupling constant. In such a way the theory is extended to include gravity and, at the same time, remains locally conformally invariant. In this paper we will consider some of the basic features of such a theory, in particular those related to inflationary and quantum cosmology.

We will see below (in section 3) that the value of the effective cosmological constant Λ in our theory is not an independent parameter, but is a function of some other parameters of the model. It is therefore at least restricted, and further investigation might discover the possibility of having it equal to zero without fine tuning the constants of the theory.

The paper is organized as follows. In the next section we present a locally conformally invariant version of a generic particle physics model coupled to gravity. In section 3 a classical cosmological scenario based on our theory is discussed. In section 4 we develop a standard quantum cosmological formalism in the Arnowitt–Deser–Misner (ADM) form for the theory considered. The general formalism will then be illustrated in section 5 on a minisuperspace cosmological model of Bianchi-IX type. In the appendix we provide the necessary geometric background including the description of the Riemann–Cartan geometric structure and local conformal transformations.

2. Conformally invariant theory

Our theory will be based on the Riemann–Cartan space U_4 (for a more detailed description and basic notation see the appendix). Riemann–Cartan structure implies the presence of the affine connection form $\omega^a{}_b$ and the metric tensor g which is covariantly constant

$$\nabla^\omega g = 0 \tag{1}$$

where the symbol ∇^ω denotes the covariant derivative specified by the affine connection $\omega^a{}_b$. The affine connection is supposed to satisfy only the metricity condition (1), hence the torsion tensor is not assumed to vanish.

In this paper we shall study a rather natural conformally invariant generalization of an ordinary field-theoretical model, with the Lagrangian of the form

$$L = \frac{1}{2}m^2(\varphi)R(\omega) + \frac{\mathrm{i}}{2}\left(\overline{\psi}\gamma^\mu\mathcal{D}_\mu\psi - \overline{(\mathcal{D}_\mu\psi)}\gamma^\mu\psi\right) - \{f\varphi\overline{\psi}\psi\}$$
$$- \frac{1}{2}|\mathcal{D}\varphi|^2 - V(\varphi) - \frac{1}{4e^2}\,\mathrm{Tr}\,F_{\mu\nu}F^{\mu\nu}. \tag{2}$$

It is constructed from a multiplet ψ of spinor fields, a multiplet φ of scalar (Higgs) fields (multiplet indices of φ and ψ will often be omitted), gauge connection field $A = A_\mu \, dx^\mu$ of some gauge group G with respect to which our theory is assumed to be invariant, metric $g = g_{\mu\nu} \, dx^\mu \otimes dx^\nu = \eta_{ab} e^a \otimes e^b$ which can be presented by the tetrad components e^a_μ or e^μ_a, and affine connection field $\omega^a{}_b = \omega^a{}_{b\mu} \, dx^\mu = \omega^a{}_{bc} e^c$. The metric signature is taken to be $(-, +, +, +)$. Greek indices are raised and lowered by the metric components $g_{\mu\nu}$ and $g^{\mu\nu}$ in the coordinate base, and Latin indices by the metric components η_{ab} and η^{ab} in the orthonormal tetrad base. $R(\omega)$ is the Riemann–Cartan curvature scalar. 'Long' derivatives $\mathcal{D}_\mu$ which enter the Lagrangian (2) are defined by the equations

$$\mathcal{D}_\mu \psi = (\partial_\mu + \omega_\mu - Q_\mu + A_\mu)\psi \tag{3}$$

$$\mathcal{D}_\mu \varphi = (\partial_\mu - \tfrac{2}{3} Q_\mu + A_\mu)\varphi \tag{4}$$

where Q_μ is the vector trace of the torsion tensor $Q_\mu = Q^\alpha{}_{\mu\alpha}$ and

$$\omega = -\tfrac{1}{8}\omega^{ab} \, [\gamma_a, \gamma_b] \tag{5}$$

is the spin connection form. γ^a are the usual constant Dirac matrices. The quantities $F_{\mu\nu}$ are the components of the curvature two-form of the gauge connection form $A = A_\mu \, dx^\mu$

$$\mathrm{i}F = \mathrm{d}A + A \wedge A. \tag{6}$$

The symbol $\{f\varphi\overline{\psi}\psi\}$ denotes the sum of various possible Yukawa couplings between the spinor and the scalar Higgs multiplets with coupling constants $\{f\}$. The values $m^2(\varphi)$ and $V(\varphi)$ are assumed to be analytic in φ, so in order to preserve conformal invariance of the action (about the conformal transformations see below) they must represent a quadratic and a quartic form of φ, respectively. We assume $m^2(\varphi)$ to be positive definite. Thus for the multiplet $\{\varphi^A, \, A = 1, \ldots, k\}$ of real scalar fields we put

$$m^2(\varphi) = \xi_{AB} \, \varphi^A \varphi^B \tag{7}$$

where $m^2(\varphi) > 0$ for $\varphi \neq 0$, and

$$V(\varphi) = \lambda_{ABCD} \, \varphi^A \varphi^B \varphi^C \varphi^D \tag{8}$$

where ξ_{AB} and λ_{ABCD} are real dimensionless constants symmetric in their indices. The functions $m^2(\varphi)$, $V(\varphi)$ and $\{f\varphi\overline{\psi}\psi\}$ are also assumed to be G-invariant. Finally, the trace in (2) is taken in the representation space of the group G.

G-invariance of the functions $m^2(\varphi)$ and $V(\varphi)$ restricts the values of the constants ξ_{AB} and λ_{ABCD} in a certain way. For example, if the scalar field multiplet $\{\varphi^A, \, A = 1, \ldots, k\}$ forms the natural representation of the group $O(k)$, and $O(k) \subset G$, then $\xi_{AB} = \xi\delta_{AB}$, $\lambda_{ABCD} = \lambda\delta_{AB}\delta_{CD}$, and ξ and λ are the only free parameters in the expressions for $m^2(\varphi)$ and $V(\varphi)$. If, however, the scalar field multiplet $\{\varphi^A\}$ consists of several sub-multiplets which under the action of the group G transform independently (as is most often the case), then there are more independent components among ξ_{AB} and λ_{ABCD}. For example, if there are two independently transformed sub-multiplets, $\{\varphi_1^a, \, a = 1, \ldots, p\}$ which is $O(p)$-invariant, and $\{\varphi_2^b, \, b = 1, \ldots, q\}$ which is $O(q)$-invariant (both $O(p)$ and $O(q)$ should be contained in G), then $m^2(\varphi) = \sum_i \xi_i m_i^2$, and $V(\varphi) = \sum_{ij} \lambda_{ij} m_i^2 m_j^2$, $i, j = 1, 2$ (here $m_1^2 = \sum_{a=1}^p \varphi_1^a \varphi_1^a$ and $m_2^2 = \sum_{b=1}^q \varphi_2^b \varphi_2^b$) are the possibilities with arbitrary ξ_i, λ_{ij}. Having in mind all such possibilities we consider the functions $m^2(\varphi)$ and $V(\varphi)$ in their general form as given by (7) and (8), respectively.

The action of the theory is written as

$$S = \int_M L\sqrt{-g}\,\mathrm{d}^4x + \int_{\partial M} m^2(\varphi)\mathcal{K}(\omega)\sqrt{h}\,\mathrm{d}^3x \tag{9}$$

where $\mathcal{K}(\omega)$ is the scalar extrinsic curvature (with respect to the connection $\omega^a{}_b$) of the boundary ∂M of the integration region M, h_{ij} $(i, j = 1, 2, 3)$ are the induced metric components on this boundary and $h = \det(h_{ij})$.

The action (9) with the Lagrangian (2) was constructed so as to differ as little as possible from the general relativity action. The only basic difference between them is the presence of the function $m^2(\varphi)$ instead of a constant $(8\pi G_\mathrm{N})^{-1}$, and the presence of torsion in the scalar curvature R and in the metric sector of the theory. One can think of this action as of the lowest terms in the expansion in field derivatives of some generic conformally invariant action based on Riemann–Cartan geometry.

As was said above, the action (9) is invariant with respect to the group of local conformal transformations

$$g(x) \to g'(x) = \exp\big(2\sigma(x)\big)g(x) \tag{10}$$

$$\varphi(x) \to \varphi'(x) = \exp\big(-\sigma(x)\big)\varphi(x) \tag{11}$$

$$\psi(x) \to \psi'(x) = \exp\big(-\tfrac{3}{2}\sigma(x)\big)\psi(x) \tag{12}$$

$$\overline{\psi}(x) \to \overline{\psi}'(x) = \exp\big(-\tfrac{3}{2}\sigma(x)\big)\overline{\psi}(x) \tag{13}$$

$$A(x) \to A'(x) = A(x) \tag{14}$$

$$\omega^a{}_b(x) \to \omega'^a{}_b(x) = \omega^a{}_b(x) \tag{15}$$

where $\sigma(x)$ is an arbitrary real function. Such an invariance is provided in particular by using the 'long' derivatives $\mathcal{D}_\mu$ as defined in (3) and (4). They involve the torsion trace vector Q_μ which, under the action of the local conformal group, transforms similarly to the Weyl gauge vector field (see the appendix)

$$Q_\mu(x) \to Q'_\mu(x) = Q_\mu(x) - \tfrac{3}{2}\partial_\mu\sigma(x). \tag{16}$$

Due to this property, 'long' derivatives (3) and (4) transform under the local conformal group just like the corresponding (spinor and scalar) fields themselves. Note that the torsion trace Q_μ in the derivative (3) of a spinor in fact drops out of the spinor kinetic term of the Lagrangian (2) due to the Hermitian form of the latter. Conformal invariance of the action (9) is also provided by the specific shape (7) and (8) of the functions $m^2(\varphi)$ and $V(\varphi)$, respectively.

Both terms of the action (9) are invariant with respect to the local conformal transformations written just above. The second, boundary, term does not affect the equations of motion. We have added it to the action in order to recover, in the natural gauge $m^2(\varphi) = $ constant, the usual Gibbons–Hawking boundary term [8] of the Hilbert–Einstein action for gravity, as we will see later.

The key difference between the theory considered here and Weyl's theory is that in the latter a special vector field is introduced as a compensating gauge field for the local conformal transformations, whereas in our theory this role is played by the torsion trace vector. We therefore use only the geometric structure already at our disposal and avoid having to introduce new entities to make our theory conformally invariant.

The equations of motion are obtained by varying the action (9) over the independent variables φ, ψ, $\overline{\psi}$, A_μ, $\omega^a{}_{b\mu}$ and e^a_μ. Varying over the affine connection components $\omega^a{}_{b\mu}$ we obtain the following equations for the torsion tensor

$$m^2(\varphi)\big(Q^\mu{}_{\nu\sigma} + \tfrac{2}{3}\delta^\mu{}_{[\nu}Q_{\sigma]}\big) = \tfrac{1}{4}\varepsilon^\mu{}_{\nu\sigma\tau}\overline{\psi}\gamma^\tau\gamma^5\psi \tag{17}$$

$$Q_\mu = \tfrac{3}{4}\partial_\mu \ln M^2(\varphi) \tag{18}$$

where $\varepsilon_{\mu\nu\sigma\tau} = e^a_\mu e^b_\nu e^c_\sigma e^d_\tau\, \varepsilon_{abcd}$, ε_{abcd} are the components of the antisymmetric tensor such that $\varepsilon_{0123} = -1$, $\gamma^\mu = e^\mu_a \gamma^a$, $\gamma^5 = (\mathrm{i}/4!)\varepsilon_{abcd}\,\gamma^a\gamma^b\gamma^c\gamma^d$, and

$$M^2(\varphi) = m^2(\varphi) + \tfrac{1}{6}\varphi^2 \qquad \varphi^2 = \sum_{A=1}^{k}\varphi^A\varphi^A. \tag{19}$$

The purely algebraic nature of equations (17), (18) for the torsion tensor components reflects the fact that torsion in our theory is non-propagating.

At this point we can substitute the expressions for the torsion from (17) and (18) back into the action (9). We will then obtain the following torsion-free effective action

$$S_{\mathrm{eff}} = \int_M L_{\mathrm{eff}}\sqrt{-g}\,\mathrm{d}^4x + \int_{\partial M} m^2(\varphi)\mathcal{K}(\Gamma)\sqrt{h}\,\mathrm{d}^3x \tag{20}$$

the Lagrangian of which is

$$\begin{aligned} L_{\mathrm{eff}} = {} & \frac{1}{2}m^2(\varphi)R(\Gamma) + \frac{\mathrm{i}}{2}\big(\overline{\psi}\gamma^\mu D_\mu\psi - \overline{(D_\mu\psi)}\gamma^\mu\psi\big) - \{f\varphi\overline{\psi}\psi\} \\ & - \frac{3}{16m^2(\varphi)}(\overline{\psi}\gamma^\mu\gamma^5\psi)(\overline{\psi}\gamma_\mu\gamma^5\psi) \\ & - \frac{1}{2}|D\varphi|^2 + 3\big(\nabla M(\varphi)\big)^2 - V(\varphi) - \frac{1}{4e^2}\,\mathrm{Tr}\,F_{\mu\nu}F^{\mu\nu} \end{aligned} \tag{21}$$

and in particular contains the term that describes a four-fermionic axial current $\times$ current interaction. Here $R(\Gamma)$ is the usual Riemannian curvature scalar and $\mathcal{K}(\Gamma)$ is the extrinsic curvature of the boundary ∂M with respect to the Riemannian connection form $\Gamma^a{}_b$. The 'long' derivatives $D_\mu\psi$ and $D_\mu\varphi$ in (21) are defined as

$$D_\mu\psi = (\partial_\mu + \Gamma_\mu + A_\mu)\psi \tag{22}$$

$$D_\mu\varphi = (\partial_\mu + A_\mu)\varphi \tag{23}$$

where

$$\Gamma_\mu = -\tfrac{1}{8}\Gamma^{ab}{}_\mu\,[\gamma^a, \gamma^b] \tag{24}$$

is the spin connection constructed from the Riemannian (torsion-free) affine connection $\Gamma^a{}_b$. The symbol ∇ will denote the usual covariant derivative with respect to the affine connection $\Gamma^a{}_b$.

The action (20) remains locally conformally invariant, although its two terms are not invariant separately, in contrast to the expression (9) for the former action. Note that in the gauge $m^2(\varphi) =$ constant, the second, boundary, term in (20) reproduces the Gibbons–Hawking boundary term [8] of the Hilbert–Einstein theory.

Equations of motion for the scalar multiplet fields φ that stem from the action (20) can be put in the following form (multiplet indices are omitted, so, for example, the equality $\varphi = 0$ denotes that all the multiplet components φ^A are zero, and $\varphi \neq 0$ means that some of the components are non-zero)

$$D_\mu^\dagger D^\mu \varphi - \frac{\nabla^2 M(\varphi)}{M(\varphi)}\varphi - \frac{\partial V(\varphi)}{\partial \varphi} - \{f\overline{\psi}\psi\}$$
$$- \frac{\xi(\varphi)}{m^2(\varphi)}\left(\varphi D_\mu^\dagger D^\mu \varphi - \frac{\nabla^2 M(\varphi)}{M(\varphi)}\varphi^2 - 4V(\varphi) - \{f\varphi\overline{\psi}\psi\}\right) = 0 \tag{25}$$

where $D_\mu^\dagger$ is the Hermitian conjugate of D_μ, $\{f\overline{\psi}\psi\}$ stands for the derivative $\partial\{f\varphi\overline{\psi}\psi\}/\partial\varphi$ and

$$\xi(\varphi) = \frac{1}{2}\frac{\partial m^2(\varphi)}{\partial \varphi} \tag{26}$$

or, writing explicitly the multiplet indices

$$\xi_A(\varphi) = \frac{1}{2}\frac{\partial m^2(\varphi)}{\partial \varphi^A} = \xi_{AB}\,\varphi^B. \tag{27}$$

Note that the left-hand side of (25) when multiplied by φ with the summation over the scalar multiplet index becomes identically zero.

For classical vacuum configurations we can set $A = 0$, $\psi = \overline{\psi} = 0$, $\varphi = C(x)\,\varphi_0$ with some scalar function $C(x)$ and $\varphi_0 =$ constant (pure conformal gauge), and one gets from (25) the equation

$$\frac{\partial V(\varphi)}{\partial \varphi} - 4\frac{\xi(\varphi)}{m^2(\varphi)}\,V(\varphi) = 0 \tag{28}$$

which is simply the extremum condition of the conformally invariant function $V(\varphi)/m^4(\varphi)$, or, in other words, of the function $V(\varphi)$ on the hypersurface given by the equation $m^2(\varphi) =$ constant in the space of φ. Due to the positivity property of the form $m^2(\varphi)$, this hypersurface is compact so (28) has non-trivial solutions. Among them there are those which minimize the potential $V(\varphi)$ on the hypersurface considered. The values φ_0 of this solution will determine in the usual way the fermionic masses through the Yukawa coupling terms in the Lagrangian and the masses of the gauge vector bosons through the gauge interactions.

Variation of the action (20) over the fermionic fields yields the following equations of motion

$$\mathrm{i}\left(\gamma^\mu(\partial_\mu + A_\mu)\psi + \frac{1}{2\sqrt{-g}}\partial_\mu(\sqrt{-g}\gamma^\mu)\psi\right) + \frac{1}{4}\varepsilon^a{}_{bcd}\,e^\mu_a\,\Gamma^{bc}{}_\mu\gamma^d\gamma^5\psi$$
$$- \frac{3}{8m^2(\varphi)}(\overline{\psi}\gamma_\mu\gamma^5\psi)\gamma^\mu\gamma^5\psi - \{f\varphi\psi\} = 0. \tag{29}$$

Variation of the action (20) over the metric with the equations of motion (29) taken into account yields

$$\begin{aligned}
m^2(\varphi)\big(R_{\mu\nu}(\Gamma) - \tfrac{1}{2}g_{\mu\nu}R(\Gamma)\big) \\
&= (\nabla_\mu\nabla_\nu - g_{\mu\nu}\nabla^2)m^2(\varphi) + D^\dagger{}_{(\mu}\varphi D_{\nu)}\varphi \\
&\quad - 6\partial_\mu M(\varphi)\partial_\nu M(\varphi) - \frac{\mathrm{i}}{2}(\overline{\psi}\gamma_{(\mu}D_{\nu)}\psi - \overline{(D_{(\mu}\psi)}\gamma_{\nu)}\psi) + \frac{1}{e^2}\operatorname{Tr} F_{\mu\sigma}F_\nu{}^\sigma \\
&\quad - \tfrac{1}{2}g_{\mu\nu}\bigg(|D\varphi|^2 - 6(\nabla M(\varphi))^2 + 2V(\varphi) \\
&\quad - \frac{3}{8m^2(\varphi)}(\overline{\psi}\gamma^\sigma\gamma^5\psi)(\overline{\psi}\gamma_\sigma\gamma^5\psi) + \frac{1}{2e^2}\operatorname{Tr} F_{\sigma\tau}F^{\sigma\tau}\bigg).
\end{aligned} \tag{30}$$

The equations obtained are invariant with respect to the local conformal transformations considered above. All the observables of the theory are also regarded as invariant. This allows one to fix the conformal gauge freedom by imposing some appropriate condition on the solutions. Two such conditions are especially convenient as can be seen from the Lagrangian (21) or from the equations of motion written just above. The first one is the gauge already mentioned, which is defined by the equation

$$m^2(\varphi) = \text{constant}. \tag{31}$$

The second is the gauge-fixing condition

$$M^2(\varphi) = \text{constant}. \tag{32}$$

3. Classical cosmology

In considering the cosmological implications of the theory proposed, we would like to start with a discussion of the physical interpretation of classical solutions. In a locally conformally invariant theory such interpretation differs somewhat from that of the standard theory of gravity.

Let us first assume that the scalar fields are already in their vacuum state, which corresponds to the present state of the universe. In standard (not conformally invariant) theory this would imply that their values are constant in space and time and minimize the effective potential. In conformally invariant theory this requirement means, according to (28), that the scalar field values minimize the function $V(\varphi)/m^4(\varphi)$. From this last condition it follows that

$$\varphi(x) = C(x)\,\varphi_0 \tag{33}$$

where $C(x)$ is an arbitrary function of spacetime coordinates, and φ_0 are some fixed non-zero constant values that minimize the function $V(\varphi)/m^4(\varphi)$. Using gauge invariance of the theory, as expressed by equations (10)–(15), we can always gauge the function $C(x)$ to a constant in any spacetime region in which it is nowhere vanishing. Our aim, however, is to see what kind of expressions we get in an arbitrary gauge.

Let us start the discussion with the question of proper time determination. In the standard theory of gravity, the interval

$$\int \mathrm{d}s = \int \sqrt{-g_{\mu\nu}\frac{\mathrm{d}x^\mu}{\mathrm{d}\tau}\frac{\mathrm{d}x^\nu}{\mathrm{d}\tau}}\,\mathrm{d}\tau \tag{34}$$

along any timelike curve $x^\mu(\tau)$ is regarded as an observable value. This is precisely the value of the proper time interval as measured by a small ideal test clock moving along the given curve. Clearly, such a value cannot correspond to any observable in a conformally invariant theory, if only because it is not conformally invariant. Let us therefore obtain the expression in the conformally invariant theory for the proper time interval measured by a test clock. To do this note that in the theory under consideration, the behaviour of a clock is to be described by the conformally invariant action. The only dimensional (in the system of units chosen $c = \hbar = 1$) parameters which will describe the clock are the masses of the particles which constitute it, and those will be proportional to the local values of the background scalar fields φ due to the Yukawa couplings in the Lagrangian (2). The frequency of the clock's internal 'oscillations' being a value of dimension mass in the system of units chosen, it will be proportional to the masses of the particles†, hence, again to the local values of φ. From all this it follows that the time shown by the clock (the number of clock internal 'oscillations') will be given (up to a constant which will depend on the specific structure of the clock) by the expression

$$\int C(s)\,\mathrm{d}s = \int C\big(x(\tau)\big)\sqrt{-g_{\mu\nu}\frac{\mathrm{d}x^\mu}{\mathrm{d}\tau}\frac{\mathrm{d}x^\nu}{\mathrm{d}\tau}}\,\mathrm{d}\tau \tag{35}$$

which is conformally invariant‡. Another way to understand the result (35) is to note that in the gauge $C(x) =$ constant the matter part of the action (20) will be the standard matter action on the curved background, with the masses of all the particles being constant. Therefore the evolution of matter will be essentially the same as in the ordinary theory on the corresponding background metric. Thus in the gauge considered the readings of a clock will be given by the expression proportional to (34). To obtain the expression for the same value in terms of the metric in some other gauge it is sufficient to perform the corresponding gauge transformation; we will obtain equations (33) and (35).

Let us now consider, from the same point of view, cosmological solutions of the conformally invariant theory. The metric line element which is described by the scale factor $a(\eta)$ then looks like

$$\mathrm{d}s^2 = a^2(\eta)(-\mathrm{d}\eta^2 + \mathrm{d}l^2) \tag{36}$$

where $\mathrm{d}l^2$ denotes the (constant in time) spatial section of the metric, and the background scalar fields are spatially homogeneous, $\varphi = \varphi(\eta) = C(\eta)\,\varphi_0$, where η is the conformal time. Let us choose any of the two time directions and ask whether the universe which is described by the metric (36) and by the homogeneous background scalar fields $\varphi(\eta) = C(\eta)\,\varphi_0$ will be observed as expanding (contracting) in this time direction. First of all we have

† For example, the Rydberg energy of the hydrogen atom in the units $\hbar = 1$ is equal to $\frac{1}{2}e^4 m$, and is proportional to the reduced mass m of the proton–electron system.

‡ In [9] it was argued that extrema of the expressions of such a type (regarded as path-functionals) will also determine conformally invariant trajectories for massive particles, provided their masses are generated dynamically by a conformal scalar field.

to specify the meaning of the universe observable expansion (contraction). To do this we must introduce comoving observers who can exchange, say, electromagnetic signals. The universe will appear as expanding (contracting) if these observers detect the signals redshifted (blueshifted). Now what one has to do is to take into account that emitters and detectors carried by the observers are described by the conformally invariant laws. Using (35) for the time interval measurements which apply to emitters and detectors as well as to clocks, one can easily conclude that the cosmological redshift (blueshift) will be given by the expression

$$1 + z = \frac{C(\eta_{\rm d})\, a(\eta_{\rm d})}{C(\eta_{\rm e})\, a(\eta_{\rm e})} \tag{37}$$

where $\eta_{\rm e}$ is the emission time of a signal by one of the observers and $\eta_{\rm d}$ is the detection time of that signal by the other observer. We thus see that the answer to the question posed will be that the universe is expanding (contracting) if the relevant conformally invariant combination $C(\eta)a(\eta)$ is increasing (decreasing) in the time direction chosen. From (37) it follows that if one chooses the gauge in which the function C is constant in time then one will obtain the familiar result for the redshift (blueshift) in terms of the scale factor. Therefore physical interpretation of cosmological solutions of the theory will be most convenient in such a gauge. On the other hand, even though by a conformal transformation we can always proceed to the conformal gauge in which the metric is static, $a(\eta) \equiv$ constant, this does not mean that the physical universe is not expanding.

So far our discussion has been concerned with the solutions with vacuum φ-configurations given by (33). Such scalar field configurations correspond to the present state of the universe. In the past, however, the scalar fields φ might have evolved in a non-trivial manner. In the standard theory of gravity, evolution of the scalar fields far from the minimum of their potential in many cases leads to inflation [10], which helps to resolve a large number of the cosmological problems of the standard scenario. It is therefore important to see whether similar evolution can take place in the conformally invariant theory. As is usual for such an analysis, we shall assume that the scalar field contribution dominates in the energy–momentum tensor, so the rest of the matter fields will be disregarded.

Consider the universe described by the Friedmann–Robertson–Walker metric

$$ds^2 = -dt^2 + a^2(t)\, dl^2 \tag{38}$$

and by homogeneous scalar fields $\varphi(t)$. The expression dl^2 as before denotes the (constant in time) homogeneous spatial section of the metric. Conformal time η is related to t by $dt = a\, d\eta$. Before studying the cosmological evolution in the conformally invariant theory, first recall the situation in standard cosmology†. In the standard theory the simplest condition for inflation is

$$|\dot{H}| \ll H^2 \tag{39}$$

where $H \equiv \dot{a}/a$ is the Hubble parameter. A sufficiently long period of universe expansion in the regime (39) provides the condition that the comoving spatial scale, which corresponds to the observable part of the universe, is well within the causally connected region [10]. It is therefore the causal structure of the universe which is of relevance to the problems of the standard cosmological model, a fact which will play a crucial role in our following

† Here and below by 'standard' we mean not conformally invariant.

reasoning. By causal structure of the universe we mean causal relations between spacio-temporal events, determined by the conformal (light-cone) structure of the metric. Note that it is these causal relations that constitute the only true observables in physics. Indeed, all our physical observations are nothing but observations of spatial and temporal coincidences or successions of different events.

In the locally conformally invariant theory the scale factor $a(t)$ is not a physically meaningful value by itself. It can be arbitrarily changed by a time-dependent conformal transformation†. We cannot, therefore, straightforwardly apply the inflationary criterion (39) to the scale factor. Even if in one of the gauges we obtain a solution for which the condition (39) is valid at some definite moment of time we can always proceed to another gauge in which this condition is violated at this moment of time. We have no other possibility than to speak about inflation in the sense of the causal structure of the universe, which may or may not be similar to the causal structure of the universe in the ordinary inflationary scenario. Note that the causal structure of the universe cannot be affected by a conformal transformation: such a transformation does not change the positions of the events on the spacetime manifold, neither does it modify the conformal structure of the metric.

It is clear that one can obtain solutions of the theory in any convenient gauge with any convenient choice of coordinates and after that proceed to any other convenient gauge and coordinates for the analysis of the causal structure. Although we cannot directly apply the criterion (39) as a criterion for inflation, we can nevertheless make use of the condition (39), if it holds for the solution in some gauge, for the analysis of the conformal structure of the metric and thereby of the causal structure of the universe. This is precisely what we shall do below.

So, let us first investigate the solutions and after that discuss their physical interpretation. One of the most convenient gauge conditions for the system of equations (25), (30) is the gauge condition (31). In this gauge the system of cosmological equations for the metric given by (38) and for the spatially homogeneous scalar fields looks like

$$H^2 + \frac{\kappa}{a^2} = \frac{1}{3m^2}\left(\frac{1}{2}\dot\varphi^2 - \frac{(\varphi\dot\varphi)^2}{12M^2} + V\right) \tag{40}$$

$$\dot H - \frac{\kappa}{a^2} = -\frac{1}{m^2}\left(\frac{1}{2}\dot\varphi^2 - \frac{(\varphi\dot\varphi)^2}{12M^2}\right) \tag{41}$$

$$\ddot\varphi + 3H\dot\varphi - \frac{\ddot M + 3H\dot M}{M}\varphi + \frac{\partial V}{\partial\varphi} - \frac{\xi(\varphi)}{m^2}\left(\varphi\ddot\varphi + 3H\varphi\dot\varphi - \frac{\ddot M + 3H\dot M}{M}\varphi^2 + 4V\right) = 0 \tag{42}$$

where $H \equiv \dot a/a$, as before, is the Hubble parameter, $\kappa = 0, \pm 1$ determines the spatial curvature of the universe, and the quantity $\xi(\varphi)$ was defined in (26). Dots denote the derivatives with respect to the cosmological time t. It is easy to see that

$$\dot\varphi^2 - \frac{(\varphi\dot\varphi)^2}{6M^2} \geqslant \dot\varphi^2 - \frac{(\varphi\dot\varphi)^2}{\varphi^2} \geqslant 0. \tag{43}$$

Hence, if we assume the scalar field potential $V(\varphi)$ to be non-negative, the right-hand side of (40) will be non-negative and the right-hand side of (41) will be non-positive.

† Note that spatially homogeneous conformal transformations in general change the range of the time coordinate t, but they do not change the range of conformal time coordinate η.

Let us discuss formal inflationary solutions to the system (40)–(42). The condition for inflation $|\dot{H}| \ll H^2$ is satisfied when

$$\dot{\varphi}^2 - \frac{(\varphi\dot{\varphi})^2}{6M^2} \ll V(\varphi). \tag{44}$$

During inflation it soon becomes possible to neglect the second time derivatives of the scalar fields (the latter begin to roll slowly down the potential $V(\varphi)$ on the hypersurface $m^2(\varphi) = \text{constant}$). From (40) and (42) we then have a very rough estimate

$$\dot{\varphi}^2 \sim \frac{V(\varphi)}{\varphi^2} m^2. \tag{45}$$

From (44) we obtain the following inflation and slow-rolling condition

$$m^2(\varphi) \ll \varphi^2 \tag{46}$$

or, taking into account the definition (7)

$$\xi_{AB} \ll 1. \tag{47}$$

This estimate is not difficult to understand. The value $m^2(\varphi)$, which is constant in the gauge chosen, determines the Planck mass M_{P} through

$$m^2(\varphi) = \frac{M_{\mathrm{P}}^2}{8\pi}. \tag{48}$$

In terms of the Planck mass the condition (47) can be written as

$$M_{\mathrm{P}}^2 \ll \varphi^2 \tag{49}$$

and looks quite familiar to those who deal with chaotic inflationary cosmology (see [10]).

Neglecting terms with the second time derivative of φ in (42) and using the condition (47) we are able to write down the following approximate equation for the scalar fields φ during the period of slow rolling

$$3H\left(\dot{\varphi} - \frac{(\varphi\dot{\varphi})}{\varphi^2}\varphi\right) + \frac{\partial V(\varphi)}{\partial\varphi} - 4\frac{\xi(\varphi)}{m^2(\varphi)}V(\varphi) = 0. \tag{50}$$

Note that if multiplied by φ with the summation over the scalar multiplet index this equation gives identically zero, as is also the case with the exact equation of motion for φ. The slow-rolling regime of the scalar field dynamics terminates as the scalar field values approach close to a (possibly local) minimum of the value $V(\varphi)/m^4(\varphi)$. After this the scalar fields start oscillating with decreasing amplitude around their stationary point. Due to various couplings between fields, such oscillations give birth to particles of various kinds. This process heats the universe in the standard manner.

Now that the solutions have been formally analysed let us turn to the question of their physical interpretation. From the above discussion it must be clear that to give such an interpretation means to reveal the causal structure which they describe. It is only this causal structure which is of importance with regard to any of the cosmological problems, including horizon and flatness problems. Our basic observation is that the causal structure of the universe described by our solutions will be quite similar to that of the inflationary universe in the standard cosmology. To see this it is sufficient to note that the spacetime coordinates (in the coordinate system $(\eta,\ x^i)$ or $(t,\ x^i)$) of all the relevant cosmological events in our model will be the same as in the standard cosmology (in the corresponding standard coordinates) and that the conformal structure of the metric in these two cases will also be the same. The second statement is quite obvious since the metric of our solution in the gauge chosen (31) is just the same as the metric in the ordinary inflationary universe scenario. The first statement is true for the following reasons. The local vacuum values (33) of the scalar fields after inflation will be constant in the gauge (31). Hence according to the arguments given at the beginning of this section, in the gauge considered the metric after inflation will represent the physical observable metric (the function $C(\eta)$ in (33), (35) and (37) will be constant). The Lagrangian (21) for the constant vacuum values of φ reduces to the usual Hilbert–Einstein Lagrangian for gravity, supplemented by the usual matter content. Therefore the evolution of matter on the cosmological background described by our solutions, in the gauge considered, will look quite similar to the evolution of matter in the standard cosmological model. The space and time coordinates of all the cosmological events of interest (helium synthesis, hydrogen recombination, galaxy and cluster formation and evolution, etc) will be of exactly the same character as in the standard cosmological model.

To check whether our solutions are free from the horizon problem it remains to see whether the comoving scale, which corresponds to the observable part of the universe, will lie within the causally connected region. Since there is no difference in the causal structure between our solutions and those of the ordinary inflationary cosmology we can use the standard analysis [10]. We know that in order for the horizon problem to be solved, our solutions must exhibit a sufficient amount ($\gtrsim 50$ e-fold) of inflation. This condition can be shown to require sufficiently small values of the constants ξ_{AB}, typically

$$\xi_{AB} \lesssim 10^{-1}. \tag{51}$$

This estimate, in general, will be model-dependent. Another condition, which is necessary for the metric to be treated quasiclassically, is

$$V(\varphi) \lesssim m^4(\varphi) \tag{52}$$

which typically implies

$$\lambda_{ABCD} \lesssim \xi_{EF}^2. \tag{53}$$

More stringent constraints on at least some of the the values of λ_{ABCD}, namely

$$\lambda_{ABCD} \sim 10^{-14} \tag{54}$$

will most probably follow from the observations of the large-scale anisotropy of the cosmic background radiation. Such constraints will be analogous to the similar constraints [10, 11] on the values of the inflation couplings in the ordinary inflationary scenario. In this paper we will not study this question.

It it worth clarifying the main circumstance which enabled us to simplify the analysis of the causal structure of the universe in the conformally invariant theory. In the case considered it was the property of the gauge condition (31), which for the vacuum configurations of the scalar fields rendered their values constant in space and time. And for the constant vacuum values of φ, the Lagrangian (21), in its turn, reduces to the familiar one, for which the analysis is well elaborated. It is clear that any gauge condition with the property mentioned immediately above will do the same job. For example, we could adopt the gauge condition $\varphi^2 =$ constant, and the analysis of the causal structure would proceed along the same lines.

We would like to emphasize once again that, although the cosmological solutions were analysed in the particular gauge $m^2(\varphi) =$ constant, the causal structure of the universe (determined by the positions of the events on the spacetime manifold and by the conformal structure of the metric) is conformally gauge-independent. In this section the conditions were found for which our model admits solutions with causal structure like that of the ordinary inflationary universe. It is quite natural to call the corresponding solutions 'inflationary'. In connection with this, note that the conditions (51)–(53), which are necessary for the solution of the horizon problem, are manifestly gauge-independent.

When illustrating the inflationary universe scenario in the context of the standard theory of gravity one often introduces hypothetical observers who exist in the inflationary stage and are capable of observing the universe expansion by exchanging signals between each other. In the standard theory of gravity it is taken for granted that such hypothetical observers are made of matter which couples to gravity in a generally covariant manner, so that the proper time interval (34) is the value which those observers actually can measure. In the conformally invariant theory we might also introduce such hypothetical observers who exist in the inflationary stage. But in doing this we are to fulfil the requirement that these observers couple to gravity in a conformally invariant manner. As a result, such observers will be incapable of measuring the conformally non-invariant value (34). Instead, their clocks will show the time which in the theory considered is expressed similarly to (35), with some scalar function $C(x)$, which will depend on the clocks' structure and transform like a scalar field under the conformal transformations. This function will somehow depend on the evolution of the scalar fields φ. The observers being *hypothetical*, this scalar function will be quite unspecified. That there is no universal relation between the readings of the clocks and the field variables of the conformally invariant theory, makes it useless to introduce hypothetical observers for illustrating inflation in this theory.

We end this section with a few comments about the cosmological constant. As it was noted in the introduction, conformal invariance of the theory forbids the usual Λ-term in the original action. However, the effective present-day Λ-term will be generated by the value V_0/m_0^4 of the function $V(\varphi)/m^4(\varphi)$ at its local minimum, as can be seen from the Lagrangian (21). The value of the effective Λ can be estimated as $\Lambda \sim (\lambda/\xi^2)M_P^4$ (here λ has the order of magnitude of the λ's in (8), and ξ has the order of magnitude of the ξ's in (7)). For this Λ-term to vanish some fine tuning of the scalar field potential parameters λ_{ABCD} seems to be necessary. If we assume the scalar field potential $V(\varphi)$ to be non-negative then to have $V_0 = 0$ at the minimum it is necessary and sufficient that a $\varphi_0 \neq 0$ exists for which $V(\varphi_0) = 0$. An interesting suggestion is that this property can somehow be provided by other symmetries of the theory (besides the local conformal one).

4. Quantum cosmology

In this section our aim is to see how the standard formalism of quantum cosmology (for reviews see [12–15]) is applied to the conformally invariant theory, which we study in this paper. For simplicity we shall restrict ourselves only to the scalar-gravitational sector of the full theory (see (20) and (21) for the action), which is described by the Lagrangian

$$L = \tfrac{1}{2}m^2(\varphi)R(\Gamma) - \tfrac{1}{2}(\nabla\varphi)^2 + 3\big(\nabla M(\varphi)\big)^2 - V(\varphi) \tag{55}$$

and by the action

$$S = \int_M L\sqrt{-g}\,\mathrm{d}^4x + \int_{\partial M} m^2(\varphi)\mathcal{K}(\Gamma)\sqrt{h}\,\mathrm{d}^3x \tag{56}$$

in which the boundary term was explained in section 2 after (21).

The metric g in the ADM form (see [12] for the best description) is written as

$$g = -(N^2 - N_iN^i)\,\mathrm{d}t \otimes \mathrm{d}t + N_i(\mathrm{d}t \otimes \mathrm{d}x^i + \mathrm{d}x^i \otimes \mathrm{d}t) + h_{ij}\,\mathrm{d}x^i \otimes \mathrm{d}x^j. \tag{57}$$

Latin indices from the middle of the alphabet run from 1 to 3. The quantities N and N_i are the lapse function and the shift vector, respectively. Substituting the metric in this form into the action (56), we obtain for a spatially closed universe

$$S = \int_M \mathcal{L}\,\mathrm{d}^3x\,\mathrm{d}t \tag{58}$$

where the Lagrangian density $\mathcal{L}$ is given by

$$\begin{aligned}\mathcal{L} = {}& \tfrac{1}{2}N\sqrt{h}\,m^2(\varphi)(\mathcal{R} + \mathcal{K}_{ij}\mathcal{K}^{ij} - \mathcal{K}^2) \\ & + N\sqrt{h}\left(\frac{2\mathcal{K}}{N}\xi_A(\varphi)\dot{\varphi}^A - \left(m^2(\varphi)\right)^{|i}{}_{|i} - \frac{2\mathcal{K}N^i}{N}\xi_A(\varphi)\partial_i\varphi^A - V(\varphi)\right) \\ & + N\sqrt{h}\,T_{AB}(\varphi)\left(\frac{1}{2N^2}\dot{\varphi}^A\dot{\varphi}^B - \frac{N^i}{N^2}\dot{\varphi}^A\partial_i\varphi^B \right.\\ & \left. - \frac{1}{2}\left(h^{ij} - \frac{N^iN^j}{N^2}\right)\partial_i\varphi^A\partial_j\varphi^B\right)\end{aligned} \tag{59}$$

and

$$\mathcal{K}_{ij} = \frac{1}{2N}(2N_{(i|j)} - \dot{h}_{ij}) \tag{60}$$

is the extrinsic curvature of the hypersurface $t = \text{constant}$ in the ADM form. Vertical lines denote covariant derivatives with respect to the metric h_{ij}, with the aid of which also small Latin indices are raised and lowered, $\mathcal{R}$ is the three-curvature scalar built of the metric h_{ij}, and

$$T_{AB}(\varphi) = \delta_{AB} - \frac{6}{M^2(\varphi)}\left(\xi_A(\varphi) + \tfrac{1}{6}\varphi^A\right)\left(\xi_B(\varphi) + \tfrac{1}{6}\varphi^B\right) \tag{61}$$

is a non-degenerate symmetric matrix of signature $(-, +, +, \ldots, +)$.

The theory described by the Lagrangian (59) is degenerate: due to invariance with respect to coordinate reparametrization the time derivatives (velocities) of the values N and N_i do not enter the Lagrangian at all; due to local conformal invariance the matrix of the second derivatives of $\mathcal{L}$ with respect to the velocities $\dot{h}_{ij}$ and $\dot{\varphi}^A$ has one zero eigenvalue. So some linear combination of the velocities $\dot{h}_{ij}$ and $\dot{\varphi}^A$ will not be expressed through the corresponding generalized momenta. As a consequence a new constraint will appear.

Let us denote the generalized momenta for the variables h_{ij} and φ^A by p^{ij} and p_A, respectively. Then, given the Lagrangian (59), we can express the velocities $\dot{\varphi}^A$ through the corresponding momenta

$$\dot{\varphi}^A = N^i \partial_i \varphi^A + \frac{N}{\sqrt{h}} T^{AB}(\varphi)\big(p_B - 2\sqrt{h}\,\mathcal{K}\xi_B(\varphi)\big) \tag{62}$$

where $T^{AB}(\varphi)$ is the matrix inverse of $T_{AB}(\varphi)$. Trying then to express the velocities $\dot{h}_{ij}$, we find that this is possible only for the traceless part of the tensor $\dot{h}_{ij}$. We have

$$\widetilde{\mathcal{K}}^{ij} = -\frac{2}{m^2(\varphi)\sqrt{h}}\widetilde{p}^{ij} \tag{63}$$

where $\widetilde{\mathcal{K}}^{ij}$ and $\widetilde{p}^{ij}$ are the traceless parts of $\mathcal{K}^{ij}$ and p^{ij}, respectively, the values $\mathcal{K}_{ij}$ being given by (60). The trace $p = p^{ij}h_{ij}$ is involved in the constraint

$$\mathcal{F} \equiv \tfrac{1}{3}\big(p^{ij}h_{ij} - \tfrac{1}{2}p_A\varphi^A\big) = 0. \tag{64}$$

The action written in the Hamiltonian form is

$$S = \int_M \big(p_A\dot{\varphi}^A + p^{ij}\dot{h}_{ij} - v\mathcal{F} - N\mathcal{H} - N^i\mathcal{H}_i\big)\,\mathrm{d}^3x\,\mathrm{d}t. \tag{65}$$

The variables v, N and N^i are Lagrange multipliers. Variations with respect to them give constraint equations. The expression for $\mathcal{F}$ is written in (64). Expressions for $\mathcal{H}$ and $\mathcal{H}_i$ are

$$\begin{aligned}\mathcal{H} = {} & \frac{2}{\sqrt{h}m^2(\varphi)}\widetilde{p}^{ij}\widetilde{p}_{ij} + \frac{1}{2\sqrt{h}}T^{AB}(\varphi)\,p_Ap_B - \frac{1}{2}m^2(\varphi)\sqrt{h}\,\mathcal{R} \\ & + \sqrt{h}\big(m^2(\varphi)\big)^{|i}{}_{|i} + \frac{1}{2}\sqrt{h}\,h^{ij}T_{AB}(\varphi)\partial_i\varphi^A\partial_j\varphi^B + \sqrt{h}\,V(\varphi)\end{aligned} \tag{66}$$

$$\mathcal{H}_i = -\tfrac{1}{3}\big(\varphi^A p_A\big)_{|i} - 2\widetilde{p}^{\,j}_{i|j} + \partial_i\varphi^A p_A. \tag{67}$$

Note that the trace of p^{ij} does not enter the constraints $\mathcal{H}$ and $\mathcal{H}_i$. To simplify the equations of the theory it is convenient to make a canonical transformation to new canonical variables related to the old ones through the following equations

$$\widetilde{h}_{ij} = h^{-1/3}h_{ij} \qquad \widetilde{\pi}^{ij} = h^{1/3}\widetilde{p}^{ij} \tag{68}$$

$$h = \det(h_{ij}) \qquad \pi_h = \frac{1}{3h}\left(p^{ij}h_{ij} - \frac{1}{2}p_A\varphi^A\right) = \frac{1}{h}\mathcal{F} \tag{69}$$

$$\chi^A = h^{1/6}\varphi^A \qquad \pi_A = h^{-1/6}p_A. \tag{70}$$

The meaning of the first line (68) in the above relations is the following: $\widetilde{h}_{ij}$ is a function of some five parameters (not specified here) which determine this symmetric matrix with unitary determinant, $\widetilde{\pi}^{ij}$ is a function of the corresponding five generalized momenta. Denoting also $h^{-1/6}N = \widetilde{N}$, $h^{1/6}\mathcal{H} = \widetilde{\mathcal{H}}$, we will have $N\mathcal{H} = \widetilde{N}\widetilde{\mathcal{H}}$, and

$$S = \int_M \left(\pi_A\dot{\chi}^A + \widetilde{\pi}^{ij}\dot{\widetilde{h}}_{ij} - \left(v - \frac{\dot{h}}{h}\right)\mathcal{F} - \widetilde{N}\widetilde{\mathcal{H}} - N^i\mathcal{H}_i\right)\mathrm{d}^3x\,\mathrm{d}t \tag{71}$$

where in terms of new variables

$$\mathcal{F} = h\pi_h \tag{72}$$

$$\widetilde{\mathcal{H}} = \frac{2}{m^2(\chi)}\widetilde{\pi}^{ij}\widetilde{\pi}_{ij} + \tfrac{1}{2}T^{AB}(\chi)\pi_A\pi_B - \tfrac{1}{2}m^2(\chi)\widetilde{\mathcal{R}} \\ + \left(m^2(\chi)\right)^{\cdot i}{}_{\cdot i} + \tfrac{1}{2}\widetilde{h}^{ij}T_{AB}(\chi)\partial_i\chi^A\partial_j\chi^B + V(\chi) \tag{73}$$

$$\mathcal{H}_i = -\tfrac{1}{3}\left(\chi^A\pi_A\right)_{\cdot i} - 2\widetilde{\pi}^{j}_{i\cdot j} + \partial_i\chi^A\pi_A. \tag{74}$$

Dots in front of small Latin indices denote covariant derivatives with respect to the metric $\widetilde{h}_{ij}$, $\widetilde{\mathcal{R}}$ is the curvature scalar built of this metric. All small Latin indices in (73), (74) and below are also lowered and raised by the metric $\widetilde{h}_{ij}$ and its inverse $\widetilde{h}^{ij}$.

From the expression (71) for the action we see that the variables h, π_h are non-dynamical. Their role becomes manifest if we write down a formal quantum path integral of the theory described by the action (71). For the wavefunction Ψ of the universe we will have

$$\Psi = \int e^{iS}\,[d\widetilde{N}]\,[dN_i]\,[dv]\,[d\pi_h]\,[dh]\,[d\widetilde{\pi}^{ij}]\,[d\widetilde{h}_{ij}]\,[d\pi_A]\,[d\chi^A]. \tag{75}$$

In this path integral it is convenient to shift the integration variable $v \to v + \dot{h}/h$ (see the expression (71) for the action). Then the integrals over the (shifted) Lagrange multiplier v and over the canonical variables π_h and h result in the overall factor

$$\int e^{-i\int v\mathcal{F}\,d^3x\,dt}\,[dv]\,[d\pi_h]\,[dh] = \int \delta[\mathcal{F}]\,[d\pi_h]\,[dh] = \int [d\sigma] \tag{76}$$

where by σ we denote the integration variable $\ln h$. In deriving (76), the expression (72) for the constraint $\mathcal{F}$ has been taken into account. The last path integral over σ is just the integral over the gauge conformal group. We see that this integral has been factorized automatically.

The quantum operator constraint $\widehat{\mathcal{F}}$ imposed on the wavefunction with the operator arrangement as written in (72) implies that the wavefunction does not depend on the variable h. Note that the remaining variables $\widetilde{h}_{ij}$, χ^A are conformally invariant. Thus the additional constraint implies that the wavefunction depends only on conformally invariant combinations of the initial variables, that is, it is conformally invariant. The constraint $\widetilde{\mathcal{H}}_i$ as usual means invariance of the wavefunction with respect to three-metric and matter field variations induced by coordinate diffeomorphisms. All the dynamical content of the theory considered is expressed by the analogue of the Wheeler–DeWitt equation

$$\widehat{\widetilde{\mathcal{H}}}\Psi = 0. \tag{77}$$

It might seem that by elimination of the conformally non-invariant variable h from the equations of the theory we have succeeded in building a consistent conformally invariant quantum gravity. This would be the case if the theory defined by the constraint equation (77) were well defined. But the latter condition is not true in our theory in the same way as the similar statement is not true in the Einstein theory of gravity. The reason is the infinity of number of the degrees of freedom together with high non-linearity of the theory. By all this we are led to the necessity of regularizations. Nevertheless, it is interesting to note that the theory developed here is no 'worse' in this respect than the usual Einstein theory of gravity. Ascribing one meaning or the other to (77) (for example, restricting it to minisuperspace) we obtain a well defined theory. Only in our case we are to respect the possible remnants of both the coordinate reparametrization invariance and local conformal invariance. In the following section we shall consider the quantum theory of minisuperspace based on our model, which is invariant with respect to spatially homogeneous conformal transformations.

The basic features of the constraint $\widetilde{\mathcal{H}}$ resemble those of the well known Wheeler–DeWitt constraint in the Einstein theory of gravity. The quadratic form in momenta in (73) is non-degenerate and has eigenvalue signs $(-, +, \ldots, +)$ and the potential part of $\widetilde{\mathcal{H}}$ is not bounded from below. In our theory, however, a negative sign comes from the matter (not metric) kinetic term in the expression (73) for $\widetilde{\mathcal{H}}$. A pseudo-Euclidean superspace metric signature enables us to have a timelike variable as in standard quantum cosmology. Such a variable can be taken to be proportional to $m(\chi)$.

To introduce a timelike variable explicitly it is convenient once again to proceed to new variables

$$\mu = \sqrt{6}\, m(\chi) \qquad \vartheta^r = \vartheta^r(\chi) \qquad r = 1, \ldots, k-1 \tag{78}$$

with corresponding conjugate momenta π_μ, π_r^ϑ, $r = 1, \ldots, k-1$. If coordinates ϑ^r are chosen so that $\vartheta^r =$ constant are rays in the χ-space which begin at the origin $\chi = 0$, then the quadratic forms in π_μ and π_r^ϑ decouple in $\widetilde{\mathcal{H}}$ and the matter part of the constraint $\widetilde{\mathcal{H}}$ takes the form

$$\begin{aligned}\widetilde{\mathcal{H}}_{\text{matter}} = &-\frac{1}{2}\pi_\mu^2 - \mu_{.i}\mu^{.i} + \frac{1}{6}(\mu^2)^{.i}{}_{.i} + \frac{1}{2\mu^2}E^{rs}(\vartheta)\pi_r^\vartheta\pi_s^\vartheta \\ &+ \frac{1}{6}\left(\frac{\mu}{m(n)}\right)^2\left(n^A_{.i}n^{.i}_B + \frac{1}{M^2(n)}m(n)_{.i}m(n)^{.i}\right) + \frac{1}{2}\mu^4 W(\vartheta)\end{aligned} \tag{79}$$

where $E^{rs}(\vartheta)$, $r, s = 1, \ldots, k-1$, is the symmetric matrix of some positive-definite form, $n^A = n^A(\vartheta)$, $A = 1, \ldots, k$ is a unit vector in the χ-space as a function of ϑ, and

$$W(\vartheta) = \frac{1}{18m^4(n(\vartheta))}V(n(\vartheta)) \tag{80}$$

is the (rescaled) scalar field potential on the hypersurface $m^2(\chi) =$ constant.

As follows from the discussion in the previous section, the elimination of the scale factor h from the theory does not eliminate *physical* cosmological expansion. The variable h is not conformally invariant and therefore cannot play the role of a physical observable, which indicates the physical expansion of the universe. As was argued in the previous section, the question of the physical expansion of the universe must be interpreted in terms of the readings of the observers' instruments. Such readings will be given in the present theory by conformally invariant expressions. Elimination of the variable h simply means choice of the gauge $h =$ constant. Needless to say that the gauge-fixing choice could be made in many different ways and that the physics is independent of this choice.

5. Minisuperspace quantum cosmology

In the previous section we were able to see how the well developed ADM formalism of quantum cosmology is applied to the conformally invariant theory described above. For simplicity we restricted ourselves to only the scalar-gravitational sector of the theory, which is described by the Lagrangian (55) with the action (56). In the previous section a theory of full superspace was considered. For many purposes, however, especially when one deals with quantum cosmology, it is desirable to perform a more simple analysis of a minisuperspace model. A minisuperspace wavefunction can be regarded as a restriction of the full wavefunction to the minisuperspace. Analysis of the minisuperspace quantum solutions will be our main task in what follows.

For definiteness we start with the homogeneous, but not necessarily isotropic, closed cosmology known as Bianchi-IX. The metric for this model can be written as (see e.g. [15])

$$\mathrm{d}s^2 = \frac{1}{2\pi^2}\left(- N^2(t)\,\mathrm{d}t \otimes \mathrm{d}t + \mathrm{e}^{2\alpha(t)}\left(\mathrm{e}^{2\beta(t)}\right)_{ij}\sigma^i \otimes \sigma^j\right) \tag{81}$$

where $N(t)$ is the lapse function, $\{\sigma^i\}$ is a homogeneous basis of one-forms on a unit spatial three-sphere, $\mathrm{e}^{\alpha(t)}$ is the scale factor of the space parametrized by the function $\alpha(t)$ while the symmetric traceless matrix $\beta_{ij}(t)$ parametrizes the anisotropy. The shift vector is absent due to spatial homogeneity of the metric. The overall factor out front is a convenient scaling. The traceless matrix β_{ij} may be chosen to be diagonal. It is convenient to write it in the form

$$\beta_{ij} = \frac{1}{\sqrt{6}}\,\mathrm{diag}\left(\beta_+ + \sqrt{3}\beta_-,\ \beta_+ - \sqrt{3}\beta_-,\ -2\beta_+\right). \tag{82}$$

For the metric (81) and for a homogeneous scalar field multiplet $\varphi(t)$ the action (56), after taking the integral over space, is written as

$$S = \int \mathcal{L}\,\mathrm{d}t \tag{83}$$

with the Lagrangian

$$\mathcal{L} = \frac{\mathrm{e}^{3\alpha}}{2N}m^2(\varphi)\left(\dot\beta_+^2 + \dot\beta_-^2 - 6\dot\alpha^2 + T_{AB}(\varphi)\dot\varphi^A\dot\varphi^B - 12\dot\alpha\xi_A(\varphi)\dot\varphi^A\right)$$
$$+ N\mathrm{e}^{\alpha}m^2(\varphi)R_{\mathrm{IX}}(\beta) - N\mathrm{e}^{3\alpha}V(\varphi) \tag{84}$$

in which the potential $V(\varphi)$ has been rescaled by $2\pi^2$ for convenience, and

$$R_{\mathrm{IX}}(\beta) = \mathrm{Tr}\left(2\mathrm{e}^{-2\beta} - \mathrm{e}^{4\beta}\right) \tag{85}$$

is half of the spatial three-curvature scalar of the unitary ($\alpha = 0$) Bianchi-IX space. The non-degenerate matrix $T_{AB}(\varphi)$ was defined in (61) and the value $\xi_A(\varphi)$ in (27).

The theory described by the Lagrangian (84) is degenerate: due to time reparametrization invariance the time derivative (velocity) of the lapse function N does not enter the Lagrangian at all; due to the remnant of the conformal invariance the matrix of the second derivatives of $\mathcal{L}$ with respect to the velocities $\dot\alpha$ and $\dot\varphi$ has one zero eigenvalue. So, as we already know, some linear combination of the velocities $\dot\alpha$ and $\dot\varphi$ will not be expressed through the corresponding generalized momenta. As a consequence a new constraint will appear.

Let us denote the generalized momenta for the variables φ^A, α, β_+ and β_- by p_A, p, p_+ and p_-, respectively. Then, given the Lagrangian (84), we can express the velocities $\dot{\varphi}^A$ and $\dot{\beta}_\pm$ through the corresponding momenta

$$\dot{\varphi}^A = T^{AB}(\varphi)\big(N\mathrm{e}^{-3\alpha}p_B + 6\dot{\alpha}\xi_B(\varphi)\big) \tag{86}$$

$$\dot{\beta}_\pm = \frac{N}{m^2(\varphi)}\mathrm{e}^{-3\alpha}p_\pm \tag{87}$$

where $T^{AB}(\varphi)$ is the matrix inverse of $T_{AB}(\varphi)$. Trying then to express the velocity $\dot{\alpha}$ through the momenta we find that this is not possible. The value p turns out to be involved in the constraint

$$\mathcal{F}(\varphi^A, p_A, \alpha, p) \equiv p - \varphi^A p_A = 0 \tag{88}$$

which is nothing but an analogue of the constraint (64) in the full superspace.

The action written in the Hamiltonian form is

$$S = \int \big(p_A\dot{\varphi}^A + p_+\dot{\beta}_+ + p_-\dot{\beta}_- + p\dot{\alpha} - v\mathcal{F} - N\mathcal{H}\big)\,\mathrm{d}t. \tag{89}$$

The variables v and N are the Lagrange multipliers. Variations with respect to them give constraint equations. The expression for $\mathcal{F}$ is written above in (88). The expression for $\mathcal{H}$ is

$$\mathcal{H} = \mathrm{e}^{-3\alpha}\left(\frac{1}{2m^2(\varphi)}(p_+^2 + p_-^2) + \frac{1}{2}T^{AB}(\varphi)p_Ap_B\right) - \mathrm{e}^{\alpha}m^2(\varphi)R_{\mathrm{IX}}(\beta) + \mathrm{e}^{3\alpha}V(\varphi). \tag{90}$$

Note that the canonical variable p does not enter the constraint $\mathcal{H}$. To simplify the equations of the theory it is convenient to make a canonical transformation from φ^A, p_A and p to new canonical variables χ^A, π_A and π_α (α, $\beta_\pm$ and $p_\pm$ being unchanged) related to the old ones through the following equations (compare with (68)–(70))

$$\chi^A = \mathrm{e}^{\alpha}\varphi^A \qquad \pi_A = \mathrm{e}^{-\alpha}p_A \qquad \pi_\alpha = \mathcal{F}(\varphi^A, p_A, \alpha, p) = p - \varphi^A p_A. \tag{91}$$

Denoting also $\mathrm{e}^{-\alpha}N = \widetilde{N}$, $\mathrm{e}^{\alpha}\mathcal{H} = \widetilde{\mathcal{H}}$, we will have $N\mathcal{H} = \widetilde{N}\widetilde{\mathcal{H}}$, and

$$S = \int \big(\pi_A\dot{\chi}^A + p_+\dot{\beta}_+ + p_-\dot{\beta}_- - (v - \dot{\alpha})\pi_\alpha - \widetilde{N}\widetilde{\mathcal{H}}\big)\,\mathrm{d}t \tag{92}$$

where in terms of new variables

$$\widetilde{\mathcal{H}} = \frac{1}{2m^2(\chi)}(p_+^2 + p_-^2) + \frac{1}{2}T^{AB}(\chi)\pi_A\pi_B - m^2(\chi)R_{\mathrm{IX}}(\beta) + V(\chi). \tag{93}$$

From the expression for the action (92) we see that the variables α and π_α are non-dynamical. Their role becomes manifest if we write down the formal quantum path integral of the theory described by the action (92). For the wavefunction $\Psi(\alpha, \beta, \chi^A)$ we have

$$\Psi = \int \mathrm{e}^{\mathrm{i}S}\,[\mathrm{d}\widetilde{N}]\,[\mathrm{d}v]\,[\mathrm{d}\pi_\alpha]\,[\mathrm{d}\alpha]\,[\mathrm{d}p_\pm]\,[\mathrm{d}\beta_\pm]\,[\mathrm{d}\pi_A]\,[\mathrm{d}\chi^A]. \tag{94}$$

After shifting the integration variable $v \to v + \dot{\alpha}$ (see the expression (92) for the action) the integrals over the (shifted) Lagrange multiplier v and over the canonical variables π_α and α lead to an overall factor

$$\int \mathrm{e}^{-\mathrm{i}\int v\pi_\alpha \mathrm{d}t}\,[\mathrm{d}v]\,[\mathrm{d}\pi_\alpha]\,[\mathrm{d}\alpha] = \int \delta[\pi_\alpha]\,[\mathrm{d}\pi_\alpha]\,[\mathrm{d}\alpha] = \int [\mathrm{d}\alpha]. \tag{95}$$

The last path integral over $\alpha(t)$ is just the (infinite) integral over the remnant of the local conformal group. It is the analogue of the factor (76) of the theory of full superspace. Again we see that this integral factorizes automatically.

The quantum operator constraint $\widehat{\pi}_\alpha$ imposed on the wavefunction of the universe implies that the wavefunction $\Psi(\alpha, \beta, \chi^A)$ does not depend on the variable α. Note that the remaining variables $\beta_\pm$, χ are conformally invariant. Thus the additional constraint implies that the wavefunction depends only on conformally invariant combinations of initial variables, that is, that it is conformally invariant. All the dynamical content of the theory is then expressed by the analogue of the Wheeler–DeWitt equation for minisuperspace (see [10, 12–15] for a review of the standard minisuperspace quantum cosmology)

$$\widehat{\widetilde{\mathcal{H}}}\Psi = 0. \tag{96}$$

In imposing constraints on the wavefunction, some particular operator ordering is to be chosen. Different choices as usual correspond to different quantum versions of the principal classical theory. We will not discuss this topic here.

Quite similar to the situation in the full superspace (see the remark at the end of the previous section), the elimination of the value α from the theory does not eliminate physical cosmological expansion. The value α being not conformally invariant, by itself it cannot represent any physical observable and cannot thus indicate physical expansion of the universe. Elimination of this variable simply means choice of the gauge $\alpha =$ constant. Physics (which was explained in section 3) is independent of this choice.

The basic features of the constraint $\widetilde{\mathcal{H}}$ were described in general in the previous section. They resemble those of the well known Wheeler–DeWitt constraint in the Einstein theory of gravity (see [12–15]). The quadratic form in momenta in (93) is non-degenerate and has eigenvalue signs $(-, +, \ldots, +)$. A negative sign comes from the matter kinetic term in the expression (93) for $\widetilde{\mathcal{H}}$ and enables us to have a timelike variable as in standard minisuperspace quantum cosmology.

To introduce a timelike variable explicitly it is convenient to proceed to new variables (78) with corresponding conjugate momenta π_μ, π_ϑ^r, $r = 1, \ldots, k-1$. If the coordinates ϑ^r are chosen as in the previous section (see the text following (78)) then the quadratic forms in π_μ and π_r^ϑ decouple in the expression for $\widetilde{\mathcal{H}}$ which then takes the form

$$\widetilde{\mathcal{H}} = \frac{1}{2\mu^2}(p_+^2 + p_-^2) - \frac{1}{2}\pi_\mu^2 + \frac{1}{2\mu^2}E^{rs}(\vartheta)\pi_r^\vartheta\pi_s^\vartheta - \frac{1}{6}\mu^2 R_{\mathrm{IX}}(\beta) + \frac{1}{2}\mu^4 W(\vartheta) \tag{97}$$

where $E^{rs}(\vartheta)$, $r, s = 1, \ldots, k-1$, is the matrix of the same positive-definite form as in (79), and $W(\vartheta)$ is given by (80).

From the description given in the final part of the previous section it is clear that $\{\vartheta\}$ can be regarded as just (arbitrary) coordinates on the unit $(k-1)$-sphere in the χ-space. This in particular has the following consequence. If μ is a good time variable (the wavefunction of the universe has WKB form in μ) then the matter field probability is distributed over a manifold with the topology of a $(k-1)$-sphere. This manifold is compact, hence the potential $W(\vartheta)$ given by (80) is bounded on it and the probability distribution may be well defined on it everywhere. To illustrate this idea, let us consider the case when the wavefunction has a WKB form in the time variable μ and, for simplicity, let us ignore the variables $\beta_\pm$, i.e. put $\beta_\pm = 0$ (this means that we turn to the de Sitter minisuperspace model). Let us also assume that the action of the third term in (97) on the wavefunction is negligible. The effective constraint operator is then

$$\mathcal{H}_{\text{eff}} = -\tfrac{1}{2}\left(\pi_\mu^2 + \mu^2 - \mu^4 W(\vartheta)\right) \tag{98}$$

and the wavefunction is written in the WKB form as

$$\Psi(\mu, \vartheta) = A(\vartheta) \exp\left(I(\mu, \vartheta)\right) \tag{99}$$

where $A(\vartheta)$ is the normalization factor. Hence ϑ (in fact $W(\vartheta)$) play the role of parameters in the equation

$$\widehat{\mathcal{H}}_{\text{eff}} \Psi = 0. \tag{100}$$

In the leading order of the WKB approximation the solution for $I(\mu, \vartheta)$ is

$$I(\mu, \vartheta) = \begin{cases} \pm \dfrac{1}{3W(\vartheta)} \left(1 - \mu^2 W(\vartheta)\right)^{3/2} & \mu^2 W(\vartheta) \leqslant 1 \\[2ex] \pm \dfrac{\mathrm{i}}{3W(\vartheta)} \left(\mu^2 W(\vartheta) - 1\right)^{3/2} & \mu^2 W(\vartheta) \geqslant 1. \end{cases} \tag{101}$$

In order to be able to neglect the action of the term $(1/2\mu^2)E^{rs}(\vartheta)\pi_r^\vartheta \pi_s^\vartheta$ on the wavefunction the normalization factor $A(\vartheta)$ is to be taken as

$$A(\vartheta) = \exp\left(\pm \frac{1}{3W(\vartheta)}\right). \tag{102}$$

The WKB solution (99) is then valid for not very large values of μ. The signs in (101) and (102) are determined by the boundary conditions. Typically the wavefunction is a linear combination of the exponents of (101) with different signs.

Let us consider, as an example, two classical choices of the boundary conditions and the corresponding wavefunctions. The tunnelling boundary condition of Vilenkin [16] demands that there be only an outgoing wave at the singular boundary $\mu \to \infty$ of the minisuperspace. The corresponding wavefunction in the leading order of the WKB approximation is given by

$$\Psi_{\mathrm{T}}(\mu, \vartheta) = \begin{cases} \exp\left(-\dfrac{1}{3W(\vartheta)}\left[1 - \left(1 - \mu^2 W(\vartheta)\right)^{3/2}\right]\right) & \mu^2 W(\vartheta) \leqslant 1 \\[2ex] \exp\left(-\dfrac{1}{3W(\vartheta)}\left[1 + \mathrm{i}\left(\mu^2 W(\vartheta) - 1\right)^{3/2}\right]\right) & \mu^2 W(\vartheta) \geqslant 1. \end{cases} \tag{103}$$

The boundary condition of Hartle and Hawking [17] is formulated in terms of a Euclidean path integral representation of the wavefunction similar to (94). The proposal is that the wavefunction is given by the path integral over compact configurations without the second boundary. Application of this proposal to the model considered here can be performed along just the same lines as it is done for the minisuperspace model based on the Einstein theory of gravity [18]. We obtain the following result

$$\Psi_{\mathrm{H}}(\mu,\vartheta)=\begin{cases}\exp\left(\frac{1}{3W(\vartheta)}\left[1-\left(1-\mu^2W(\vartheta)\right)^{3/2}\right]\right) & \mu^2W(\vartheta)\leqslant 1\\ \exp\left(\frac{1}{3W(\vartheta)}\right)\cos\left[\frac{1}{3W(\vartheta)}\left(\mu^2W(\vartheta)-1\right)^{3/2}-\frac{\pi}{4}\right] & \mu^2W(\vartheta)\geqslant 1.\end{cases} \tag{104}$$

According to the widespread interpretation of the wavefunction [19] the probability density is to be defined through the probability flux vector on any of the hypersurfaces in the minisuperspace which is crossed by the probability flux lines from the same side everywhere. In our case such a hypersurface is most conveniently chosen as $\mu=$ constant. The probability density is then defined in the space of variables $\{\vartheta\}$ and is given by

$$\mathrm{d}P(\vartheta)=J^{(\mu)}(\vartheta)\,\mathrm{d}\Sigma_{(\mu)}(\vartheta) \tag{105}$$

where

$$J^{(\mu)}=\mathrm{Im}\left(\Psi\frac{\partial}{\partial\mu}\Psi^*\right) \tag{106}$$

is the probability flux vector component in the direction of μ, and $\mathrm{d}\Sigma_{(\mu)}(\vartheta)$ is the surface volume element of the surface $\mu=$ constant. This volume element is easily calculable to be

$$\mathrm{d}\Sigma_{(\mu)}=\frac{\mu^{k-1}}{m^k(n(\vartheta))}\sqrt{\frac{m^2(n(\vartheta))+6\xi^2(n(\vartheta))}{m^2(n(\vartheta))+\frac{1}{6}}}\,\mathrm{d}S^{k-1}(\vartheta) \tag{107}$$

where $\xi^2=\sum_A\xi_A\xi_A$, k is the dimension of the χ-space, and $\mathrm{d}S^{k-1}(\vartheta)$ is the surface volume element of the unit sphere S^{k-1} in the χ-space. Calculating the fluxes $J_{\mathrm{T}}{}^{(\mu)}$ and $J_{\mathrm{H}}{}^{(\mu)}$ correspondingly for the tunnelling wavefunction of Vilenkin (103) and for the expanding-universe component of the wavefunction of Hartle and Hawking (104) we obtain the expressions for the probability densities in the corresponding cases ($\mu^2W>1$), to the leading WKB order

$$\mathrm{d}P_{\mathrm{T}}(\vartheta)=C_{\mathrm{T}}\exp\left(-\frac{2}{3W(\vartheta)}\right)\mathrm{d}\Sigma_{(\mu)} \tag{108}$$

$$\mathrm{d}P_{\mathrm{H}}(\vartheta)=C_{\mathrm{H}}\exp\left(\frac{2}{3W(\vartheta)}\right)\mathrm{d}\Sigma_{(\mu)} \tag{109}$$

where C_{T} and C_{H} are corresponding normalization constants. The expressions obtained differ only in the signs in the powers of the leading exponent factors. Thus for the tunnelling boundary conditions the probability of having a large potential $W(\vartheta)$ at the onset of classical universe evolution is exponentially large whereas for the boundary conditions of Hartle and Hawking such probability is exponentially suppressed. This means a high probability of inflation in the first case and a low one in the second case.

To conclude this section we wish to stress once again that the probabilities (108) and (109) are distributed on a compact manifold with the topology of a $(k-1)$-sphere in the space of the fields $\{\chi\}$. On this manifold the field potential $W(\vartheta)$ is bounded, and if it is also smooth enough then the expressions like (108), (109) may be applicable everywhere on this manifold. This situation is very different from what we have in the usual cases when probability distributions are defined typically on spaces like R^n with the scalar field potential unbounded.

6. Discussion

In this paper we have presented a generalization of a realistic field theory model which incorporated gravity and which was invariant under the group of local conformal transformations. The model was based on Riemann–Cartan geometry and the vector trace of torsion played the role of the gauge vector potential for the conformal group. We have considered inflationary universe dynamics based on our model and found that an inflationary stage is allowable provided the couplings satisfy some natural constraints. We also developed standard quantum gravitational formalism for the scalar-gravitational sector of the model. We have seen that conformally non-invariant dynamical variables can be eliminated from the equations of the theory so that the wavefunction of the universe turns out to be independent of them. Although our treatment was rather formal, nevertheless, it may indicate that it is possible to construct and operate with a conformally invariant quantum theory of gravity.

We also considered a simple minisuperspace formulation of the theory under discussion. It has been illustrated that invariance with respect to local conformal transformations (to be precise, their spatially homogeneous remnant) can be consistently implemented into a quantum theory of minisuperspace by an appropriate operator arrangement choice. Then this symmetry becomes manifest as conformal invariance of the wavefunction, i.e. it depends only on conformally invariant variables.

An interesting issue, which remained unresolved, concerns the possibility of having zero cosmological constant at the present cosmological epoch. In our model, the cosmological constant, although in certain sense constrained, is not automatically vanishing (see the end of section 3). Whether or not it can be made equal to zero (exactly or approximately) without unnatural fine tuning of the parameters is an open question.

Acknowledgments

The authors enjoyed valuable discussions with the members of the Department of Astrophysics and Elementary Particles of the Bogolyubov Institute. One of us (YS) would like to express gratitude to Brown University for hospitality, and especially to thank Professor Robert Brandenberger for his invitation to visit Brown University, where the final version of this paper was written. The work of YS was also supported in part by the National Research Council under the COBASE program.

Appendix

Our theory is based on the Riemann–Cartan space U_4 (for a good detailed description see, e.g. [20]). This space naturally arises within the framework of the Poincaré gauge theory of gravity. The Riemann–Cartan structure of U_4 implies the presence of the affine connection form and the metric tensor, which is covariantly constant. Let $\{e_a(x)\}$ be an arbitrary field of bases in the tangent space of U_4 at each point x, and $\{e^a(x)\}$ be the field of their dual bases, that is

$$\langle e^a(x), e_b(x)\rangle = \delta^a_b. \tag{A1}$$

Here and below Latin indices run from 0 to 3. The affine connection form $\omega^a{}_b(x)$ referred to the basis $\{e^a(x)\}$ defines the covariant derivative (denoted by ∇^{ω}) of tensors in U_4.

The metric tensor g can be developed as follows

$$g = g_{ab}e^a \otimes e^b = g_{\mu\nu}\,\mathrm{d}x^\mu \otimes \mathrm{d}x^\nu \tag{A2}$$

where $g_{ab}(x)$ and $g_{\mu\nu}(x)$ are symmetric components. In the Riemann–Cartan space U_4 the affine connection form $\omega^a{}_b$ satisfies the metricity condition

$$\nabla^\omega g = 0. \tag{A3}$$

The metric components g_{ab} ($g_{\mu\nu}$), and their inverse (in the sense of matrices) g^{ab} ($g^{\mu\nu}$), allow one to raise and lower Latin (Greek) indices. The metricity condition (A3) written in terms of the components g^{ab} of the metric reads

$$\omega^{(ab)} \equiv \tfrac{1}{2}(\omega^{ab} + \omega^{ba}) = \tfrac{1}{2}\,\mathrm{d}g^{ab}. \tag{A4}$$

Torsion tensor components $Q^a{}_{bc} = Q^a{}_{[bc]} \equiv \frac{1}{2}(Q^a{}_{bc} - Q^a{}_{cb})$ are defined by the Cartan structural equations as

$$\mathrm{d}e^a + \omega^a{}_b \wedge e^b = -Q^a{}_{bc}e^b \wedge e^c. \tag{A5}$$

From (A3) it follows that

$$\omega^a{}_{bc} = \Gamma^a{}_{bc} + Q^a{}_{bc} + Q_{bc}{}^a + Q_{cb}{}^a \tag{A6}$$

where $\Gamma^a{}_{bc}$ are the components of the Riemannian connection form (which is constructed in the usual way from the metric tensor) written in the basis $\{e_a\}$, $\{e^a\}$.

In order to describe spinors one has to choose an orthonormal vector basis (tetrad) $\{e_a\}$, in which

$$g_{ab} = g(e_a, e_b) = \eta_{ab} = \mathrm{diag}(-1, 1, 1, 1). \tag{A7}$$

In this basis the metricity condition (A4) becomes

$$\omega^{(ab)} = 0 \qquad \text{or} \qquad \omega^{ab} = \omega^{[ab]}. \tag{A8}$$

Henceforth, all formulae will refer to an arbitrary orthonormal basis.

The spin connection form, which determines parallel transport and covariant derivatives of four-spinors, is defined by

$$\omega = -\tfrac{1}{8}\omega^{ab}\,[\gamma_a, \gamma_b] \tag{A9}$$

where γ^a are the usual constant Dirac γ-matrices, which satisfy

$$\{\gamma^a, \gamma^b\} = -2\eta^{ab}. \tag{A10}$$

The covariant derivative of a spinor is then defined as

$$\nabla^\omega \psi = \mathrm{d}\psi + \omega\psi. \tag{A11}$$

It transforms like a spinor under the action of the local Lorentz group L_6.

For any group G of intrinsic gauge transformations, we will denote its connection form by $A = A_\mu \, dx^\mu$. Then both locally Lorentz and G-invariant derivatives of any field multiplet $f(x)$ will be

$$Df = \nabla^\omega f + Af \tag{A12}$$

where A implies the matrix of the corresponding representation of the algebra of the group G.

Let us now consider the group of local conformal transformations whose action on the metric g is

$$g(x) \to g'(x) = \exp(2\sigma(x))\, g(x) \tag{A13}$$

where $\sigma(x)$ is an arbitrary smooth function. Equation (A13) means that the metric g has conformal weight two†

$$w(g) = 2. \tag{A14}$$

From (A1) and (A7) we immediately obtain conformal weights of the tetrad basis vectors and their dual one-forms

$$w(e_a) = -1 \qquad w(e^a) = 1. \tag{A15}$$

For the affine and gauge connection forms we set

$$w(\omega^a{}_b) = 0 \qquad w(A) = 0 \tag{A16}$$

and conformal weights of scalar and spinor fields are, as usual

$$w(\varphi) = -1 \qquad w(\psi) = -\tfrac{3}{2}. \tag{A17}$$

The second group of Cartan structural equations define the Riemann–Cartan curvature two-form

$$R^a{}_b(\omega) = d\omega^a{}_b + \omega^a{}_c \wedge \omega^c{}_b \tag{A18}$$

which is conformally invariant and whose components relative to the coordinate basis $\{dx^\mu\}$

$$R^a{}_{b\mu\nu}(\omega) = \partial_\mu \omega^a{}_{b\nu} - \partial_\nu \omega^a{}_{b\mu} + \omega^a{}_{c\mu}\omega^c{}_{b\nu} - \omega^a{}_{c\nu}\omega^c{}_{b\mu} \tag{A19}$$

constitute those of the curvature tensor. From this equation one constructs the scalar curvature

$$R(\omega) = R^{ab}{}_{\mu\nu}(\omega) e^\mu_a e^\nu_b \tag{A20}$$

† A field $f(x)$ will be said to have conformal weight $w(f)$ if under the action of local conformal group it transforms as

$$f(x) \to f'(x) = \exp\left(w(f)\sigma(x)\right) f(x).$$

whose conformal weight is

$$w(R(\omega)) = -2. \tag{A21}$$

From (A5) the transformation law for the torsion tensor components follows, namely

$$Q^a{}_{bc}(x) \to Q'^a{}_{bc}(x) = \exp(-\sigma(x))\big(Q^a{}_{bc}(x) + \delta^a{}_{[b}\langle \mathrm{d}\sigma(x),\, e_{c]}(x)\rangle\big). \tag{A22}$$

As can be shown, these transformations affect only the vector trace part of the torsion tensor

$$Q_a = Q^b{}_{ab} = Q_\mu e^\mu_a \tag{A23}$$

so that

$$Q_\mu(x) \to Q'_\mu(x) = Q_\mu(x) - \tfrac{3}{2}\partial_\mu\sigma(x). \tag{A24}$$

The locally Lorentz and G-invariant derivative Df defined by the expression (A12) does not preserve the conformal properties of the fields. Using the transformation law (A24) of the torsion trace vector and taking into account (A16) and (A17) we can construct a new derivative

$$\mathcal{D}f = Df + \tfrac{2}{3}w(f)Qf \tag{A25}$$

with the desired property

$$w(\mathcal{D}f) = w(f) \tag{A26}$$

where Q is the one-form of the torsion trace

$$Q = Q_a e^a = Q_\mu \,\mathrm{d}x^\mu. \tag{A27}$$

The derivative $\mathcal{D}$ is used in our paper in constructing locally conformally invariant actions.

References

[1] Weyl H 1918 *Sitzung. Preuss. Akad. Wiss.* 465; 1920 *Raum. Zeit. Matterie.* (Berlin: Springer)

[2] Dirac P A M 1973 *Proc. R. Soc.* A **333** 403
Utiyama R 1973 *Prog. Theor. Phys.* **50** 2080; 1975 *Prog. Theor. Phys.* **53** 565
Freund P G O 1974 *Ann. Phys., NY* **84** 440
Hayashi K, Kasuya M and Shirafugi T 1977 *Prog. Theor. Phys.* **57** 431
Hayashi K and Kugo T 1979 *Prog. Theor. Phys.* **61** 334
Cheng H 1988 *Phys. Rev. Lett.* **61** 2182
Wheeler J T 1990 *Phys. Rev.* D **41** 431
Hochberg D and Plunien G 1991 *Phys. Rev.* D **43** 3358

[3] Mannheim P D and Kazanas D 1989 *Astrophys. J.* **342** 635; 1991 *Phys. Rev.* D **44** 417
Mannheim P D 1990 *Gen. Rel. Grav.* **22** 289; 1992 *Astrophys. J.* **391** 429; 1993 *Astrophys. J.* **419** 150

[4] Zhytnikov V V 1993 *Int. J. Mod. Phys.* **8A** 5141

[5] Buchbinder I L, Odintsov S D and Shapiro I L 1992 *Effective Action in Quantum Gravity* (Bristol: IOP Publishing)
Elizalde E and Odintsov S D 1993 *Mod. Phys. Lett.* **8A** 3325
Bytsenko A A, Elizalde E and Odintsov S D 1993 *Prog. Theor. Phys.* **90** 677

[6] Nieh H T and Yan M L 1982 *Ann. Phys., NY* **138** 237

[7] Obukhov Yu N 1982 *Phys. Lett.* **90A** 13
Dereli T and Tucker R W 1982 *Phys. Lett.* **110B** 206
[8] Gibbons G W and Hawking S W 1977 *Phys. Rev.* D **15** 2752
[9] Mannheim P D 1993 *Gen. Rel. Grav.* **25** 697
[10] Linde A D 1990 *Particle Physics and Inflationary Cosmology* (Chur: Harwood Academic)
[11] Peebles P J E 1993 *Principles of Physical Cosmology* (Princeton, NJ: Princeton University Press)
[12] Misner C N, Thorn K S and Wheeler J A 1973 *Gravitation* (San Francisco, CA: Freeman)
[13] Alvarez E 1989 *Rev. Mod. Phys.* **61** 561
[14] Halliwell J J 1990 Introductory lectures on quantum cosmology *Proc. Jerusalem Winter School on Quantum Cosmology and Baby Universes* ed T Piran
[15] Duncan M J 1990 Quantum geometrodynamics (Lectures presented at TASI summer school) *Preprint* UMN-TH-916/90
[16] Vilenkin A 1984 *Phys. Rev.* D **30** 509; 1986 *Phys. Rev.* D **33** 3560; 1988 *Phys. Rev.* D **37** 888
[17] Hawking S 1982 *Astrophysical Cosmology: Proc. Study Week on Cosmology and Fundamental Physics* ed H A Bruck, G V Coyne and M S Longair (Vatican: Pontificae Academiae Scientarium Scripta Varia); 1983 *Relativity, Groups and Topology II* ed B S DeWitt and R Stora (Amsterdam: North Holland); 1984 *Nucl. Phys.* B **239** 257
Hartle J B and Hawking S W 1983 *Phys. Rev.* D **28** 2960
[18] Halliwell J J and Louko J 1989 *Phys. Rev.* D **39** 2206
[19] Vilenkin A 1989 *Phys. Rev.* D **39** 1116
[20] Hehl F W, Von der Heyde P, Kerlick G D and Nester J M 1976 *Rev. Mod. Phys.* **48** 393

Class. Quantum Grav. **11** (1994) 2483–2497. Printed in the UK

Quantum cosmology and higher-order Lagrangian theories

Henk van Elst†§, James E Lidsey‡‖ and Reza Tavakol†¶

† School of Mathematical Sciences, Queen Mary and Westfield College Mile End Road, London E1 4NS, UK

‡ NASA/Fermilab Astrophysics Center, Fermi National Accelerator Laboratory, Batavia IL 60510, USA

Received 28 March 1994, in final form 14 July 1994

Abstract. In this paper the quantum cosmological consequences of introducing a term cubic in the Ricci curvature scalar R into the Einstein–Hilbert action are investigated. It is argued that this term represents a more generic perturbation to the action than the quadratic correction usually considered. A qualitative argument suggests that there exists a region of parameter space in which neither the tunnelling nor the no-boundary boundary conditions predict an epoch of inflation that can solve the horizon and flatness problems of the standard big bang model. This is in contrast to the R^2-theory.

PACS numbers: 9880H, 0450, 0460K, 9880C

1. Introduction

An important motivation for the development of the quantum cosmology programme has been to explain the initial conditions for the emergence of the universe as a dynamical object, which can be described in terms of a classical theory of gravity, Einstein's theory of general relativity. In quantum cosmology, in principle, one must find the form of the wavefunction Ψ satisfying the Wheeler–DeWitt equation [1]. This equation describes the annihilation of the wavefunction by the Hamiltonian operator and since it admits an infinite number of solutions, one must also choose the boundary conditions in order to specify the wavefunction uniquely. Such boundary conditions must be viewed as an additional physical law since, by definition, there is nothing external to the universe. In practice, one assumes, at least implicitly, that a finite subset of all possible boundary conditions is favoured by cosmological observations, in the sense that the wavefunctions corresponding to such boundary conditions predict outcomes which are compatible with observations. For example, if one believes in the inflationary scenario, the requirement that sufficient inflation occurred, in order to solve the assorted problems of the standard big bang model can, in principle, restrict the number of plausible boundary conditions.

Among the set of all possible choices the *tunnelling from nothing* boundary condition proposed by Linde [2] and Vilenkin [3] and the *no-boundary* boundary condition proposed by Hartle and Hawking [4] have been the subject of intense discussion. Given the non-uniqueness of such conditions, the question arises as to the consequences of choosing different boundary conditions for the resulting wavefunction of the universe and its

§ E-mail address: hve@maths.qmw.ac.uk
‖ E-mail address: jim@fnas09.fnal.gov
¶ E-mail address: reza@maths.qmw.ac.uk

corresponding probability measures. An important study in this regard is due to Vilenkin [5], who considered the effects of the above boundary conditions within the context of Einstein gravity minimally coupled to a self-interacting scalar field. He restricted his analysis to the minisuperspace corresponding to the spatially closed, isotropic and homogeneous Friedmann–Lemaître–Robertson–Walker (FLRW) model of the universe and showed that the tunnelling wavefunction predicts initial states that are likely to lead to sufficient inflation, whereas the no-boundary wavefunction does not.

It is sometimes argued that this result indicates that observations favour the tunnelling as opposed to the no-boundary boundary condition. However, the precise relation between the boundary conditions and the observations is determined by the specific models employed and since such models always involve idealizations in the form of a set of simplifying assumptions, it follows that the above conclusion cannot be made *a priori*. Indeed it only makes sense, in general, if the correspondence between the observations and the boundary conditions is robust under physically motivated perturbations to the underlying quantum cosmological model.

Consequently, it is important to consider the 'stability' of the above conclusions. In particular, are the conclusions robust under higher-order perturbations to the Einstein–Hilbert action? Quadratic and higher-order terms in the Riemann curvature tensor and its traces appear in the low-energy limit of superstrings [6] and they also arise when the usual perturbation expansion is applied to general relativity [7, 8]. Such terms diverge as the initial singularity is approached, but can, in principle, be eliminated if higher-order corrections are included in the action. In four-dimensional spacetimes the Hirzebrucht signature and Euler number imply that the most general, four-dimensional gravitational action to quadratic order is

$$S = \mathrm{d}^4x\, \sqrt{-g}\, \frac{1}{2\kappa^2}\left[R - \gamma\, C_{\alpha\beta\gamma\delta}\, C^{\alpha\beta\gamma\delta} + \epsilon_1\, R^2\right] \qquad (1)$$

where R is the Ricci curvature scalar of the spacetime with metric tensor $g_{\mu\nu}$, $g = \det g_{\mu\nu}$, $C_{\alpha\beta\gamma\delta}$ is the Weyl conformal curvature tensor, κ^2 is the gravitational coupling constant and ϵ_1 and γ are coupling constants of dimension $(\mathrm{length})^2$.

These higher-order terms lead to important local effects. If $\epsilon_1 < 0$, the vacuum can become unstable and the theory admits tachyonic solutions [9, 10]. Consequently this case is generally dismissed as unphysical. The presence of the γ-contribution also leads to local instabilities [9]. However, the classical action simplifies further for spatially homogeneous and isotropic four-geometries, since the conformal flatness of these spacetimes implies that the Weyl tensor vanishes. The effects of including quadratic terms have been investigated in [11–14]. In particular Mijić *et al* [13] studied the effects of such perturbations on Vilenkin's result [5] and found that those results remain robust in the sense that the inflationary scenario still favours the tunnelling boundary condition in the presence of quadratic terms in the action. On the other hand, Biswas and Guha have recently arrived at the opposite conclusion [11].

The renormalization of higher-loop contributions introduces terms into the effective action that are higher than quadratic order. Consequently it is important to also study the effects of these additional terms. In this paper we shall investigate what happens to the wavefunction if an R^3-contribution is present. By employing the conformal equivalence of higher-order gravity theories with Einstein gravity coupled to matter fields, we argue that this term represents a more general perturbation to the Einstein–Hilbert action than the R^2-correction, at least within the context of four-dimensional FLRW spacetimes. We then consider the conditional probability that an inflationary epoch of sufficient duration can occur. We estimate how the qualitative behaviour of this quantity changes when higher-order perturbations to the action are included. Our main result is that for the R^3-theory there exists a finite region of parameter space in which neither of the boundary conditions discussed above predict an epoch of inflationary expansion that leads to the observed universe. We use (dimensionless) Planckian units defined by $\hbar = c = G = 1$ throughout and define $\kappa^2 = 8\pi$.

2. Higher-order Lagrangians as Einstein gravity plus matter

The wavefunction of the universe in higher-order Lagrangian theories can be determined in one of two ways. It is well known that theories with a Lagrangian given by a differentiable function of the Ricci curvature scalar are conformally equivalent to Einstein gravity with a matter sector containing a minimally coupled, self-interacting scalar field [15, 16]. The precise form of the self-interaction is uniquely determined by the higher-derivative metric terms in the field equations. It follows that one can start either from the original action or the conformal action and derive the corresponding Wheeler–DeWitt equation [16]. One takes the related Lagrangian as the defining feature of the theory and then applies the canonical quantization rules. The advantage of the conformal transformation is that it allows the known results from Einstein gravity to be carried over to the higher-order examples and we shall follow such an approach in this paper. It is important to emphasise, however, that the conformal transformation is nonlinear and in general quantizing the transformed theory is not equivalent to quantizing the original theory.

Consider the general, D-dimensional, vacuum theory

$$S = \mathrm{d}^D x\, \sqrt{-g_D}\, \left[f(R)\right] \tag{2}$$

where the Lagrangian $f(R)$ is some arbitrary differentiable function of the Ricci curvature scalar satisfying $\{f(R), \mathrm{d}f(R)/\mathrm{d}R\} > 0$ and g_D is the determinant of the D-dimensional spacetime metric $g_{D\,\mu\nu}$. If we perform the conformal transformation [15]

$$\tilde{g}_{D\,\mu\nu} = \Omega^2\, g_{D\,\mu\nu} \qquad \Omega^2 = \left(2\kappa^2\, \frac{\mathrm{d}f(R)}{\mathrm{d}R}\right)^{2/(D-2)} \tag{3}$$

and define a new scalar field

$$\kappa\,\bar{\phi} \equiv \left(\frac{D-1}{D-2}\right)^{1/2} \ln\left[2\kappa^2\, \frac{\mathrm{d}f(R)}{\mathrm{d}R}\right] \tag{4}$$

the conformally transformed action takes the Einstein–Hilbert form

$$S = \mathrm{d}^D x\, \sqrt{-\tilde{g}_D}\left[\frac{\tilde{R}}{2\kappa^2} - \tfrac{1}{2}\,\tilde{\nabla}_\mu\bar{\phi}\,\tilde{\nabla}^\mu\bar{\phi} - U(\bar{\phi})\right] \tag{5}$$

where the self-interaction potential is given by

$$U(\bar{\phi}) \equiv \left(2\kappa^2\, \frac{\mathrm{d}f[\,R(\bar{\phi})\,]}{\mathrm{d}R}\right)^{-D/(D-2)} \left(R(\bar{\phi})\, \frac{\mathrm{d}f[\,R(\bar{\phi})\,]}{\mathrm{d}R} - f[\,R(\bar{\phi})\,]\right). \tag{6}$$

Definition (4) yields a correspondence between the values of the Ricci curvature scalar R and the values of the scalar field $\bar{\phi}$. We shall consider the quadratic and cubic Lagrangians

$$f_2(R) = \frac{1}{2\kappa^2}\left(R + \epsilon_1\, R^2\right) \tag{7}$$

$$f_3(R) = \frac{1}{2\kappa^2}\left(R + \epsilon_1\, R^2 + \epsilon_2\, R^3\right) \tag{8}$$

in four dimensions, where the parameters ϵ_1 and ϵ_2 have dimensions (length)^2 and (length)^4, respectively, before the introduction of Planckian units. The corresponding potentials as defined in (6) then are given by [13, 18]

$$U_{f_2}(\bar{\phi}) = \frac{1}{8\kappa^2\,\epsilon_1}\left[1 - \exp(-\sqrt{2/3}\,\kappa\,\bar{\phi})\right]^2 \tag{9}$$

$$U_{f_3}(\bar{\phi}) = \frac{{\epsilon_1}^3}{27\kappa^2\,{\epsilon_2}^2}\exp(-2\sqrt{2/3}\,\kappa\,\bar{\phi})\left[-1 + \frac{9\epsilon_2}{2{\epsilon_1}^2}\left[1 - \exp\left(\sqrt{2/3}\,\kappa\,\bar{\phi}\right)\right]\right.$$
$$\left. + \left(1 - \frac{3\epsilon_2}{{\epsilon_1}^2}\left[1 - \exp\left(\sqrt{2/3}\,\kappa\,\bar{\phi}\right)\right]\right)^{3/2}\right]. \tag{10}$$

It follows from (4) and $\mathrm{d}f(R)/\mathrm{d}R > 0$ that in the R^3-theory

$$\exp(\sqrt{2/3}\,\kappa\,\bar{\phi}) = 1 + 2\epsilon_1\,R + 3\epsilon_2\,R^2 \tag{11}$$

and if $\epsilon_2 < 0$, the scalar field is not well defined in the limit when $|R|$ diverges. This can also be seen in that the potential $U_{f_3}(\bar{\phi})$ in (10) will become imaginary for a certain value $\bar{\phi} > 0$. However, since the high curvature limit is precisely the region where quantum cosmological effects should be important, we assume $\epsilon_2 > 0$ in the following. The region $\mathrm{d}f(R)/\mathrm{d}R < 0$ should be excluded, because in this case the universe cannot evolve into Minkowski space [17]. For positive ϵ_1 and ϵ_2 the potentials (9) and (10) are positive semi-definite for all values of $\bar{\phi}$.

In the classical R^3-theory the requirement that the inflationary epoch lasts sufficiently long implies that the coupling constants must satisfy $|\epsilon_2| \ll \epsilon_1^2$ [17]. Moreover, the observed isotropy of the cosmic microwave background radiation requires that $\epsilon_1 \approx 10^{11}$ [13]. In view of these constraints we specify $\epsilon_1 = 10^{11}$ in the subsequent numerical calculations. Figures 1(*a*) and (*b*) illustrate the behaviour of the potentials (9) and (10) for $\epsilon_1 \approx 10^{11}$ and $\epsilon_2 \approx 10^{20}$. The effect of decreasing the value of the parameter ϵ_1 is to increase the height of the plateau and the relative maximum of the potentials in the quadratic and cubic cases, respectively. This reflects the fact that decreasing this parameter is equivalent to increasing the energy scales involved. In this sense there exists no continuous transformation from an R^2-theory to the ordinary Einstein–Hilbert action as this parameter approaches zero. A similar problem arises when quadratic terms containing the Ricci and the Riemann tensors are also included [18]. In the neighbourhood of the origin of $\bar{\phi}$ corresponding to smaller values of R the quadratic term in the action dominates and the potentials in this region are equivalent. This can be seen by expanding the last of the three terms in the square brackets of (10). The first-order contribution cancels the remaining terms in $U_{f_3}(\bar{\phi})$ and the second-order term reduces the form of $U_{f_3}(\bar{\phi})$ to that of $U_{f_2}(\bar{\phi})$. Hence the two potentials are effectively identical if the third- and higher-order terms in the expansion can be neglected. It is straightforward to show that this is a consistent approximation if

$$\kappa\,\bar{\phi} \ll \kappa\,\bar{\phi}_{\mathrm{limit}} \equiv \sqrt{3/2}\,\ln\left(\frac{2{\epsilon_1}^2}{\epsilon_2}\right). \tag{12}$$

For polynomial Lagrangians with $f(R) = \left(\sum_{k=1}^{n} \epsilon_{k-1}\,R^k\right)/2\kappa^2$, the detailed form of the corresponding potential $U(\bar{\phi})$ is extremely complicated and generally not expressible in an analytically closed form. Nevertheless, one can determine the qualitative behaviour of the potential at small and large $\bar{\phi}$. Close to the origin the quadratic term in the action again dominates and the potential in this region is therefore similar to (9). The asymptotic behaviour at infinity, however, depends critically upon the combination of the highest degree n of the polynomial and the dimensionality D of the spacetime [15]. More precisely, for

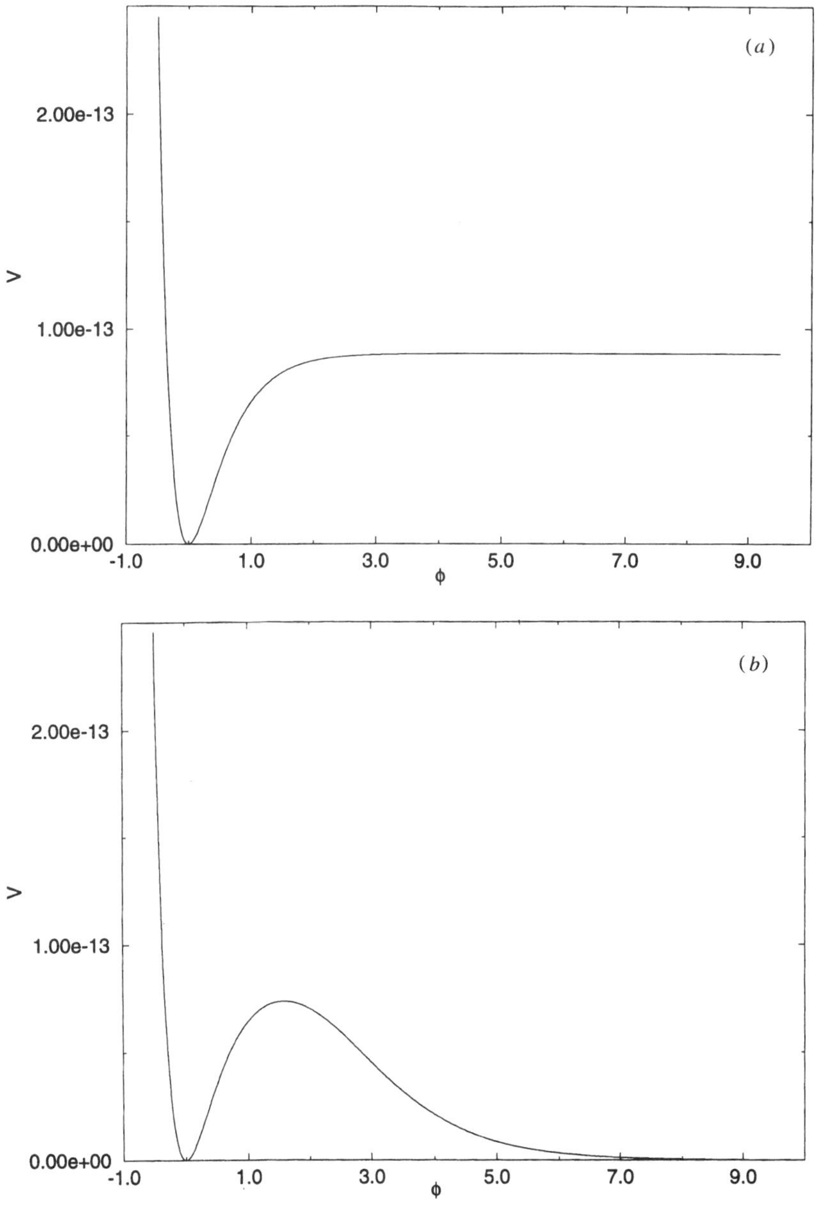

Figure 1. (a) The effective self-interaction potential (9) corresponding to the R^2-theory with $\epsilon_1 = 10^{11}$. The scalar field and magnitude of the potential have been rescaled via (13) to enable easy comparison with the results of section 4. (b) The rescaled effective self-interaction potential (10) corresponding to the R^3-theory with $\epsilon_1 = 10^{11}$ and $\epsilon_2 = 10^{20}$.

$D > 2n$ the potential is unbounded from above, for $D = 2n$ it flattens into a plateau and for $D < 2n$ the potential has an exponentially decaying tail [17]. In particular, if $D < 2n$ the effective scalar field potential $U(\bar{\phi})$ is qualitatively equivalent to the cubic potential (10). As a result, when $D = 4$ the qualitative behaviour of $U(\bar{\phi})$ does not change relative to the cubic case as terms with $n > 3$ are considered, although the relative position of the maximum of $U(\bar{\phi})$ will be n-dependent. This implies that the $n = 2$ contribution is rather special in four dimensions, whereas the R^3-term is in fact a more generic perturbation. Thus, it is instructive to consider this case further.

3. Behaviour of the wavefunction

Within the context of the spatially closed FLRW minisuperspace models, the Wheeler–DeWitt equation derived from theory (5) has been solved for an arbitrary potential, subject to the condition that the momentum operator for the scalar field can be neglected [3, 5]. This is self-consistent if $|\mathrm{d}V/\mathrm{d}\phi| \ll \max\{|V|, a^{-2}\}$, where a represents the cosmological scale factor and

$$V \equiv \frac{16}{9} U \qquad \phi \equiv \sqrt{\frac{4\pi}{3}}\,\bar{\phi}\,. \tag{13}$$

The WKB approximations of the wavefunctions satisfying the quantum tunnelling boundary condition (Ψ_{T}) and the no-boundary proposal (Ψ_{NB}) then take the forms [5]

$$\Psi_{\mathrm{T}} = (1 - a^2\,V)^{-1/4}\,\exp\left[\frac{(1 - a^2\,V)^{3/2} - 1}{3V}\right] \tag{14}$$

$$\Psi_{\mathrm{NB}} = (1 - a^2\,V)^{-1/4}\,\exp\left[\frac{1 - (1 - a^2\,V)^{3/2}}{3V}\right] \tag{15}$$

in the classically forbidden (Euclidian signature) region defined by $a^2\,V < 1$, and

$$\Psi_{\mathrm{T}} = \mathrm{e}^{\mathrm{i}\pi/4}\,(a^2\,V - 1)^{-1/4}\,\exp\left[-\frac{1}{3V}\right]\,\exp\left[-\mathrm{i}\,\frac{(a^2\,V - 1)^{3/2}}{3V}\right] \tag{16}$$

$$\Psi_{\mathrm{NB}} = 2\,(a^2\,V - 1)^{-1/4}\,\exp\left[\frac{1}{3V}\right]\,\cos\left[\frac{(a^2\,V - 1)^{3/2}}{3V} - \frac{\pi}{4}\right] \tag{17}$$

in the classically allowed (Lorentzian signature) region $a^2\,V > 1$. Substituting for $V(\phi)$ from the potentials of the quadratic and cubic Lagrangians of section 2, it can readily be seen that the wavefunctions corresponding to the quadratic and cubic theories have very different types of behaviour, at least for large ϕ. In the quadratic case both Ψ_{T} and Ψ_{NB} remain bounded. However, for the cubic case Ψ_{NB} becomes divergent in the classically allowed region whilst Ψ_{T} remains regular. In this sense then the qualitative behaviour of the wavefunction satisfying the no-boundary proposal is fragile with respect to cubic perturbations to the action. This is significant because often the quadratic corrections to the action are taken as representative of higher-order perturbations.

To proceed it is important to ensure that for the regimes under consideration the conformal transformation (3) remains non-singular. This is the case if the condition $\mathrm{d}f(R)/\mathrm{d}R \neq 0$ is valid for all values of R. The conformal transformation is singular at the point

$$R = -\frac{1}{2\epsilon_1} \tag{18}$$

in the R^2-theory and at the point

$$R = -\frac{\epsilon_1}{3\epsilon_2}\left[1 \pm \sqrt{1 - \frac{3\epsilon_2}{\epsilon_1^2}}\right] \tag{19}$$

for the R^3-theory. Since ϵ_1 and ϵ_2 are taken to be positive, these conditions imply that in both cases the problematic values of R lie in the region $R < 0$. However, for a classical, spatially closed FLRW model, the Ricci curvature scalar is given by

$$R = 6(1-q)\left(\frac{\dot{a}}{a}\right)^2 + \frac{6}{a^2} \tag{20}$$

where $q \equiv -\ddot{a}\,a/\dot{a}^2$ defines the deceleration parameter and a dot denotes differentiation with respect to cosmic proper time. Now if, as is generally assumed, the universe tunnels into the Lorentzian region in an inflationary phase ($q < 0$), it follows that R will be positive-definite. Thus, the conformal transformation is self-consistent in these theories.

4. Interpretation of the wavefunction

In the previous section we saw that the wavefunctions corresponding to the tunnelling and the no-boundary boundary conditions have qualitatively different modes of behaviour for the quadratic and cubic theories. To see what predictive effects such changes might have, we employ the notion of a probability density ρ as is usually done. For the cases of the tunnelling and the no-boundary boundary conditions, respectively, ρ takes the form [5]

$$\rho_{\mathrm{T}}(a,\phi) = C_{\mathrm{T}} \exp\left[-\frac{2}{3V(\phi)}\right] \tag{21}$$

$$\rho_{\mathrm{NB}}(a,\phi) = C_{\mathrm{NB}} \exp\left[\frac{2}{3V(\phi)}\right] \tag{22}$$

on surfaces of constant scale factor in the classically allowed region of minisuperspace, where the normalization constants C_{T} and C_{NB} are given by

$$C_{\mathrm{T}}^{-1} = \int_{V(\phi)>0} \mathrm{d}\phi\, \exp\left[-\frac{2}{3V(\phi)}\right] \quad C_{\mathrm{NB}}^{-1} = \int_{V(\phi)>0} \mathrm{d}\phi\, \exp\left[\frac{2}{3V(\phi)}\right]. \tag{23}$$

Since $\rho(\phi)$ is usually not normalizable, the common practice is to employ the notion of a conditional probability [16]. One argues that the initial values of the scalar field must lie in the range $\phi_{\mathrm{min}} < \phi_{\mathrm{i}} < \phi_{\mathrm{P}}$. The lower limit ϕ_{min} follows from the requirement that the universe expands at least until the formation of large-scale structure and the upper bound follows from the condition that $V(\phi_{\mathrm{P}}) \approx 1$, since the minisuperspace approximation is unlikely to be valid when the potential energy of the matter sector exceeds the Planck density. However, in a chaotic inflationary scenario there is a critical value of the scalar field, ϕ_{suf}, and sufficient inflation occurs if $\phi_{\mathrm{i}} > \phi_{\mathrm{suf}}$ but not for $\phi_{\mathrm{i}} < \phi_{\mathrm{suf}}$. We must therefore calculate the conditional probability that sufficient inflation occurs given that ϕ_{i} is bounded by ϕ_{min} and ϕ_{P}. This quantity takes the form [16]

$$P(\phi_{\mathrm{i}} > \phi_{\mathrm{suf}} \,|\, \phi_{\mathrm{min}} < \phi_{\mathrm{i}} < \phi_{\mathrm{P}}) = \frac{\int_{\phi_{\mathrm{suf}}}^{\phi_{\mathrm{P}}} \rho(\phi)\,\mathrm{d}\phi}{\int_{\phi_{\mathrm{min}}}^{\phi_{\mathrm{P}}} \rho(\phi)\,\mathrm{d}\phi} \tag{24}$$

and allows us to determine which of the two boundary conditions considered here 'naturally' predicts a phase of sufficiently long inflationary expansion. Sufficient inflation is a prediction of a theory if $P \approx 1$, whereas it is not if $P \ll 1$.

For standard reheating the minimum amount of inflation that solves the horizon problem is determined by the condition $N \equiv \ln(a_{\mathrm{f}}/a_{\mathrm{i}}) \approx 65$, where subscripts i and f denote the values of the scale factor at the onset and end of inflation, respectively [19]. It is then straightforward to deduce from the classical field equations that

$$N \approx 65 \approx 6 \int_{\phi_f}^{\phi_{suf}} V(\phi) \left(\frac{dV(\phi)}{d\phi} \right)^{-1} d\phi \qquad (25)$$

where the value of the scalar field at the end of inflation, ϕ_f, is computed from the relation

$$\frac{1}{12} \left[V^{-1}(\phi) \frac{dV(\phi)}{d\phi} \right]^2_{\phi=\phi_f} = 1\,. \qquad (26)$$

This condition corresponds to the breakdown of the slow-roll approximation [20]†. Once ϕ_f is known, the value of ϕ_{suf} can be determined numerically by evaluating the integral in (25).

To understand how the probability densities (21) and (22) change in the quadratic and cubic cases, we shall consider them in turn. Since (21) and (22) are usually *not* normalizable (unless the range of values that ϕ can take is bounded), we set the 'normalization constants' equal to one as is the common practice.

4.1. The quadratic case

To begin with, we note that the shape of $V(\phi)$ does not qualitatively change with changes in the coupling constant ϵ_1. This parameter only fixes the height of the plateau and as a result leaves the shapes of the two probability densities unchanged. Consequently the qualitative behaviours of the probability densities are robust with respect to changes in ϵ_1. Figure 2(*a*) gives a plot of ρ_T showing that it starts at zero when $\phi = 0$ and asymptotically approaches a constant value. On the other hand, as can be seen from figure 2(*b*), ρ_{NB} decreases from infinity and asymptotically approaches a constant value. We should emphasise here that since the probability distribution functions (21) and (22) typically take values of the order $\exp(\pm 10^{14})$, we, for the sake of graphical representation, applied nonlinear scalings of the kinds $\tilde{\rho}_T = {\rho_T}^{1/C}$ and $\tilde{\rho}_{NB} = \ln\left({\rho_{NB}}^{1/C}\right)$, respectively (where C is a constant) to the two probability distribution functions. Note, however, that the values of the argument ϕ remain uneffected by this scaling.

Contrary to the claim of Biswas and Guha [11], the two probability distribution functions reveal *no* qualitative changes as compared to the case of 'chaotic'-type potentials (e.g. $V(\phi) = m^2\phi^2/2$) as discussed by Vilenkin [5] and Halliwell [16]. This means that the tunnelling wavefunction has its maximum nucleation probability for the universe coming into existence somewhere on the plateau of the potential $V(\phi)$, whereas the no-boundary wavefunction peaks near the true minimum of the potential at $\phi = 0$. Translated into initial values of the Ricci curvature scalar, this means that the tunnelling wavefunction prefers values of R_i near the Planck scale, whereas the no-boundary wavefunction favours a universe of large initial size, i.e. small R_i [13].

We now consider the conditional probability (24). The range of values of ϕ_i is specified by the range of initial values R_i. In Planckian units, where $R_P = 1$, we deduce that $\phi_P = 13.0$. The value of ϕ_f is calculated from (26) to be $\phi_f = 0.38$ and condition (25) is therefore satisfied for $\phi_{suf} = 2.27$. Since the conditional probability measure (24)

† Strictly speaking, conditions (25) and (26) are only valid in spatially flat FLRW models, but we are considering spatially closed cases in this work. However, during inflation the curvature term in the Friedmann equation is redshifted to zero within one Hubble expansion time and the universe effectively becomes spatially flat at an exponentially fast rate. For our purposes, therefore, these expressions remain valid.

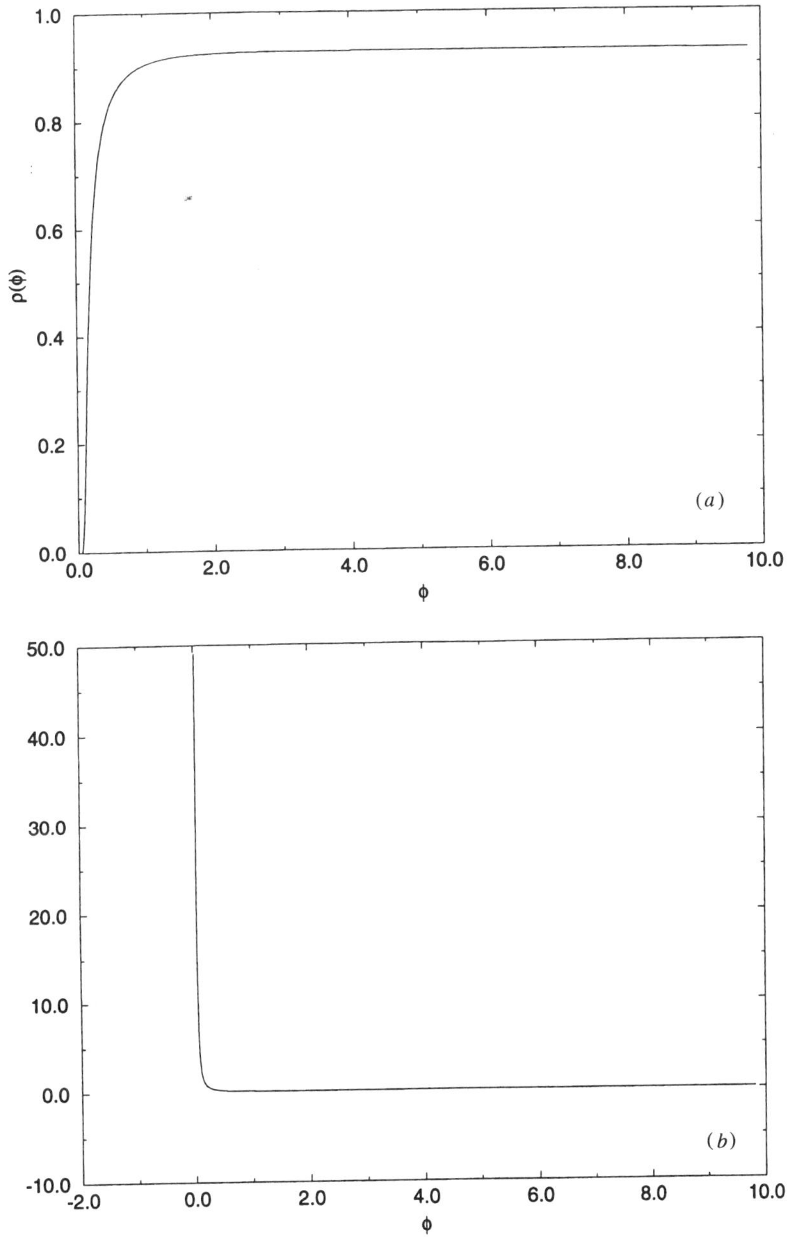

Figure 2. (*a*) The probability distribution $\rho_T(\phi)$ associated with the tunnelling boundary condition for the R^2-theory with a rescaling $\rho_T(\phi) = [\exp(-2/3V)]^{10^{-14}}$; (*b*) The probability distribution $\rho_{NB}(\phi)$ associated with the no-boundary boundary condition for the R^2-theory with a rescaling $\rho_{NB}(\phi) = \ln[\exp(2/3V)]^{10^{-14}}$. We choose these particular rescaled values of $\rho(\phi)$ in order to obtain easily interpretable plots from our numerical program ($\epsilon_1 = 10^{11}$).

essentially amounts to a comparison of areas between the $\rho(\phi)$ curve and the positive ϕ-axis in figures 2(*a*) and (*b*), it seems obvious that the tunnelling wavefunction leads to sufficient inflation whereas the no-boundary wavefunction does not. This is in line with the conclusions of Vilenkin [5] and Mijić *et al* [13] and in contrast to what is claimed by Biswas and Guha [11].

4.2. The cubic case

We now consider the effects of adding a cubic term to the action. In general, ρ_T is peaked around the maximum of $V(\phi)$ at ϕ_{max} and falls off to zero on both sides. In contrast ρ_{NB} decreases from infinity near $\phi = 0$ to a minimum at ϕ_{max} and diverges again as $\phi \to \infty$. In this sense the presence of the cubic term drastically alters the shapes of the two probability distributions. This qualitative behaviour is illustrated in figures 3(a) and (b) for $\epsilon_1 = 10^{11}$ and $\epsilon_2 = 10^{20}$.

Now, regarding the location of the maximum nucleation probability, the tunnelling case is unambiguous since there is only a single peak in the probability distribution function. Note, however, that in the cubic case this wavefunction favours *smaller* values of the initial curvature R_i (namely ϕ_i) as compared to those in the quadratic case, where they are of Planckian order. On the other hand, the case of the no-boundary boundary condition is ambiguous because of the presence of two peaks in the probability distribution function, corresponding respectively to low and high values of R_i.

From a practical point of view, the question arises as to whether the tunnelling wavefunction still predicts a phase of sufficiently long inflationary expansion immediately after tunnelling into the Lorentzian signature region. To investigate this, we confined ourselves to the region on the left of the maximum in the potential (10), i.e. $\phi \leqslant \phi_{max}$. Although inflation occurs on both sides of the turning point, there is no end to the superluminal expansion if the field rolls down the right-hand side and consequently there is no reasonable mechanism of reheating [17]. On the basis of these physical considerations it is therefore more appropriate to *identify* the upper limit ϕ_P of the integrals in (24) with ϕ_{max} rather than with the Planck limit.

The specific value of the conditional probability depends on the magnitude of ϵ_2 and it is therefore necessary to determine the relevant range of values for this parameter. We noted in section 2 that ϵ_2 is bounded from above by the condition $\epsilon_2 \ll {\epsilon_1}^2$. As ϵ_2 is decreased relative to a *fixed* ϵ_1, the location of the maximum is shifted to larger values of ϕ and eventually beyond the Planck limit ϕ_P. This follows since the model reduces to the R^2-theory for which the potential exhibits a plateau, i.e. the maximum is effectively located at infinity in this case. However, according to condition (12) the region over which the cubic and quadratic potentials are equivalent also increases as ϵ_2 decreases. The question then is whether ϕ_{limit} grows faster or slower than ϕ_{max}. By explicitly calculating the values of ϕ_{max} and ϕ_{limit} it is found that ϕ_{limit} exceeds ϕ_{max} for all parameter values $\epsilon_2 \leqslant 10^{20}$. This implies that the R^2- and R^3-theories are equivalent for $\phi < \phi_{max}$ in this range. Hence the results in subsection 4.1 for R^2-theory may be carried over directly to the cubic case in this region of the variable ϕ, although there is the important difference that the upper bound on ϕ_i is now identified with ϕ_{max} and not ϕ_P.

For any given ϵ_2 the end of inflation occurs at $\phi_f = 0.38$ as in the R^2-case, since the R^3-contribution is negligible at very small ϕ. Unfortunately, a direct numerical integration of (24) cannot be performed, because the integrands are typically of the orders of of $\exp(\pm 10^{14})$. However, since the probability density ρ is a single valued, positive-definite function of ϕ, it follows that a handle on the qualitative behaviour of the conditional probability can be obtained by investigating how the area under the $\rho(\phi)$ curve changes as ϵ_2 changes. The problem then reduces to determining how the limits of the integrals in the numerator and denominator vary as the parameters of the theory are altered.

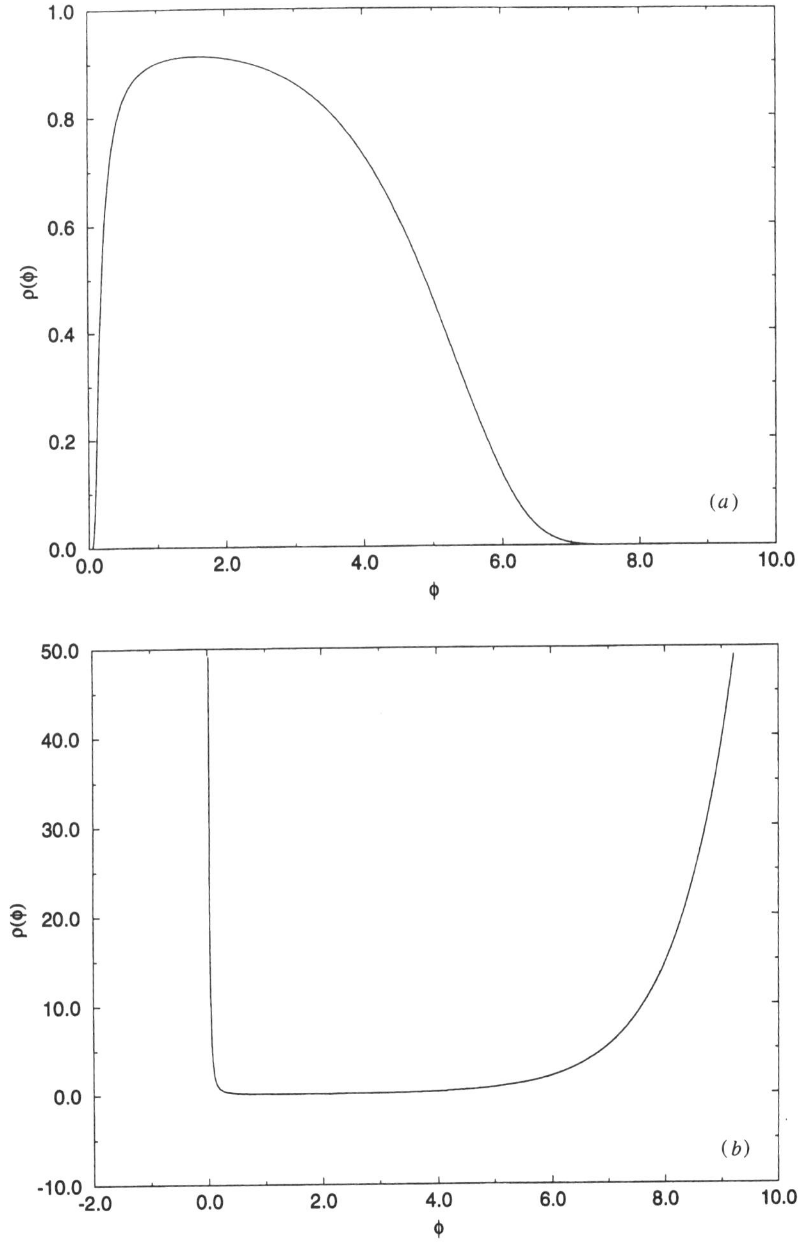

Figure 3. (*a*) The tunnelling probability distribution $\rho_T(\phi)$ for the R^3-theory with the same rescaling as for figure 2(*a*); (*b*) The no-boundary probability distribution $\rho_{NB}(\phi)$ for the R^3-theory with the same rescaling as for figure 2(*b*). ($\epsilon_1 = 10^{11}$, $\epsilon_2 = 10^{20}$).

The dependences of the parameters of interest on ϵ_2 are summarized in table 1. We find that ϕ_{suf} for the potential (10) settles at the same value as in the quadratic case when ϵ_2 is of order 10^{18} or smaller. We also find that ϕ_{max} rapidly approaches ϕ_{suf} in the region $10^{18} \leqslant \epsilon_2 \leqslant 10^{20}$. This implies that the integral in the numerator of the conditional probability (24) becomes *much smaller* than the term in the denominator for $\epsilon_2 \geqslant 10^{18}$.

Table 1. Summarizing, for different values of ϵ_2, the values of the scalar field corresponding to $R_P = 1$ (ϕ_P), the limit of ϕ below which the R^2- and R^3-potentials are equivalent ($\phi_{\rm limit}$), the location of the maximum in the potential ($\phi_{\rm max}$) and the values of the field that just lead to sufficient inflation ($\phi_{\rm suf}$). We specify $\epsilon_1 = 10^{11}$ throughout due to microwave background considerations. As ϵ_2 increases to order of 10^{18}, the magnitudes of the quantities $\phi_{\rm max}$ and $\phi_{\rm suf}$ become comparable to one another and this implies that the numerator in the conditional probability approaches zero. This suggests that the conditional probability will become significantly smaller than unity for values of $\epsilon_2 \geqslant 10^{18}$.

ϵ_2	10^{20}	10^{18}	10^{16}	10^{14}	10^{12}	10^{10}	10^{8}	10^{6}
ϕ_P	23.6	21.3	19.0	16.7	14.4	13.1	13.0	13.0
$\phi_{\rm limit}$	2.65	4.95	7.25	9.56	11.9	14.2	16.5	18.8
$\phi_{\rm max}$	1.59	2.68	3.78	4.94	7.06	9.34	11.7	13.0
$\phi_{\rm suf}$	1.59	2.24	2.27	2.27	2.27	2.27	2.27	2.27

Consequently the tunnelling approach does not predict a phase of sufficiently long inflation in this region, contrary to the results for the R^2-model. We further note that for the same range of initial values of ϕ, the no-boundary wavefunction shows no qualitative change from the quadratic case. Consequently, it appears that neither boundary condition predicts inflation for this choice of the parameters ϵ_1 and ϵ_2. This behaviour occurs because the presence of the cubic perturbation severely restricts the range of initial field values ϕ_i for which a phase of sufficiently long inflationary expansion is likely.

Including the full range of values of ϕ_i up to the Planck limit ϕ_P would not significantly improve this result in the tunnelling scheme. In the no-boundary case, however, the integral in the numerator of (24) would have a large contribution from the second peak in ρ_{NB}. However, this range of ϕ_i was excluded, as discussed above, in order to avoid the problem of exiting the inflationary expansion.

Even though the conditional probability P of (24) cannot be estimated numerically in this case, nevertheless, we present a set of values of 'scaled conditional probabilities' in the appendix, which are obtained by applying a nonlinear scaling to the probability distribution functions as discussed in subsection 4.1. These values, which may be treated as qualitative indicators of P, also support the conclusions given in this section.

5. Discussion and conclusions

In this paper we have investigated how the probability of realising sufficient inflation from quantum cosmology is altered when higher-order corrections to the Einstein–Hilbert action are introduced. Our results confirm that the addition of quadratic terms to the action does not reverse the conclusions of Vilenkin [5] regarding the effects of boundary conditions on the likelihood of sufficient inflation, in contrast to some recent claims [11]. Our feeling is that the difference in conclusions arises because we assume, in agreement with other analyses [5, 13, 14], that tunnelling to the Lorentzian region can occur at *any* point on the plateau of the potential, whereas Biswas and Guha assume *a priori* that the initial state is around $\phi \approx 0$. This seems to us to go against the spirit of quantum cosmology. If one specifies the boundary conditions in such a way, that the universe is in the Euclidian regime for small scale factors, the role of quantum cosmology is to provide a probability distribution for the initial conditions at the point of tunnelling, when the universe enters the classical Lorentzian regime. In our minisuperspace model this yields a measure of the most likely value of the scalar field ϕ_i. One should therefore consider all possible values of ϕ_i, if the most probable one is to be determined. Negative values of ϕ_i can be excluded in this model, since they may lead to super-Planck energy densities, so the range of interest is therefore $0 \leqslant \phi_i < \phi_P$.

We found that cubic perturbations can produce qualitative changes to the nature of the probability distribution function $\rho(\phi)$. From a physical point of view one is confined to consider initial values of the scalar field that allow an exit from the inflationary expansion. As a result the important physical (as opposed to purely mathematical) consequences of cubic perturbations are that they restrict the measure of allowed initial field values ϕ_i that lead to sufficient inflation. This is in agreement with the classical arguments [17]. By considering the conditional probability (24) (see also the appendix) we have argued that if the coupling constant ϵ_2, which determines the strength of the R^3-contribution to the Lagrangian, exceeds a critical value, neither the tunnelling nor the no-boundary boundary conditions predict an epoch of sufficient inflation, in the sense that the conditional probability is significantly less than unity in both cases.

Our results appear to exhibit some generality in four dimensions. As discussed in section 2, the qualitative shape of the self-interaction potential $V(\phi)$ remains unaltered if general polynomial perturbations with a highest-order term $\epsilon_{n-1} R^n$ are considered. In general this result is true when $D < 2n$. This immediately implies that neither of the two probability distributions ρ_T and ρ_{NB} for the $n = 3$ case will be qualitatively affected under $n > 3$ perturbations. The qualitative conclusions drawn for the case of cubic perturbations in subsection 4.2 therefore remain robust under higher-order perturbations to the action, although of course the details of what happens will depend on how the precise location of the maximum in the potential $V(\phi)$ is related to the highest-order term.

However, the consequences of the quadratic and the cubic perturbations (as well as those of general polynomial types) depend crucially on the values of the free parameters of the system, namely ϵ_k $(k = 1, \ldots, n-1)$, D, n, as well as on the initial field values ϕ_i. In particular, the dimensionality D of the spacetime is crucial in deciding the maximum degree n of perturbations allowed ($D < 2n$ say) above which the perturbations would be qualitatively inconsequential, i.e. the system would be robust.

Finally we remark that inflation is possible, at least at the classical level, if the field is initially placed to the right of the maximum in (10) and given sufficient kinetic energy to travel over the hill towards $\phi = 0$. Unfortunately, our analysis cannot consider this possibility since the scalar field momentum operator in the Wheeler–DeWitt equation then becomes important and the solutions (21) and (22) are no longer valid. Furthermore, if one is prepared to include the effects of the R^3-contribution in the action, the cubic term $R\Box R$ should also be considered. In this case the effective theory resembles Einstein gravity minimally coupled to two scalar fields after a suitable conformal transformation on the metric [17] and in principle a similar analysis to the one presented here can be followed for this more general case. We shall return to some of these questions in future.

Acknowledgments

We thank Jonathan J Halliwell and Andrei D Linde for helpful remarks. HvE is supported by a Grant from the Drapers' Society at QMW. JEL is supported by the Particle Physics and Astronomy Research Council (PPARC), UK, and is supported at Fermilab by the DOE and NASA under grant no NAGW-2381. RT is supported by the PPARC, UK, under grant no H09454.

Appendix

As was pointed out in subsection 4.1, the integrands involved in the definition of conditional probability typically have magnitudes of order $\exp(\pm 10^{14})$, which makes the numerical calculation of the integrals not possible in practice. Now due to the nature of these numbers no linear scaling of the probability function ρ can bypass this difficulty. The question then arises as to whether appropriate nonlinear scalings exist which keep the conditional probability P invariant. To see that there do *not*, recall that the only scalings that leave the Wheeler–DeWitt equation, $H\,\Psi = 0$, of the D-dimensional minisuperspace models of quantum cosmology invariant are given by $\tilde{H} = \Omega^{-2}\,H$, $\tilde{\Psi} = \Omega^{\gamma}\,\Psi \rightarrow \tilde{H}\,\tilde{\Psi} = \Omega^{\gamma-2}\,H\,\Psi = 0$, ($\Omega(q)$ is an arbitrary function of the minisuperspace coordinates q) provided γ and ξ (a free parameter in the Wheeler–DeWitt equation) are given by $\gamma = (2-D)/2$ and $\xi = -(D-2)/8(D-1)$ respectively [21]. Effectively this amounts to a redefinition of the potential $U(q)$ and the DeWitt metric of minisuperspace $f^{\alpha\beta}(q)$, which occur in the Hamilton operator H. More importantly, under such scale transformations the conserved probability current density j^{α} defined from Ψ remains unchanged. This freedom, however, is not of much use in bypassing the numerical difficulty mentioned above in order to obtain quantitative values for P. Nevertheless, if we confine ourselves to qualitative information, we may choose nonlinear (but monotonic) scalings of ρ, which, while violating the invariance properties of the model, would nevertheless supply us with a qualitative indicator of P. This is not dissimilar to the way nonlinear scalings of functions are employed for the purpose of graphical representation.

To calculate a qualitative indicator of P we define the *nonlinearly scaled conditional probability* $\tilde{P}$ as

$$\tilde{P}(\phi_{\rm i} > \phi_{\rm suf}\,|\,\phi_{\rm min} < \phi_{\rm i} < \phi_{\rm P}) \equiv \frac{\int_{\phi_{\rm suf}}^{\phi_{\rm P}} \rho^{1/C}(\phi)\,\mathrm{d}\phi}{\int_{\phi_{\rm min}}^{\phi_{\rm P}} \rho^{1/C}(\phi)\,\mathrm{d}\phi} \tag{A1}$$

where C is the index of nonlinear scaling. Clearly such a scaling will not change the qualitative behaviour of $\rho_{\rm T}$ and the values of its argument ϕ, and therefore the values of the boundaries of the integrals occurring in (24) (as listed in table 1) remain the same. Furthermore, such scalings leave P invariant in the limiting cases where $P = 0$ and $P = 1$.

Table A1. Behaviour of the nonlinearly scaled conditional probability distribution $\tilde{P}$ for the tunnelling wavefunction $\Psi_{\rm T}$ in the cubic model of subsection 4.2. We specify $\epsilon_1 = 10^{11}$ throughout.

ϵ_2	10^{20}	10^{18}	10^{16}	10^{14}	10^{12}	10^{10}	10^{8}	10^{6}
$\tilde{P}$	0.00	0.20	0.45	0.59	0.72	0.79	0.84	0.85

Here we chose $C = 10^{14}$. Table A1 gives the values of $\tilde{P}$ as a function of ϵ_2 for the tunnelling wavefunction in the case of the R^3-model, calculated for $\epsilon_1 = 10^{11}$ and the boundary values of ϕ given in table 1. We approximated $\phi_{\rm min}$ by $\phi_{\rm f} = 0.38$. For the corresponding value of $\tilde{P}$ for the Vilenkin model in the R^2-case of subsection 4.1 we found $\tilde{P} = 0.85$.

As can be seen from table A1, the behaviour of $\tilde{P}$ supports the conclusions drawn in subsection 4.2 on the basis of qualitative analysis.

Note added in proof. After this paper was submitted, related papers by Coule and Biswas and Guha [22] appeared in the literature.

References

[1] DeWitt B S 1967 *Phys. Rev.* **160** 1113
Wheeler J A 1968 *Battelle Rencontres* ed C DeWitt and J A Wheeler (New York: Benjamin)
[2] Linde A D 1984 *Zh. Eksp. Teor. Fiz.* **87** 369 (*Sov. Phys.–JETP* **60** 211); 1984 *Lett. Nuovo Cimento* **39** 401; 1984 *Rep. Prog. Phys.* **47** 925
[3] Vilenkin A 1984 *Phys. Rev.* D **30** 509
[4] Hartle J B and Hawking S W 1983 *Phys. Rev.* D **28** 2960
Hawking S W 1984 *Nucl. Phys.* B **239** 257
[5] Vilenkin A 1988 *Phys. Rev.* D **37** 888
[6] Candelas P, Horowitz G T, Strominger A and Witten E 1985 *Nucl. Phys.* B **258** 46
[7] Barth N H and Christensen S M 1983 *Phys. Rev.* D **28** 1876
[8] Antoniadis I and Tomboulis E T 1986 *Phys. Rev.* D **33** 2756
[9] Stelle K S 1978 *Gen. Rel. Grav.* **9** 353
[10] Horowitz G T 1980 *Phys. Rev.* D **21** 1445
Teyssandier P and Tourrenc Ph 1983 *J. Math. Phys.* **24** 2793
[11] Biswas S and Guha J 1993 *Class. Quantum Grav.* **10** L123
[12] Boulware D G 1984 *Quantum Theory of Gravity* ed S M Christensen (Bristol: Hilger)
Hawking S W and Luttrell J C 1984 *Nucl. Phys.* B **247** 250
Vilenkin A 1985 *Phys. Rev.* D **32** 2511
Horowitz G T 1985 *Phys. Rev.* D **31** 1169
Hawking S W and Wu Z C 1985 *Phys. Lett.* **151B** 15
Odintsou S R and Shapiro I L 1992 *Class. Quantum Grav.* **9** 873
Kasper U 1993 *Class. Quantum Grav.* **10** 869
[13] Mijić M B, Morris M S and Suen W-M 1989 *Phys. Rev.* D **39** 1496
[14] Bicknell G V 1974 *J. Phys. A: Math. Gen.* **7** 1061
Whitt B 1984 *Phys. Lett.* **145B** 176
[15] Barrow J D and Cotsakis S 1988 *Phys. Lett.* **214B** 515
Maeda K 1989 *Phys. Rev.* D **39** 3159
[16] Halliwell J J 1991 *Quantum Cosmology and Baby Universes* ed S Coleman, J B Hartle, T Piran and S Weinberg (Singapore: World Scientific)
[17] Berkin A L and Maeda K 1990 *Phys. Lett.* **245B** 348
Barrow J D and Cotsakis S 1991 *Phys. Lett.* **258B** 299
[18] Simon J Z 1992 *Phys. Rev.* D **45** 1953
[19] Guth A H 1981 *Phys. Rev.* D **23** 347
[20] Steinhardt P J and Turner M S 1984 *Phys. Rev.* D **29** 2162
[21] Halliwell J J 1988 *Phys. Rev.* D **38** 2468
[22] Coule D H 1994 *Class. Quantum Grav.* **11** 1365
Biswas S and Guha J 1994 *Class. Quantum Grav.* **11** 1369

Class. Quantum Grav. 11 (1994) 2499–2505. Printed in the UK

Flux-confinement in dilatonic cosmic strings

G W Gibbons† and C G Wells‡

DAMTP, University of Cambridge, Silver Street, Cambridge CB3 9EW, UK

Received 14 December 1993

Abstract. We study dilaton electrodynamics in flat spacetime and exhibit a set of global cosmic-string-like solutions in which the magnetic flux is confined. These solutions continue to exist for a small enough dilaton mass but cease to exist above a critical value depending on the magnetic flux. There also exist domain wall and Dirac monopole solutions. We discuss a mechanism whereby magnetic monopoles might have been confined by dilaton cosmic strings during an epoch in the early universe during which the dilaton was massless.

PACS numbers: 0420, 0440N, 1127

1. Introduction

There has been considerable interest recently in four-dimensional dilaton-Einstein–Maxwell and dilaton-Einstein–Yang–Mills theory. In the Abelian case, single and multi static and stationary black hole solutions with electric and magnetic charges have been extensively studied. Another class of interesting solutions describe magnetic fields with or without black holes. In the case of the Einstein–Maxwell theory, these are based on the Melvin solution which represents a sort of super-massive cosmic string [1, 2]. The Melvin solution may be generalized to include a coupling to the dilaton [3]. Related black hole metrics have been obtained and their relevance as instantons for the Schwinger production of black holes discussed [4]. Recently Maki and Shiraishi [5] have obtained some interesting time-dependent solutions with a dilaton potential.

In this paper, in order to gain some physical insight into dilaton electrodynamics and its non-Abelian generalization, we study the simpler flat-space version in which the effects of gravity are ignored. The action is

$$\int \mathrm{d}^4x\left(- \tfrac{1}{4}\exp(-2\tilde{\kappa}\phi)F_{\mu\nu}F^{\mu\nu} - \tfrac{1}{2}\eta^{\mu\nu}\partial_\mu\phi\partial_\nu\phi\right) \tag{1.1}$$

where $\eta^{\mu\nu} = \mathrm{diag}\,(-1, +1, +1, +1)$ and the dimensionful quantity $\tilde{\kappa}$ in the action can be changed by a suitable rescaling of the variables. Hence, its numerical value (as long as it does not vanish) has no physical significance in the purely classical theory that we study here. The action (1.1) can be obtained from the standard gravitational case in the Einstein conformal gauge in which the Maxwell field is multiplied by $\exp(-g\sqrt{4\pi G}\phi)$, by letting Newton's constant G tend to zero while at the same time $g\sqrt{4\pi G}$ tends to a finite limit

† E-mail address GWG1@AMTP.CAM.AC.UK
‡ E-mail address CGW11@AMTP.CAM.AC.UK

$\tilde{\kappa}$.† The field $F_{\mu\nu}$ may be Abelian or non-Abelian. In the latter case there remains, again in the classical theory, sufficient freedom to scale the Yang–Mills connection A so that $F = \mathrm{d}A + A \wedge A$. Note that had we taken the above limit of the gravitational Lagrangian expressed in string conformal gauge (i.e in terms of the metric $\exp(2\tilde{\kappa}\phi)g_{\mu\nu}$) we would have obtained action different from (1.1)).

The flat space version (1.1) has been studied by Lavrelashvili and Maison [7] and also by Bizon [8] who have obtained sphaleron-type solutions in which the Yang–Mills field is confined by the attractive forces exerted by the dilaton, which replace the attractive forces due to gravity in the Bartnik–McKinnon solution [9]. In the Abelian case it is easy to obtain the general static spherically symmetric purely magnetic solution representing a Dirac monopole coupled to the dilaton. Bizon has pointed out a special case of the spherically symmetric Dirac monopoles that may be generalized to give multi-Dirac-monopole solutions. These are Bogomol'nyi-type solutions and may be regarded as a limiting case of the multi-black-hole solutions as we shall show in section 4. It is also completely straightforward to obtain plane-wave solutions in which the dilaton and the photon are travelling parallel to one another.

The main point of this paper is to derive cosmic string-like solutions.

2. Permeabilities and permittivities

It follows immediately from (1.1) that the equations of motion for the field F in the presence of the dilaton field are those for a medium in which the electric permittivity ϵ is given by

$$\epsilon = \exp(-2\tilde{\kappa}\phi) \tag{2.1}$$

and the magnetic permeability μ is given by

$$\mu = \exp(2\tilde{\kappa}\phi). \tag{2.2}$$

The product $\epsilon\mu$ is unity and so the velocity of light remains one everywhere. Thus regions of spacetime for which $\phi < 0$ are diamagnetic while regions with $\phi > 0$ are paramagnetic. One does not usually encounter permittivities ϵ which are less than unity. One exception is the case of vacuum polarization in quantum chromodynamics which is 'anti-screening'rather than 'screening'as for all conventional material media. In the non-Abelian theory $\exp(\tilde{\kappa}\phi)$ plays the role of spacetime-dependent gauge coupling 'constant' and in string perturbation theory its expectation value plays the role of a variable coupling constant. Thus weak coupling corresponds to a diamagnetic phase and strong coupling to a paramagnetic phase. The action (1.1) is invariant under the simultaneous change of the sign of the dilaton field ϕ and the replacement of the Maxwell field F by its Hodge dual. This symmetry therefore interchanges the weak and the strong coupling phases.

The equation for ϕ in the static case is

$$\nabla^2\phi = -\tilde{\kappa}\exp(-2\tilde{\kappa}\phi)\big(\boldsymbol{B}^2 - \boldsymbol{E}^2\big) \tag{2.3}$$

where $\boldsymbol{E}$ and $\boldsymbol{B}$ have their usual meaning. It follows from (2.3) that ϕ can have no maximum in a purely magnetic field and no minimum in a purely electric field. Thus if ϕ is taken

† The sign of g we are using in this paper is consistent with the conventions of the majority of string theorists though it differs from that used in our recent preprint [6].

to be zero at infinity then magnetic regions tend to be paramagnetically polarized ($\mu > 1$) and electric regions tend to be dielectrically polarized ($\mu < 1$). Intuitively, magnetic flux ($\int \boldsymbol{B}\, \mathrm{d}\boldsymbol{S}$) tends to get self-trapped in strong coupling domains and electric flux ($\int \boldsymbol{D}\, \mathrm{d}\boldsymbol{S}$) in weak coupling domains, where $\boldsymbol{D} = \epsilon \boldsymbol{E}$ is the electric displacement and (for later use) $\boldsymbol{H} = \mu^{-1}\boldsymbol{B}$ is the magnetic induction. These observations are borne out by the particular solutions mentioned above. Thus for the Maison–Lavrelashvili–Bizon sphalerons μ has a maximum at the centre and decreases monotonically to unity at infinity. For magnetic-black-hole solutions μ increases monotonically inwards from unity at infinity. In the extreme case it becomes infinitely large and positive as one approaches the horizon. From the string point of view the infinitely long throat is a strong coupling region. In the electrically charged case the opposite is true. There is a parallel here with monopoles in the Yang–Mills theory and vortices in the Abelian Higgs theory. In the case of $SU(2)$ Yang–Mills theory with the Higgs in the adjoint representation, one may take the components of the Higgs field as a triplet of permittivities. The parallel is not completely precise but it is the case for the 't Hooft–Polyakov monopole that the associated permeabilities monotonically increase as one moves radially inwards. Similarly for the Nileson–Olesen vortex, one may think of the magnetic flux as being confined inside a core of high permeability where the Higgs field has a smaller magnitude than it does at infinity. We shall see similar features arising for dilaton electrodynamics.

The paramagnetic behaviour described above and the existence of the dilaton-Melvin solution strongly suggest that there may be non-singular static cosmic string-type solutions in which a finite amount of flux is trapped. This is indeed the case, as we shall show in the next section.

3. Dilaton cosmic strings and domain walls

In the static case the dilatonic Maxwell equations are readily seen to be satisfied if the magnetic induction $\boldsymbol{H} = (0, 0, H)$, where H is a constant, as long as the dilaton satisfies the two-dimensional Liouville equation:

$$\nabla^2\phi = -\tilde{\kappa} H^2 \exp(2\tilde{\kappa}\phi) \ . \tag{3.1}$$

Following Liouville [10], the general solution of (3.1) is

$$\mu = \exp(2\tilde{\kappa}\phi) = \frac{4}{\tilde{\kappa}^2 H^2}\frac{f'(\zeta)g'(\bar{\zeta})}{\left(1 + f(\zeta)g(\bar{\zeta})\right)^2} \tag{3.2}$$

where f and g are locally holomorphic functions and $\zeta = x + \mathrm{i}y$. If ϕ is real then

$$\overline{g(\bar{\zeta})}/f(\zeta)$$

must be a real-valued holomorphic function of ζ and hence constant. With no loss of generality, this constant may be taken to be unity and therefore the solution we require is

$$\mu = \exp(2\tilde{\kappa}\phi) = \frac{4}{\tilde{\kappa}^2 H^2}\frac{|f'(\zeta)|^2}{\left(1 + |f(\zeta)|^2\right)^2}. \tag{3.3}$$

Note that f and $1/f$ give the same solution ϕ.

Choosing different functions f gives different types of solution. For example, choosing $f(\zeta) = \zeta$ gives a cylindrically symmetric solution with finite total magnetic flux:

$$\Phi = \int_{\mathbb{R}^2} B \, \mathrm{d}S = \frac{4\pi}{\tilde{\kappa}^2 H} \quad . \tag{3.4}$$

This solution may be obtained as the limit of the dilaton-Melvin solution.

The magnetic contribution to the total energy per unit length of our solution is $\frac{1}{2}\Phi H = 2\pi/\tilde{\kappa}^2$, which is independent of the magnetic field H. The dilaton, however, contributes a logarithmically divergent energy per unit length because of its logarithmic dependence on the radius. In this respect our solution resembles a global rather than a local string. However, it should be pointed out that the solution, in common with its gravitational version, breaks neither electromagnetic gauge invariance nor any compact internal symmetry.

In addition to the single-string solution there are multi-string solutions. Thus choosing a rational function of Brouwer degree k, i.e. the ratio of two polynomials of order p and q, gives a solution with finite total magnetic flux

$$\Phi = k\frac{4\pi}{\tilde{\kappa}^2 H}. \tag{3.5}$$

Note that the permeability decreases to zero at infinity as $(x^2 + y^2)^{-(|p-q|+1)}$. Thus the weak coupling region at infinity is strongly diamagnetic and confines the magnetic flux Φ. The multi-string solutions are not axisymmetric. In flat-space Maxwell theory the only regular solution is the uniform magnetic field which is necessarily axisymmetric. When gravity is included, this goes over into the Melvin solution which also has axial symmetry. If one insists that the metric be boost-invariant then the axisymmetry, and hence uniqueness, follow by a version of Birkhoff's theorem [11]. However, the proof given in [11] does not go through in the presence of a scalar field. This suggests that there may exist static non-axisymmetric multi-dilaton-Melvin solutions.

If we take $f(\zeta) = \exp(\zeta)$ then

$$\mu = \exp(2\tilde{\kappa}\phi) = \frac{1}{\tilde{\kappa}^2 H^2}\frac{1}{\cosh^2 x}. \tag{3.6}$$

This solution describes a sheet or membrane confining an amount of flux per unit length $2/(\tilde{\kappa}^2 H)$. It may be thought of as a sort of domain wall separating two weak coupling domains.

4. Monopoles and Bogomol'nyi solutions

In this section we shall consider a general static magnetic solution. Since

$$\boldsymbol{\nabla} \times \boldsymbol{H} = 0 \tag{4.1}$$

we may locally introduce a magnetic potential χ by

$$\boldsymbol{H} = \boldsymbol{\nabla}\chi. \tag{4.2}$$

If we make the *ansatz*

$$\frac{1}{\tilde{\kappa}} \exp(-\tilde{\kappa}\phi) = \chi \tag{4.3}$$

then all the equations will be satisfied if in addition

$$\nabla^2 \exp(\tilde{\kappa}\phi) = 0. \tag{4.4}$$

These solutions are the same as those mentioned by Bizon and may be obtained from the multi-black-hole solutions by a limiting procedure. Our cosmic-string solutions are not a special case.

In addition to these Bogomol'nyi solutions, it is easy to find the general spherically symmetric monopole solution. If one insists that ϕ does not blow up at finite non-zero radius or at infinity one finds that

$$\exp(\tilde{\kappa}\phi) = \frac{1}{\alpha} \sinh \alpha\tilde{\kappa} P \left(\frac{1}{r} + \frac{1}{b}\right) \tag{4.5}$$

and

$$\boldsymbol{B} = \frac{P}{r^3}\boldsymbol{r} \tag{4.6}$$

where the constant of integration b is chosen so that $\phi = 0$ at infinity, P is the total magnetic charge and α is an arbitrary dimensionless constant. Just as in the case of magnetic black holes and 't Hooft–Polyakov monopoles, we find that the magnetic permeability increases monotonically inwards. The Bogomol'nyi solution (4.4) is obtained from the general solution (4.5) and (4.6) by letting α go to zero.

5. Massive dilatons

It is widely believed by string theorists that the dilaton acquires a mass due to non-perturbative effects connected with the breaking of supersymmetry. It is therefore of interest to ask what the effect of a mass term for the field ϕ might have on our solutions. In this section we simply add in a mass term 'by hand'. Thus we add to the Lagrangian in (1.1) a term $-\frac{1}{2}m^2\phi^2$. One checks that the work above goes through as long as one replaces ∇^2 by $\nabla^2 - m^2$ in (2.3) and (3.1). However, the solution (3.2) is no longer valid and we do not know how to solve the new version of (3.1) exactly. The circularly symmetric solutions may be treated by regarding the radial coordinate r as a fictitious time variable. The equation for ϕ becomes that of a particle with position ϕ moving on a line experiencing a time-dependent frictional force, $-\frac{1}{r}\frac{\mathrm{d}\phi}{\mathrm{d}r}$, and subject to a potential $U(\phi)$:

$$\begin{aligned} &\frac{\mathrm{d}^2\phi}{\mathrm{d}r^2} + \frac{1}{r}\frac{\mathrm{d}\phi}{\mathrm{d}r} = \frac{\mathrm{d}U(\phi)}{\mathrm{d}\phi} \\ &U(\phi) = -\tfrac{1}{2}m^2\phi^2 + \tfrac{1}{2}H^2\exp(2\tilde{\kappa}\phi). \end{aligned} \tag{5.1}$$

Regularity at the origin implies that the radial derivative of ϕ vanishes there. A solution exists for each value of ϕ at the origin. If $m = 0$, one has those solutions given in section 3 in which ϕ decreases monotonically to minus infinity at infinity as

$$\tilde{\kappa}\phi \sim -2\log r. \tag{5.2}$$

For non-vanishing mass m, the behaviour depends on the ratio $m^2/(2\mathrm{e}\tilde{\kappa}^2 H^2)$, where e is the base of natural logarithms. If $0 < m^2/(2\mathrm{e}\tilde{\kappa}^2 H^2) < 1$, the potential $U(\phi)$ is a monotonic increasing function of ϕ and ϕ decreases monotonically with r for all values of $\phi(0)$, and tends to minus infinity at infinity,

$$\phi \sim -c \exp(mr) \tag{5.3}$$

where c is a positive constant.

However, if $m^2/(2\mathrm{e}\tilde{\kappa}^2 H^2)$ is greater than unity then the potential $U(\phi)$ has a local minimum and a maximum. The behaviour of the solutions depends upon $\phi(0)$. If $\tilde{\kappa}\phi(0)$ is positive and sufficiently large, then ϕ decreases monotonically to minus infinity as before. However, if $\tilde{\kappa}\phi(0)$ lies in a finite interval bounded below by the smaller solution x of the equation

$$x = \frac{\tilde{\kappa}^2 H^2}{m^2} \exp(2x) \tag{5.4}$$

then the solutions oscillate about the minimum with an amplitude which decreases to zero as r tends to infinity. Finally if $\tilde{\kappa}\phi(0) < x$, the solutions decrease monotonically to minus infinity.

Thus for given magnetic field H there always exist solutions with finite total flux. If, however, the mass is large enough, one has solutions in which ϕ tends a minimum value dependent upon H of the potential $U(\phi)$.

We have repeated these calculations for the monopole solutions of section 4 in the case of a massive dilaton. We find that the qualitative behaviour of the dilaton is very similar to that in the massless case.

One might imagine a cosmological scenario in which the dilaton is initially massless at some high temperature and acquires a mass during a cosmological phase transition at a lower temperature. If cosmic strings of the type we have described, confining a finite flux, were initially present and the mass were large enough, it seems from our calculation that provided $\phi(0)$ took suitable values the flux would become unconfined. If this were true it might have important consequences for magnetic monopoles. If flux was confined by strings at early times then one might expect magnetic monopoles, of the sort described in section 4, to be found at the ends of flux tubes. These flux tubes should pull the monopoles together and cause their rapid annihilation. At late times magnetic fields would become unconfined. In this way one might have a natural solution to the monopole problem. Clearly more work needs to be done to establish whether this picture is really viable.

It is interesting to note that dilaton electrodynamics with an effective mass term has already been invoked [12] to account for a possible primordial magnetic field. It would be interesting to investigate the relation between that work and the monopoles and vortices described in this paper.

6. Conclusions

We have investigated cosmic string solutions to dilaton electrodynamics. In terms of the effective magnetic permeability in the massless case, they are perfectly diamagnetic at infinity, i.e. the flux is confined. Furthermore, we have found that when the dilaton acquires a mass, cosmic string solutions exist which do not share this confinement property. We have also exhibited a domain-wall solution in this theory and investigated the Dirac monopole analogue for the dilaton which is both massless and massive. In addition, we have proposed a mechanism which may solve the monopole problem by using these cosmic strings to catalyse the annihilation of primordial magnetic monopoles.

Acknowledgments

We would like to thank Paul Shellard for helpful discussions and suggestions.

References

[1] Bonnor W B 1954 *Proc. Phys. Soc.* A **67** 225
[2] Melvin M A 1964 *Phys. Lett.* **8** 65
[3] Gibbons G W and Maeda K 1988 *Nucl. Phys.* B **298** 741–75
[4] Dowker F, Gauntlett J P, Kastor D A and Traschen J 1993 hep-th 9309075
[5] Maki T and Shiraishi K 1993 *Preprint* AJC HEP-16 Akita Junior College
[6] Gibbons G W and Wells C G 1993 *DAMTP preprint* R93/25 gr-qc 9310002
[7] Lavrelashvili G and Maison D 1992 *Phys. Lett.* **295B** 67
[8] Bizon P 1993 *Phys. Rev.* D **47** 1656–63
[9] Bartnik R and McKinnon J 1988 *Phys. Rev. Lett.* **61** 41
[10] Liouville J 1853 *J. Math.* **18** 71–2
[11] Gibbons G W 1991 *The Physical Universe* Lecture Notes in Physics **383** ed J D Barrow, A B Henriques, M T V T Lago and M S Longair (Berlin: Springer)
[12] Ratra B 1992 *Astrophys. J.* **391** L1–4

Class. Quantum Grav. **11** (1994) 2507–2515. Printed in the UK

A non-traditional cosmological fluid

Caroline Lewis†
Physics Department, Furman University, Greenville, SC 29613, USA

Received 8 February 1994, in final form 6 June 1994

Abstract. Non-Newtonian corrections to an imperfect cosmological fluid result in a density contrast growing as a conformal time to the tenth power during a first order cosmological phase transition. Further questions of dissipative processes in non-equilibrium cosmology are discussed.

PACS numbers: 9880H, 0440N, 6460

1. Introduction

Given the present interest in the large scale structure of the universe, it is surprising that comparatively little exploration of the matter side of the Einstein equation has been made. In this paper I investigate the effect on large scale structure in the universe of adding the simplest non-Newtonian correction to the usual stress–energy tensor. This can be done by making the viscosity term in the stress energy tensor for an imperfect cosmological fluid proportional to a function of the velocity gradient in the fluid. One then computes the solutions to the density perturbation evolution equations that result from using the gauge-invariant, cosmological perturbation formalism set up by Bardeen [1]. The non-Newtonian terms cause a non-linear amplification of the density perturbations, which can then act as seeds for galaxy formation.

The motivation for this unusual viscosity treatment comes from the Einstein result [2, 3] that the effective viscosity for a suspension of rigid spheres in an otherwise uniform fluid is proportional to the concentration of the spheres. This work has been extended by Schowalter *et al* [4] to the case of Newtonian droplets immersed in a fluid, where all of the shear viscosity, bulk viscosity and thermal conduction terms (for the fluid) need to be corrected for the concentration of the drops, which will in turn change as the velocity gradient changes with time.

The early universe will be treated as a two-phase system during one of the now popular first-order cosmological phase transitions: clumps (rough globules) of matter immersed in an otherwise uniform fluid medium. The effective viscosity of the imperfect cosmological fluid during this phase transition is calculated using bubble nucleation theory; one can then examine the behaviour of density perturbations using the Bardeen formalism with non-Newtonian corrections.

The standard Hot Big Bang theory maintains that $\delta\rho/\rho$ grows as $t^{2/3}$. There appears to be an amplification in the growing solution for the density contrast over that of the standard Hot Big Bang model because of the non-linear dependence of stress on velocity gradients which occurs for non-Newtonian fluids.

† Present address: Center for Relativity, Physics Dept, University of Texas at Austin, Austin, TX 78712, USA.

0264-9381/94/102507+09$19.50

This approach gives a conceptually different approach to the problem of large scale structure formation in cosmology; it puts more emphasis on the right hand side (matter) of what is after all the Einstein *equation*. Most treatments of relativistic cosmology focus on the geometry side of the Einstein equation. In section 2, the density contrast formalism is set up. Section 3 gives the bubble nucleation theory and the main results; in section 4 there is the discussion and conclusions including reference to the idea of dissipative structure. I use the conventions of Bardeen, with Latin indices ranging from 0 to 3, Greek indices from 1 to 3. The Einstein summation convention is used throughout and units such that $c = 8\pi G = 1$ are used.

2. Density contrast formalism

This section is a review of the Bardeen relativistic, gauge-invariant formalism as used to treat density perturbations. This tool (as opposed to the much simpler Newtonian approximation cosmology) is used because perturbations on scales of cosmological interest are superhorizon-sized in the early universe, that is, the physical wavelength λ_{phy} is greater than the Hubble radius H^{-1}.

The background metric tensor is Friedmann–Robertson–Walker (FRW) with line element

$$ds^2 = S^2(\tau)(-d\tau^2 + {}^3 g_{\alpha\beta}\, dx^\alpha\, dx^\beta) \tag{2.1}$$

where S is the cosmic scale factor, τ is the conformal time, and ${}^3g_{\alpha\beta}$ is the metric of 3-space with uniform spatial curvature K. The perturbed metric tensor is given by $g_{\alpha\beta}$:

$$\begin{aligned}
g_{00} &= -S^2(\tau)[1 + 2A(\tau)Q^{(0)}(x^\mu)] \\
g_{0\alpha} &= -S^2 B^{(0)}(\tau) Q^{(0)}_\alpha(x^\mu) \\
g_{\alpha\beta} &= S^2\{[1 + 2H_L(\tau)Q^{(0)}(x^\mu)]^3 g_{\alpha\beta} + 2H^{(0)}_T(\tau)Q^{(0)}_{\alpha\beta}(x^\mu)\}
\end{aligned} \tag{2.2}$$

where the scalar harmonics describing spatial wave motion are solutions of the scalar Helmholtz equation $Q^{(0)\uparrow\alpha}_{\uparrow\alpha} + k^2 Q^{(0)}$, where the arrow represents the covariant derivative with respect to the 3 metric. Define $Q^{(0)}_\alpha = -(1/k)Q^{(0)}_{\uparrow\alpha}$ and the traceless, symmetric $Q^{(0)}_{\alpha\beta} = k^{-2}Q^{(0)}_{\uparrow\alpha\beta} + \frac{1}{3}{}^3g_{\alpha\beta}Q^{(0)}$ where k is the comoving wavenumber. The stress–energy tensor, $T_{\alpha\beta}$, is given by:

$$\begin{aligned}
E &= -T^0_0 = E_0[1 + \delta(\tau)Q^{(0)}(x^\mu)] \\
P &= \tfrac{1}{3}T^\alpha_\alpha = P_0(\tau) + P_0(\tau)\pi_L(\tau)Q^{(0)}(x^\mu) \\
T^\alpha_\beta &= P_0[1 + \pi_L Q^{(0)}]\delta^\alpha_\beta + P_0\pi_T(\tau)Q^{(0)\alpha}_\beta \\
T^0_\alpha &= (E_0 + P_0)(v^{(0)} - B^{(0)})Q^{(0)}_\alpha \\
T^\alpha_0 &= -(E_0 + P_0)v^{(0)}Q^{(0)\alpha}
\end{aligned} \tag{2.3}$$

using the notation of Bardeen. Especially note the matter 3-velocity $v^{(0)}(\tau)Q^{(0)\alpha}(x^\mu)$ (the 4-velocity is defined with respect to the energy frame, where the energy flux vanishes, so that no heat conduction term appears) and the anisotropic stress perturbation coefficient π_T. E_0 and P_0 are the background energy density and pressure, respectively. For later reference, define the entropy perturbation, η, as a measure of the difference between the background equation of state and the equation of state in the physical spacetime:

$$\eta(\tau)Q^{(0)} = \left(\pi_L - \frac{E_0}{P_0}\frac{\mathrm{d}P_0}{\mathrm{d}E_0}\delta\right)Q^{(0)}. \tag{2.4}$$

The condition $\eta(\tau) = 0$ characterizes an adiabatic perturbation.

This matter formulation is sufficient for studying the broad features of dissipative behaviour, a more complete treatment would use relativistic kinetic theory [5, 6].

Gauge-invariant metric potentials are constructed out of the metric perturbation coefficients while gauge-invariant density contrasts and velocities are derived from the matter perturbation coefficients. I choose to work on a uniform-Hubble-constant hypersurface, meaning that the extrinsic curvature, K, is spatially uniform on each hypersurface. Then the gauge-invariant metric potentials are

$$\phi_h = \bar{H}_L + \tfrac{1}{3}\bar{H}_T \qquad \alpha_h = \bar{A} \tag{2.5}$$

where the bars represent the uniform-Hubble-constant hypersurface condition.

The gauge-invariant density contrast is

$$\varepsilon = [1 + \tfrac{3}{2}k^2(E_0 + P_0)S^2]^{-1}[\varepsilon_m + 3(1+w)\phi_m] \tag{2.6}$$

where

$$\begin{aligned} \varepsilon_m &= \delta + 3(1+w)\frac{1}{k}\frac{\dot{S}}{S}(v^{(0)} - B^{(0)}) \\ \phi_m &= \bar{H}_L + \tfrac{1}{3}\bar{H}_T \\ w &= E_0/P_0 \end{aligned} \tag{2.7}$$

with the bars in the ϕ_m case representing the comoving hypersurface condition.

To calculate the time variation of ε I utilized the not very well known section V of the Bardeen paper in which a relationship to non-linear dynamics is set up. This section is specifically in the language of the Arnowitt–Deser–Misner (ADM) $(3+1)$-split of spacetime and ends up with a convenient expression for an adiabatic gauge-invariant density perturbation in terms of a double integral over the dissipative source term. Recall that in the ADM approach there is an intrinsic metric tensor $h_{\alpha\beta}$ describing the spatial hypersurface and an intrinsic curvature tensor $K_{\alpha\beta}$ which describes the imbedding of the spatial hypersurface in spacetime. Proper spacetime slicing between the hypersurfaces is done using the lapse function, N. Dynamic equations will evolve the $h_{\alpha\beta}$, $K_{\alpha\beta}$ from hypersurface to hypersurface. Each hypersurface is labelled by τ and n_α, the unit future-directed 4-vector normal to the hypersurface.

$$n^0 = N^{-1} \qquad n^\alpha = N^{-1}N^\alpha \tag{2.8}$$

where the shift vector N^α equals the coordinate 3-velocity to the worldline. N^α describes how spatial coordinates are propagated from one slice to another. The Einstein equation is written in terms of the trace and traceless parts of the extrinsic curvature tensor and the momentum and energy initial value equations as well as the geometry evolution equations are calculated. This is especially convenient to write out for the uniform-Hubble-constant hypersurface where one can specify $K_{,\alpha} = 0$, $K_{,0} = 0$, to simplify the evolution equations down to second order. Geometrical evolution reduces to a second-order linear differential equation for the geometric perturbation, with a dissipative matter source term. Defining $x = k\tau$ and $\beta = 2/(3w + 1)$, Bardeen gets that for the superhorizon-sized fluctuation condition $x/\beta \ll 1$,

$$\varepsilon \approx \tfrac{2}{3}\beta^{-2}x^2\phi_h \qquad \alpha_h \approx \frac{w}{1+w}\eta. \tag{2.9}$$

The metric potential is given by the evolution equation

$$g' + 2(\beta/x)g \approx (\beta/x)^2 w\pi_T \tag{2.10}$$

where

$$g = \phi_h' - (\beta/x)\alpha_h \approx \phi_h' + \frac{\beta w x^{-1}\eta}{1+w} \tag{2.11}$$

with $' = \mathrm{d}/\mathrm{d}x$. (See Bardeen for further details on the evolution equations for metric potentials as related to dissipative terms and matter velocities.)

For no perturbation at $x = 0$, the solution is

$$g = \beta^2 x^{-2\beta} \int_0^x w\pi_T y^{2\beta-2}\,\mathrm{d}y \tag{2.12}$$

and then the main result of this section is that the gauge-invariant metric potential is given by a straightforward integral over the dissipative terms,

$$\phi_h = \int_0^x g(y)\,\mathrm{d}y - \frac{\beta}{1+w}\int_0^x w\eta y^{-1}\,\mathrm{d}y. \tag{2.13}$$

This is quite simple.

I will be dealing only with anisotropic stress perturbations in this paper, meaning that the entropy perturbation η is zero, i.e. these are adiabatic perturbations. So

$$\phi_h = \int_0^x \beta^2 y^{-2\beta}\,\mathrm{d}y \int_0^y w\pi_T z^{2\beta-2}\,\mathrm{d}z \tag{2.14}$$

which is then simply related to ε for superhorizon-sized fluctuations.

Now the question is: what phenomenological theory gives an expression for π_T? The next section of this paper, on non-equilibrium thermodynamics and statistical mechanics, answers this question.

3. First-order cosmological phase transitions

The anisotropic stress perturbations, π_T, will be calculated in this section on the basis of cosmological phase transition theory. This is one way to implement a non-Newtonian correction to a cosmological fluid: bubbles in a fluid (due to first order phase transition) cause an effective viscosity which can then act as a source for the growth of density perturbations as seen in the last section.

First, I will give some general comments on phase transitions in the early universe [7]. Exotic quantum fields flourish in the high temperatures after the (Hot) Big Bang. I will consider a generic quantum field, ϕ. During a phase transition (normalized) ϕ evolves from a high temperature, symmetric minimum at $\phi = 0$ (false vacuum) to a low temperature, symmetry-breaking minimum at $\phi = \sigma$ (true vacuum). If there is a potential barrier separating these two minima the phase transition is a first-order one, otherwise it is second order. In a first-order phase transition the potential barrier can be penetrated by either quantum or thermal tunnelling, in which bubbles of true vacuum nucleate and then expand, in the first case at near the speed of light and in the second case at about $c/\sqrt{3}$ [8]. The latter, lower speed is due to the fact that in the case of thermal tunnelling not all of the false vacuum energy goes into the kinetic energy of the bubble walls, some goes into an ambient plasma associated with the quantum fluid. I will be considering a first-order phase transition proceeding through thermal tunnelling in the so-called small supercooling (SSC) approximation, which says that the temperature at the end of the phase transition, T_e, is only slightly less than the temperature at the start of the phase transition,

$$T_s : \frac{T_s - T_e}{T_s} \ll 1.$$

This approximation allows classical bubble nucleation theory to be used (pair production is not possible under the SSC approximation). Effectively, the only difference here between quantum and thermal tunnelling is the speed of bubble expansion. Examples of possible first-order cosmological phase transitions are the electroweak transition, the QCD phase transition and some inflationary scenarios. Extended inflation [9] would be the best choice for a first-order cosmological phase transition here because the high energy scales associated with the Hubble constant for this era will offset the large time delays for re-entry of superhorizon-sized fluctuations so as to produce a non-negligible density contrast ($\delta\rho/\rho \approx H\tau$ where τ, the time delay between first and second horizon-crossing, is dimensionally sensible). I will say more about extended inflation at the end of this section.

Bubble nucleation theory will be needed to calculate the effective viscosity related to π_T. The Einstein statistical mechanics result for spheres suspended in a fluid leads to the following estimate for the effective viscosity

$$\eta_{\text{eff}} = \eta_0[1 + \tfrac{5}{2}F(t)] \tag{3.1}$$

where η_0 is the original viscosity of the fluid, $F(t)$ is the volume fraction of, in this case, bubbles of true vacuum nucleated out of the now surrounding false vacuum. The increased dissipation is due to the collisions between the interfaces of the different vacuua. This does lead to a non-Newtonian correction for the stress–energy tensor where the specific dependence on velocity gradient will depend on the microphysics of the particular phase transition. The volume fraction (that is, the concentration) is calculated using classical bubble nucleation theory to be

$$F(t) = \int_{t_c}^{t} \mathrm{d}t'\, p(t')V(t, t') \tag{3.2}$$

where the volume of the bubble $V(t, t') = \frac{4}{3}\pi v_b^3(t - t')^3$ at time t of the bubble nucleated at time t'. t_c is the critical time at the start of the phase transition which I will normally take as the start of the phase transition at time t_s and v_b is the physical velocity of the bubble wall expansion in an expanding spacetime ($c/\sqrt{3}$ in this case). $p(t)$ is the nucleation rate, that is, the number of bubbles nucleated per unit volume and time. $p(t)$ is given by [10] in an approach to equilibrium formalism. It is an energetically difficult process to form the interface of a new phase in a fluid: bubbles are nucleated and then fluctuate until they reach a critical size, a_c, at which time they quickly expand until they coalesce and fill the space.

A size distribution function f is defined such that $f\,da$ is the number of nuclei per unit volume with sizes in the range da. The following Fokker–Planck equation holds

$$\frac{\partial f}{\partial t} = -\frac{\partial p}{\partial a} \tag{3.3}$$

where the nucleation rate

$$p = -B\frac{\partial f}{\partial a} + Af. \tag{3.4}$$

B is a size diffusion coefficient and A is a constant determined by using the condition that $p = 0$ at $f = f_0$, the equilibrium value of the distribution function. Reference [10] gives that

$$p \cong 2\sqrt{\frac{\alpha}{T}}B(a_{cr})f_0(a_{cr}) \tag{3.5}$$

where α is the surface tension coefficient (work needed to form a unit surface area) and T is the temperature. Above the critical temperature $f \ll f_0$ is approximately constant (the nuclei are growing steadily larger, with no 'reversals') so that

$$p \approx Af \qquad A \approx (\mathrm{d}a/\mathrm{d}t)_{\mathrm{macroscopic}}. \tag{3.6}$$

Then

$$B(a)\,|_{a\to a_{cr}} = \frac{T}{8\pi\alpha(a-a_{cr})}\left(\frac{\mathrm{d}a}{\mathrm{d}t}\right)_{\mathrm{macroscopic}} \tag{3.7}$$

so that finally,

$$p \cong 2\sqrt{\frac{\alpha}{T}}\frac{T}{8\pi\alpha}(t-t_c)^{-1}f_0(a_{cr}) \tag{3.8}$$

where $f_0 = \mathrm{const} \times \exp[-(4\pi\alpha a_{cr}^2/3T)]$, with the constant determined microscopically for the particular scenario. (The constant is of the order of a_{cr}^2/V^2 where V is the typical volume of the increment being added on to the nucleus.)

To simplify the calculation I will make the reasonable assumptions that for the duration of the phase transition

$$a - \mathrm{const} \qquad \alpha - \mathrm{const} \qquad v_b \approx c/\sqrt{3} \qquad T \approx T_c \approx \mathrm{const} \tag{3.9}$$

(for the latter, averaging over the fluctuations inherent in a non-equilibrium process such as a phase transition.) and I will also use that $w = P_0/E_0 = \frac{1}{3}$ (for radiation or highly relativistic fluid behaviour, referring to the background FRW universe). $p(t)$ is the number of nuclei formed past the critical range in stationary conditions per unit time and unit volume.

I can now integrate out for the volume fraction $F(t)$ then and convert to $F(\tau)$ using

$$S(\tau)\,\mathrm{d}\tau = \mathrm{d}t \qquad F(t) = c_1(t-t_\mathrm{c})^3 \tag{3.10}$$

where

$$c_1 = \frac{c}{9}\sqrt{\frac{T}{\alpha}}v_\mathrm{b}^3 \exp\left[\frac{4\pi\alpha a_\mathrm{c}^2}{3T}\right] \tag{3.11}$$

and

$$t = \frac{S_0}{\tau_0}\frac{\tau^{\beta+1}}{(\beta+1)} \tag{3.12}$$

where the subscript '0' represents the value of the quantity at the time of radiation–matter equality; subscript 'c' represents values at the critical time of the phase transition. Now,

$$\eta_\mathrm{eff} \approx \eta_0[1+\tfrac{5}{2}F(\tau)]. \tag{3.13}$$

Dimensionally,

$$\eta_\mathrm{eff} \approx P_0\pi_T\tau_\mathrm{mft} \tag{3.14}$$

where the mean free time $\tau_\mathrm{mft} \approx (t-t_\mathrm{c})$ is reasonable. Therefore $P_0\pi_T(t-t_\mathrm{c}) \approx \eta_\mathrm{eff}$. Use $P_0 = wE_0$ and $E_0 = 3\beta^2\tau^{-2}/S^2$ where $\beta = 2/(3w+1) = 1$, here. Then

$$w\pi_T \approx w\eta_\mathrm{eff}/P_0(t-t_\mathrm{c}) = \frac{\eta_0 S_0^2}{3\beta^2\tau_0^2}\left\{\frac{\tau^{\beta+1}}{[S_0/\tau_0(\beta+1) - c\tau^{-(\beta+1)}]} + \frac{5c_1\tau^{(2\beta+1)}[(S_0\tau^{\beta+1}/\tau_0(\beta+1)) - c]^2}{6}\right\} \tag{3.15}$$

where $c = t_\mathrm{c}$. Finally the main result of this paper comes from integrating out

$$\phi_h = \int \beta^2 x^{-2\beta}\,\mathrm{d}x \int_0^x w\pi_T y^{2\beta-2}\,\mathrm{d}y \tag{3.16}$$

to give

$$\phi_h = \frac{A\tau^2}{6B} + \frac{AC^2D\tau^4}{24} - \frac{5ABCD\tau^6}{126} + \frac{5AB^2D\tau^8}{432} + \frac{AC^{3/2}\tanh^{-1}\left[\left(\sqrt{B}/\sqrt{C}\right)\tau\right]}{B^{5/2}\tau} + \frac{AC\log(C-B\tau^2)}{2B^2} \tag{3.17}$$

where

$$A = \frac{\eta_0 S_0^2}{3\tau_0} \qquad B = \frac{S_0}{2\tau_0} \qquad C = t_\mathrm{c} \qquad D = c_1. \tag{3.18}$$

Therefore the dominant term for a given phase transition will cause the gauge-invariant density contrast ε to grow as τ^{10}:

$$\varepsilon \approx \tfrac{2}{3}\beta^{-2}x^2\phi_h \approx \left(\frac{10AB^2D}{1296}\right)k^2\tau^{10}. \tag{3.19}$$

This could be useful for galaxy formation.

This result is good for a generic first-order phase cosmological phase transition of approximate parameters and such that $x/\beta \ll 1$, i.e. at times $\tau \ll 1/k$ ($\beta = 1$ here).

For comoving wavenumbers of cosmological interest this suggests that extended inflation (EI) [9] would be the best choice of phase transition. (See Hwang [9] for effective fluid treatments of generalized gravity theories such as EI.) While a specific fit of EI to the present work is the subject of a future paper, it is worthwhile to briefly discuss the constraints on bubble formation due to the cosmic microwave background radiation (CMBR) and large scale structure (LSS). The history of a thermal bubble [9] is such that there is a flow of false vacuum energy from the kinetics of the bubble wall expansion to the formation of massive scalar particles which will decay into relativistic particles which eventually thermalize (hopefully) to a near isotropic and homogeneous radiation-dominated universe. For this to work in EI depends upon the competition between the expansion rate H and the bubble nucleation rate per unit volume Γ. One must worry about so called 'big bubbles', i.e., bubbles produced early in the inflationary phase which later reach astrophysically interesting sizes. These could overly disturb the CMBR. Suppression of these 'big bubbles' requires that at early times $\Gamma/H^4 \leqslant 10^{-4}$; however, to complete the phase transition $\Gamma/H^4 \geqslant 9/4\pi$ at late times is required [9, La, Steinhardt and Bertschinger reference].

The density perturbations formed in the non-Newtonian scenario of this paper may be formed in addition to those formed from quantum fluctuations of the inflation field. The phase transition parameters will have to be tuned in order to ensure that the CMBR temperature fluctuation limits prevail. This is a challenging question for any work on LSS formation in the early universe.

4. Discussion and conclusions

We have seen that for superhorizon-sized density fluctuations generated during a first-order cosmological phase transition there can be a time rate of growth going as the conformal time to the tenth power for the duration of the phase transition. This striking growth rate is best discussed in the context of the Carr and Silk work [11] on 'grains' in the early universe (these are the same as what I have called bubbles in this paper). The question is: how can a density perturbation with significant amplitude on a superhorizon scale result from zero initial perturbation at the start of the phase transition without violating energy conservation or causality? Carr and Silk have tracked the energetics of this process. Inherent randomness in the bubble nucleation process gives rise to statistical ($\sqrt{N}$) fluctuations (where N is the number of grains) in the pressure because of the differing equations of state in the two phases. If these pressure perturbations persist longer than one expansion time they can generate growing density perturbations on the superhorizon scale which scale as $N^{-1/2}$ when they come within the horizon. Another statistical fluctuation is associated with the peculiar velocities of the grains which are bound to develop through non-linear effects. In this case the additional step in the energetics process is that the peculiar velocities cause $\sqrt{N}$ pressure perturbations which in turn lead to density fluctuations growing as $N^{-1/2}$ on re-entry.

These spontaneous fluctuations can generate structure on galactic or larger scales only if the fractional pressure perturbation at early times is of order 10^{-2} or greater on the same scales [12]. This may be feasible for some scenarios of extended inflation. The non-Newtonian corrections will be considered as a dissipative enhancement of this effect in a forthcoming paper [13], which also considers the power spectrum of the fluctuations. Attention also is paid to the speed of the dissipative signals (subluminal speeds are ensured in the extended thermodynamic framework of Israel and Stewart [14, 15]).

The non-Newtonian structure formation scenario indicated in this paper is to be regarded as an alternative to structure formation through quantum fluctuations during an inflationary epoch or through 'Bubble Universe' mechanisms where the bubbles in the cosmological phase transition directly cause LSS formation. Using viscosity (a measure of dissipation) to build up structure is a promising approach to the LSS problem of the early universe. A fundamental direction of this work is to explore all types of non-equilibrium processes in the early universe and to investigate whether the idea of dissipative structure discussed by Prigogine and Nicolis [16] can be applied to relativistic cosmology.

Acknowledgments

I would like to acknowledge the receipt of a grant from the Knight Foundation which helped in the completion of this work. I would also like to thank Bernard Carr, Ethan Vishniac and Harry King for valuable discussions as well as the Astronomy Unit of the Queen Mary and Westfield College of the University of London, the Center for Relativity of the University of Texas at Austin and King, Matkowski and Associates for kind hospitality.

References

[1] Bardeen J M 1980 *Phys. Rev.* D **22** 1882
[2] Einstein A 1906 *Ann. Phys., Lpz.* **19** 289
[3] Landau L D and Lifschitz E M 1959 *Fluid Mechanics* (Oxford: Pergamon)
[4] Schowalter W R 1978 *Mechanics of Non-Newtonian Fluids* (Oxford: Pergamon)
[5] Jou D, Casas-Vasquez J and Lebon G 1993 *Extended Irreversible Thermodynamics* (Berlin: Springer)
[6] de Groot S R, van Leeuwen W A and van Weert C 1980 *Relativistic Kinetic Theory: Principles and Applications* (Amsterdam: North-Holland)
[7] Kolb E W and Turner M S 1990 *The Early Universe* (Reading, MA: Addison-Wesley)
[8] Weinberg E Private communication
[9] La D and Steinhardt P J 1989 *Phys. Rev. Lett.* **62** 376
Hwang J 1990 *Phys. Rev.* D **42** 2601
La D, Steinhardt P and Bertschinger E 1989 *Phys. Lett.* **231B** 231
Turner M, Weinberg E and Widrow L 1992 *Phys. Rev.* D **46** 2384
[10] Lifschitz E M and Pitaevskii L P 1982 *Physical Kinetics* (Oxford: Pergamon)
[11] Carr B J and Silk J 1983 *Astrophys. J.* **268** 1
[12] Press W H and Vishniac E T 1980 *Astrophys. J.* **239** 1
[13] Lewis C M 1994 to be published
[14] Israel W 1976 *Ann. Phys.* **100** 310
[15] Israel W and Stewart J M 1979 *Ann. Phys., NY* **118** 341
[16] Nicolis G and Prigogine I 1977 *Self-Organization in Nonequilibrium Systems* (New York: Wiley)

Class. Quantum Grav. **11** (1994) 2517–2521. Printed in the UK

On spherically symmetric cosmology in Riemann–Cartan spacetime with spin-density

Larry L Smalley† and Jean P Krisch‡

†Department of Physics, University of Alabama in Huntsville, Huntsville, AL 35899, USA

‡Department of Physics, University of Michigan, Ann Arbor, MI 48109, USA

Received 13 July 1993, in final form 2 July 1994

Abstract. We develop the average effective energy–momentum tensor for a spherically symmetric cosmology with randomly oriented spin using the improved energy–momentum tensor in the framework of the self-consistent Einstein–Cartan theory with spin-density.

PACS numbers: 9530S, 9880C, 4775, 0340G, 0450

It has been known for nearly two decades in the investigations of Einstein–Cartan (EC) theories that the spin-density of matter heavily influences [1] and could dominate the behaviour of spin aligned fluids for extremely large matter density which may occur at either early or later stages of the universe [2–4]. It was then shown by Hehl, von der Heyde and Kerlick [5] that the general conclusions reached above for aligned spins would also hold for randomly oriented spins which are locally isotropic, i.e. spherically symmetric. Except for Kopczyński [2], however, the spins in these theories were generally associated with the quantum mechanical spin of particles. We feel, however, that it is not necessary to identify the spin-density of a fluid with the spin of an elementary particle. Our approach involving an improved energy–momentum tensor for spinning fluids [6] is similar to the concept of Israel [7], Bailey and Israel [8], Bailey [9] or Kopczyński [5, 10], where the cosmological fluid particles, which may be galaxies or clusters of galaxies, have intrinsic but classical spin. Although one could investigate within the context of general relativity (GR) [11, 12] the effect of spin on important cosmological issues such as inflation in the early universe [13], the natural arena should be within an EC theory in Riemann-Cartan (RC) spacetime. In fact, recent investigations by Martin *et al* [14] have indicated that consistency relations [15, 16] restrict most interesting spinning fluid configurations in GR even though their work seems to require further analysis. We will find from this work that the lack of spin-squared terms, which do not occur in the GR calculation, limits the range of examples, but we will leave the proof to a future work. *A fortiori*, in the extensive review of the EC theory, Hehl *et al* [17] have shown that the field equations and conservation laws for the metric and the angular momentum arise *naturally* from variational considerations within the EC theory. In RC spaces a fundamental field quantity is the torsion, which is identified with the antisymmetric part of the RC connection $S_{ij}{}^{k} = \Gamma_{[ij]}{}^{k}$. The relationship between torsion and physical spin has been clarified and made more preicse [18]. We now know that the spin is directly identified with a natual object in the geometry, namely the trace-free torsion, $\tilde{S}_{ij}{}^{k} = S_{ij}{}^{k} + \delta^{k}_{[i}\zeta_{j]}$, where the torsion vector $\zeta_i = \frac{2}{3}S_{ix}{}^{x}$. Within the context of a Lagrangian-based self-consistent EC theory [6], the torsion vector enters the theory depending on the form of the fluid constraint. For the case of particle number conservation in the fluid frame, the constraint is given by $\overset{*}{\nabla}_i(\rho u^i)$ where the 'star' derivative $\overset{*}{\nabla}_i \equiv \nabla_i + 2S_{ix}{}^x$. The torsion field equation then shows that the torsion vector vanishes. For the discussion of stars, gas clouds and galaxies, this type of constraint is probably sufficient; however, for cosmological models, this may not be the case. If the constraint is given by $\nabla_i(\rho u^i)$ (which is the direct minimal coupling generalization of the fluid constraint to RC [19]), then the torsion vector remains active; however, its interpretation in the theory is somewhat incomplete, since the torsion field equation can only resolve it in terms of the Lagrange multiplier for the fluid constraint [18] (as will be seen later on) without a complete solution (model) of the field equations. One interpretation obtained by applying Gauss' law for a region between two spacelike hypersurfaces enclosing all matter leads to mass 'creation' (or annihilation), which may have cosmological significance [6]. This interpretation nevertheless does seem controversial. Since this work involves cosmological questions, we necessarily have used the constraint which keeps the torsion vector active.

Our investigations seem to indicate that the effects of spin-density in cosmological problems such as inflation, final collapse, or even collapsing objects, should be investigated in an RC spacetime [20]. Our averaging treatment of the proper torsion represents an older variation of the microscopic viewpoint of torsion [21]. In fact, using the improved energy–momentum tensor with spin-density, Gasperini [22] has demonstrated spin-dominated inflation in the EC theory for a spinning fluid with randomly oriented spin. This result seems most reasonable in comparison with similar conclusions obtained from an earlier *ad hoc* description of spinning fluids [5] obtained by just adding to the energy–momentum tensor a spin-density contribution based upon a generalization of the special relativistic treatment of Halbwachs [23]. We point out that the improved energy–momentum tensor for a spinning fluid within an EC theory is also a generalization of Halbwachs' classical treatment of spin to RC spacetime, but here we depart from others through the use of a self-consistent, Lagrangian formulation which also treats the thermodynamic properties of the fluid [19]. Thus with these comments in mind, the purpose of this paper is to discuss the changes that occur in the 'average' contribution to the energy–momentum tensor for randomly oriented spins when the self-consistent description of a spinning fluid is used. An expression for the average energy–momentum tensor is then obtained.

The results of the self-consistent EC description of a spinning fluid are the field equations [6, 18,]†

$$G^{(ij)}(\Gamma) - 2\overset{*}{\nabla}_k T^{k(ij)} = \kappa T^{ij} \tag{1}$$

$$\widehat{S}_{\ell m}{}^k = \tfrac{1}{2}\kappa\rho s_{\ell m} u^k \tag{2}$$

$$\zeta_\ell = -\kappa\rho\lambda_2 u_\ell \tag{3}$$

where $G^{(ij)}(\Gamma)$ are the symmetric components of the Einstein tensor in RC space;time with torsion $S_{ij}{}^k = \Gamma_{[ij]}{}^k$, where the modified torsion is $T_{ij}{}^k \equiv \widehat{S}_{ij}{}^k - 2\zeta_{[i}\delta^k_{j]}$, s_{ij} is the spin/particle, ρ is the mass density, $\kappa = 8\pi G$, and λ_2 is the Lagrange multiplier for the fluid constraint. The improved energy–momentum tensor is

$$T^{ij} = T_F{}^{ij} + T_S{}^{ij} \tag{4}$$

† Our basic notation follows that of [25], except that we use latin letters for spacetime indices.

where

$$T_{\mathrm{F}}{}^{ij} = [\rho(1+\epsilon)]u^i u^j + pg^{ij} \tag{5}$$

$$T_{\mathrm{S}}{}^{ij} = 2\rho u^{(i}s^{j)k}\dot{u}_k + \overset{*}{\nabla}_\ell[\rho u^{(i}s^{j)\ell}] - \rho\omega_k{}^{(i}s^{j)k}. \tag{6}$$

The LHS of equation (1) can be written as the Riemannian Einstein tensor plus torsion-squared correction terms defining the 'spin' correction tensor $-\tau^{ij}$ such that [5, 25]

$$\kappa\tau^{ij} = -4T^{ik}{}_{[\ell}T^{j\ell}{}_{k]} - 2T^{ik\ell}T^{j}{}_{k\ell} + T^{k\ell i}T_{k\ell}{}^{j} + \tfrac{1}{2}g^{ij}\left[4T_k{}^p{}_{[m}T^{km}{}_{p]} + T^{kpm}T_{kpm}\right]. \tag{7}$$

We then combine equation (7) with (4) to obtain the effective Riemannian energy–momentum tensor

$$T_{\mathrm{R}}{}^{ij} = T^{ij} + \tau^{ij} \tag{8}$$

and then find its average for a randomly oriented, spherically symmetric spinning fluid.

The average of the effective energy–momentum tensor is a two-step process. Since a general spherically symmetric tensor takes the form $au^i u^j + bg^{ij}$, we can rewrite it in this form for convenience. The average is then easily found. We note that the fluid part is already in this form. The spin parts are then

$$T_{\mathrm{S}}{}^{ij} = -\tfrac{1}{3}[4\Omega_{k\ell}S^{k\ell} + 4\kappa S^2 - \omega_{k\ell}S^{k\ell}]u^i u^j - \tfrac{1}{3}[\Omega_{k\ell}S^{k\ell} + \kappa S^2 - \omega_{\kappa\ell}S^{k\ell}]g^{ij} \tag{9}$$

and

$$\kappa\tau^{ij} = [\tfrac{5}{6}\kappa^2 S^2 - 6\zeta^2]u^i u^j + [\tfrac{1}{12}\kappa^2 S^2 - 3\zeta^2]g^{ij} \tag{10}$$

where the spin-density tensor $S^{ij} = \rho s^{ij}$, and $S_{ij}S^{ij} = 2S_k S^k \equiv 2S^2$ where S_k is the spin vector, $\zeta^k\zeta_k = \zeta^2$, and $\Omega_{k\ell}$ is the vorticity tensor in GR [26]†. In taking the average of a spherically symmetric, isotropic system of randomly oriented spins, then for the spin average itself $\langle S^{ij}\rangle = 0$, but for the spin-squared terms $\langle S_{ij}S^{ij}\rangle \neq 0$. First we note that if we ignore the vorticity, angular velocity and torsion vector terms, we obtain the old EC results for the average of the effective energy–momentum tensor [5, 27]‡

$$\langle T_{\mathrm{R}}{}^{ij}\rangle_{\mathrm{NI}} = [\rho(1+\epsilon) + p - \tfrac{1}{2}\kappa S^2]u^i u^j + [p - \tfrac{1}{4}\kappa S^2]g^{ij} \tag{11}$$

where the subscript NI refers to the results obtained from the non-improved effective energy–momentum tensor. We have also set $\langle S^2\rangle = S^2$ for convenience.

Another way of looking at the results given by equation (11) is to assume that on average there is no correlation between the randomly oriented spins and the GR vorticity of the congruences associated with 4-velocity, i.e. we have $\langle\Omega_{ij}S^{ij}\rangle = 0$. At first one might

† See, for example, [26].

‡ These are the results for the spin-squared terms obtained by [27], but in order to obtain these results they defined the torsion $S_{ij}{}^k = 2\Gamma_{[ij]}{}^k$, used the *ad hoc* classical Weyssenhoff relation between torsion and spin $S_{ij}{}^k = S_{ij}u^k$, and used $S_{ij}S^{ij} = 2S^2$. On the other hand, [5] also used the classical Weyssenhoff relation, but defined the torsion as in this paper, and defined $2S_{ij}S^{ij} = S^2$. Thus in reality, these results are very different from one another because of conflicting definitions. On the surface, the compensating factors of 2 in [27] gives the same look-alike results as the exact calculation of the EC limit found in this paper, but we can not rectify an additional factor of 2 in [5].

want to extend this conclusion to the $\omega_{ij}S^{ij}$ terms, but this would be incorrect, since we can write

$$\omega_{ij}S^{ij} = \omega_{ij}^{(\mathrm{GR})}S^{ij} - \kappa S^2 \tag{12}$$

where we have separated a RC term into a general relativistic term plus torsion related terms. In the form of equation (12) we easily note that the average is non-zero, because of the presence of the spin-squared term. We further argue that the average of the general relativistic term vanishes (similarly to the vorticity-spin term), since it can be written in terms of products of the uncorrelated tetrads (or, equivalently, metric terms) without any torsion terms present. Thus

$$\langle \omega_{ij}S^{ij} \rangle = -\kappa S^2. \tag{13}$$

Combining this with equations (9) and (10), we find in the self-consistent EC theory the average of the effective energy–momentum tensor

$$\langle \kappa T_{\mathrm{R}}{}^{ij} \rangle_{\mathrm{I}} = \left[\kappa\rho \left(1 + \epsilon + \frac{p}{\rho} \right) - \tfrac{5}{6}\kappa^2 S^2 - 6\zeta^2 \right] u^i u^j + [\kappa p - \tfrac{7}{12}\kappa^2 S^2 - 3\zeta^2] g^{ij} \tag{14}$$

where the subscript I refers to the improved results. Note this does not reduce to the old EC results in equation (11), because of the contribution of the angular velocity term in the improved energy–momentum tensor given by equation (6). The effect of the negative contribution of the spin-density to the energy-density is about twice as large as the old EC results. However, because of the torsion vector field equation (3), the torsion vector contributes positive terms to the energy–momentum density. This is not entirely conclusive, since one needs to introduce source fields (e.g. a vector field Lagrangian sensitive to the torsion vector) and then include the contribution of those fields in the energy–momentum tensor as well. The Lagrange multiplier can be removed from equation (3) by solving the constraint field equation from the variation of the Lagrangian [6]. This gives an equation of motion for λ_2

$$\kappa\rho\dot{\lambda}_2 = \kappa \left[\rho \left(1 + \epsilon + \frac{p}{\rho} \right) + T_{\mathrm{S}} \right] + 3(\kappa\rho\lambda_2)^2 \tag{15}$$

where T_{S} is the spin energy density. Equation (15) for λ_2 can be solved within the context of some cosmological model. For the example of constant density ρ and uniform total energy density, the contribution of the torsion vector terms to the average energy–momentum equation (14) is a monotonically increasing function of proper time. Although an unlikely model, it does serve to show that the torsion vector terms and the spin terms act oppositely to one another.

Thus, although we agree with the general conclusion of Gasperini concerning spin-dominated inflation [22], his result is based upon equation (11), which neglects the angular velocity contributions which would give a much stronger spin effect, as is seen in the average for the self-consistent EC theory given in equation (14).

In conclusion, equation (14) describes the average effective energy–momentum tensor due to the effects of torsion in an RC spacetime. The averaging process could be thought of as the transition from the microscopic to the macroscopic description of a spin field.

In a different approach to cosmological models, Smalley has shown how one can obtain a spacetime-dependent cosmological 'function' in a Weyl spacetime [28]. It has been known for some time that in some affine geometries the torsion and Weyl vectors may be proportional [29]. One way to see this is to note that the fluid constraint in an affine geometry will contain an independent connection that can have both torsion and non-metricity, the trace of the non-metricity being the Weyl vector. The field equations for the torsion vector and the Weyl vector then give the proportionality. The similarities of the results of spacetime geometries with torsion vector or Weyl vector have been pointed out earlier [30]. The linking of the torsion vector to a cosmological function would be interesting. However, the relationship to this work is only suggestive, since one first needs a Lagrangian theory with all the appropriate elements, including source fields as well. We leave this for future work.

Finally, we mention that other approaches to the torsion vector have not been included in this work, such as the possible relationship of torsion and electromagnetism by Hammond [31] or that of Smalley and Krisch [32].

References

[1] Hehl F W and von der Heyde P 1973 *Ann. Inst. H Poincaré* A **19** 179
[2] Kopczyński W 1972 *Phys. Lett.* **39A** 219; 1973 *Phys. Lett.* **43A** 63
[3] Trautman A 1973 *Nat. Phys. Sci.* **242** 7
[4] Stewart J and Hájícek P 1973 *Nat. Phys. Sci.* **244** 96
[5] Hehl F W, von der Heyde P and Kerlick G D 1974 *Phys. Rev.* D **10** 1066
Hehl F W 1974 *Gen. Rel. Grav.* **5** 491
[6] Ray J R and Smalley L L 1982 *Phys. Rev. Lett.* **49** 1059; 1983 *Phys. Rev. Lett.* **50** 626E; *Phys. Rev.* D **27** 1383
[7] Israel W 1973 *Nuovo Cimento* **7** 860
[8] Bailey I and Israel W 1975 *Commun. Math. Phys.* **42** 65
[9] Bailey I 1979 *Ann. Phys., NY* **119** 76
[10] Kopczyński W 1986 *Phys. Rev.* D **34** 352
[11] Ray J R and Smalley L L 1982 *Phys. Rev.* D **26** 2619
[12] Ray J R, Smalley L L and Krisch J P 1987 *Phys. Rev.* D **35** 3261
[13] Bedran M L and Vasconcellos-Vaidya 1984 *Nuovo Cimento* **41** 73
[14] Martin M A P, Vasconcellos-Vaidya E P and Som M M 1991 *Class. Quantum Grav.* **8** 2225
[15] Amorim R 1984 *Phys. Lett.* **104A** 259
[16] Ray J R and Smalley L L 1986 *Phys. Rev.* D **34** 3268
[17] Hehl F W, von der Heyde P, Kerlick G D and Nester J M 1976 *Rev. Mod. Phys.* **48** 393
[18] Smalley L L and Ray J R 1986 *Gen. Rel. Grav.* **18** 549
[19] Ray J R 1972 *J. Math. Phys.* **13** 1451
[20] Arkuszewski W, Kopczyński W and Ponomariev V N 1974 *Ann. Inst. H Poincaré* **21** 89; 1975 *Commun. Math. Phys.* **45** 183
[21] Hehl F W, McCrea J D, Mielke E W and Ne'eman Y 1989 *Found. Phys.* **19** 1075
[22] Gasperini M 1986 *Phys. Rev. Lett.* **56** 2873
[23] Halbwachs F 1960 *Théorie relativiste des fluides à spin* (Paris: Gauthier–Villars)
[24] Hehle F W 1974 *Gen. Rel. Grav.* **5** 491
[25] Schouten J A 1954 *Ricci Calculus* (Berlin: Springer)
[26] Ehlers J and Kundt W 1962 *Gravitation: An Introduction to Current Research* ed L Witten (New York: Wiley) pp 49–101
[27] Arkuszewski W, Kopczyński W and Ponomariev V N 1974 *Ann. Inst. H Poincaré* **21** 89
[28] Smalley L L 1993 *Class. Quantum Grav.* **10** 1179
[29] Hehl F W, Lord E A and Smalley L L 1978 *Gen. Rel. Grav.* **9** 691
[30] Smalley L L 1986 *Phys. Rev.* D **33** 3590
[31] Hammond R T 1988 *Gen. Rel. Grav.* **20** 813
[32] Smalley L L and Krisch J P 1991 *Class. Quantum Grav.* **8** 1889; 1991 *J. Math. Phys.* **33** 1073; 1992 *Int. J. Theor. Phys.* **31** 1253

Class. Quantum Grav. **11** (1994) 2523–2536. Printed in the UK

Kleinian signature change

L J Alty†

Department of Applied Mathematics and Theoretical Physics, University of Cambridge, Cambridge CB3 9EW, UK

Received 9 May 1994, in final form 10 August 1994

Abstract. We study the propagation of plane-waves in a fixed flat background spacetime whose metric changes from a Lorentzian signature $(+---)$ to a Kleinian signature $(++--)$. When the change of signature is discontinuous there is an apparent difference in wave behaviour depending upon whether the wave describes matter with integer or half-integer spin. A more detailed study of the Dirac equation shows that, while it is possible to decrease this disparity, the two matter classes still do not behave identically. A more consistent picture finally emerges when the same problem is attempted with a metric which changes signature continuously. This leads to the prediction of some surprising physical phenomena.

PACS numbers: 0420G, 0365P

1. Introduction

The idea that the signature of the metric of spacetime may have changed from Euclidean to Lorentzian in the early universe, developed from the well known 'no boundary' proposal of Hartle and Hawking [1]. The recent revival of interest in signature change phenomena is due to Dray *et al* [2, 3] who studied the massless scalar field, and Ellis *et al* [4] who looked for cosmological solutions of the classical Einstein field equations in this context. Hayward [5] clarified much of this work, in particular, he re-emphasized a point made by Gibbons and Hartle [6] that, in order to avoid distributional contributions to the field equations, the second fundamental form of the surface at which the signature changes must be zero. There is some argument as to whether discontinuous or continuous changes in signature provide a better description of the effect, indeed both descriptions suffer from some problems [4]. In this paper, both continuous and discontinuous signature change are considered.

A metric on a 4-manifold M will be said to be of *Kleinian signature* if it has signature $(++--)$. If one accepts that it is possible for the signature of the metric of spacetime to change from Euclidean to Lorentzian, then one must also accept that it is possible for the signature of the metric of spacetime to change from Lorentzian to Kleinian. Indeed, this idea of more general signature transitions was postulated by Sakharov [7].

If particles are free to move between Lorentzian and Kleinian regions then, due to the structure of the Kleinian light cone, dramatic physical effects would be observable. For example, it may be possible to use a Kleinian region for time travel (with respect to time in the Lorentzian region). It may also be possible to extract energy from a Kleinian region using a method similar to that for extracting energy from rotating black holes. If this latter effect is possible, then the amount of energy that could be extracted is unlimited.

† E-mail address: lja13@amtp.cam.ac.uk

0264-9381/94/102523+14$19.50

In this paper, we examine solutions of the wave equations for various matter types on a fixed flat background spacetime which changes from a Lorentzian signature $(+---)$ to a Kleinian signature $(++--)$. In section 2 we solve the wave equation for the scalar, Dirac, and electromagnetic fields, using the discontinuous description of signature change. The results appear to be dependent on whether the matter has integer or half-integer spin. In section 3 we examine the Dirac equation in further detail to try to resolve this apparent discrepancy. In section 4 we solve the wave equation for the scalar and Dirac fields using the continuous description of signature change. These final results suggest a resolution of the discrepancy found in the discontinuous case, but at the same time predict some surprising physical phenomena.

2. Discontinuous signature change

Here we solve the wave equations for different matter types propagating on a background spacetime with metric given by

$$ds^2 = g_{ab}\,dx^a dx^b = \begin{cases} dt^2 - dx^2 - dy^2 - dz^2 & z < 0 \\ dt^2 - dx^2 - dy^2 + dz^2 & z > 0 . \end{cases} \tag{2.1}$$

Thus the background spacetime is flat with discontinuous signature change from Lorentzian to Kleinian across the surface $z = 0$ (see figure 1). The exact nature of the surface $z = 0$ is not specified as it is unimportant. However, since the change of signature is discontinuous, it need not be the case that the metric is degenerate at $z = 0$.

For $z < 0$ we take the standard plane-wave solutions, and for $z > 0$ we seek solutions which decay exponentially in z as $z \to \infty$. In order that the components of the energy–momentum tensor remain bounded, we cannot admit solutions which grow exponentially in z as $z \to \infty$. The inclusion of such solutions would violate our assumption that the background spacetime is fixed and flat.

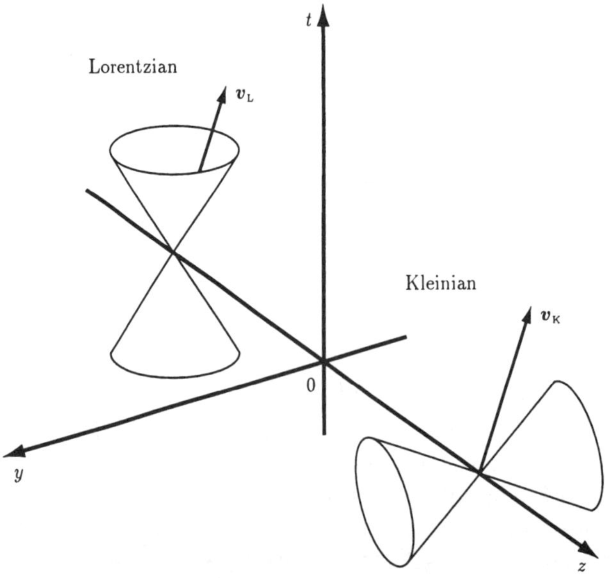

Figure 1. Light cones in Lorentzian and Kleinian regions of spacetime. The vectors v_L and v_K show typical time-like vectors.

Let ϕ be a linear combination of allowable solutions, then at $z = 0$ we impose the following natural boundary condition:

(i) ϕ is continuous.

If the wave equation for ϕ is second-order then we impose the additional boundary condition

(ii) $\partial\phi/\partial z$ is continuous.

The scalar, Dirac and electromagnetic fields are treated in this manner.

2.1. *Scalar field*

The wave equation for a massive scalar field ϕ is the Klein–Gordon equation

$$(\Box + m^2)\phi = 0 \tag{2.2}$$

where $\Box = g^{ab}(\partial/\partial x^a)(\partial/\partial x^b)$. Thus for $z < 0$, $\Box$ is the standard d'Alembertian operator. ϕ has position $\boldsymbol{x} = (x, y, z)$ and momentum $\boldsymbol{p} = (p_x, p_y, p_z)$. Throughout this paper we will use the following notation:

$$\boldsymbol{y} = (x, y) \tag{2.3}$$

$$\boldsymbol{k} = (p_x, p_y) \tag{2.4}$$

$$p = p_z\,. \tag{2.5}$$

Solving (2.2) for $z < 0$ we find

$$\phi = \mathrm{e}^{\mathrm{i}(\boldsymbol{k}\cdot\boldsymbol{y}+pz-\omega t)} \tag{2.6}$$

where $\omega^2 = \boldsymbol{k}^2 + p^2 + m^2$. This is the standard plane-wave solution.

For $z > 0$ we require solutions of (2.2) that decay as $z \to \infty$. We find

$$\phi = \mathrm{e}^{\mathrm{i}(\boldsymbol{k}\cdot\boldsymbol{y}-\omega t)-pz} \tag{2.7}$$

where, once again, $\omega^2 = \boldsymbol{k}^2 + p^2 + m^2$.

Set

$$\phi = \mathrm{e}^{\mathrm{i}(\boldsymbol{k}\cdot\boldsymbol{y}-\omega t)} \begin{cases} a\mathrm{e}^{\mathrm{i}pz} + b\mathrm{e}^{-\mathrm{i}pz} & z < 0 \\ c\mathrm{e}^{-pz} & z > 0 \end{cases} \tag{2.8}$$

where $p > 0$.

Applying the boundary conditions discussed above we find that $b = -\mathrm{i}a$ and $c = (1 - \mathrm{i})a$, and since

$$\left|\frac{\text{incident amplitude}}{\text{reflected amplitude}}\right| = \left|\frac{a}{b}\right| = 1 \tag{2.9}$$

we see that the scalar field is completely reflected at $z = 0$.

This result is typical of wave equation solutions with integer spin. Upon encountering a region of discontinuous Kleinian signature change, the incoming wave is totally reflected with a phase shift (see figure 2).

2.2. *Dirac field*

The metric of spacetime g_{ab} is given by (2.1). For $z < 0$ we define the usual gamma matrices γ_{L}^a by $\{\gamma_{\mathrm{L}}^a, \gamma_{\mathrm{L}}^b\} = 2g^{ab}$, and similarly, for $z > 0$ we define the Kleinian gamma matrices γ_{K}^a by $\{\gamma_{\mathrm{K}}^a, \gamma_{\mathrm{K}}^b\} = 2g^{ab}$.

We wish to solve the following (Dirac) equations:

$$(\gamma_{\mathrm{L}}^a\partial_a + \mathrm{i}m)\Psi = 0 \qquad z < 0 \tag{2.10}$$

$$(\gamma_{\mathrm{K}}^a\partial_a + \mathrm{i}m)\Psi = 0 \qquad z > 0\,. \tag{2.11}$$

Choosing $\gamma_{\mathrm{K}}^0 = \gamma_{\mathrm{L}}^0$, $\gamma_{\mathrm{K}}^1 = \gamma_{\mathrm{L}}^1$, and $\gamma_{\mathrm{K}}^2 = \gamma_{\mathrm{L}}^2$, there are two cases given by $\gamma_{\mathrm{K}}^3 = \pm\mathrm{i}\gamma_{\mathrm{L}}^3$.

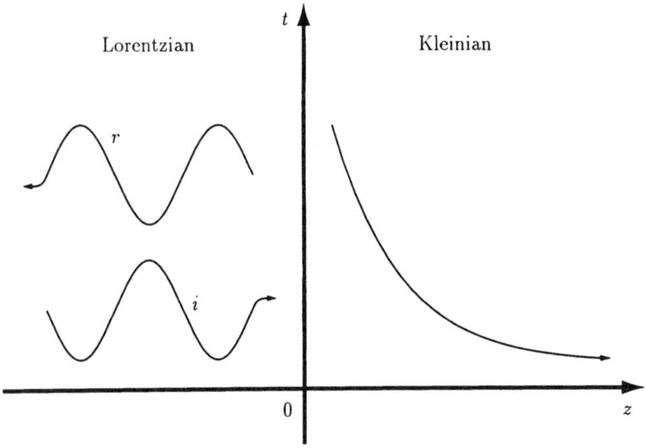

Figure 2. One type of behaviour of wave equation solutions in a spacetime that changes signature from Lorentzian to Kleinian. The incident and reflected waves are labelled i and r respectively. The incident wave is completely reflected with a phase shift.

Case 1. $\gamma_K^3 = -i\gamma_L^3$.

Let the Dirac field Ψ have the form

$$\Psi = e^{i(\boldsymbol{k}\cdot\boldsymbol{y}-\omega t)}\chi(z) \tag{2.12}$$

then for $z < 0$ we find using (2.10) that

$$\frac{\partial\chi}{\partial z} = -i\gamma_L^3(\gamma_L^0\omega - \boldsymbol{\gamma}_L\cdot\boldsymbol{k} - m)\chi \tag{2.13}$$

where $\boldsymbol{\gamma}_L\cdot\boldsymbol{k} = \gamma_L^1 p_x + \gamma_L^2 p_y$.

Let $\chi = V_+e^{ipz}$ then, on defining the operator

$$\mathcal{D} = -\gamma_L^3(\gamma_L^0\omega - \boldsymbol{\gamma}_L\cdot\boldsymbol{k} - m) \tag{2.14}$$

equation (2.13) gives the eigenvector equation

$$\mathcal{D}V_+ = pV_+ . \tag{2.15}$$

Similarly, if $\chi = V_-e^{-ipz}$ we get

$$\mathcal{D}V_- = -pV_- . \tag{2.16}$$

For $z > 0$ we find using (2.11) that

$$\frac{\partial\chi}{\partial z} = \gamma_L^3(\gamma_L^0\omega - \boldsymbol{\gamma}_L\cdot\boldsymbol{k} - m)\chi . \tag{2.17}$$

Let $\chi = Ve^{-pz}$ then (2.17) gives the eigenvector equation

$$\mathcal{D}V = pV . \tag{2.18}$$

Comparing (2.15), (2.16), and (2.18) we conclude that V_+ and V_- are linearly independent, and that we may take $V = V_+$. Thus if χ has the form

$$\chi = \begin{cases} aV_+e^{ipz} + bV_-e^{-ipz} & z < 0 \\ cVe^{-pz} & z > 0 \end{cases} \tag{2.19}$$

then on applying the boundary conditions at $z = 0$ we are forced to set $a = c$ and $b = 0$. Thus the incoming wave is totally absorbed by the Kleinian region (see figure 3).

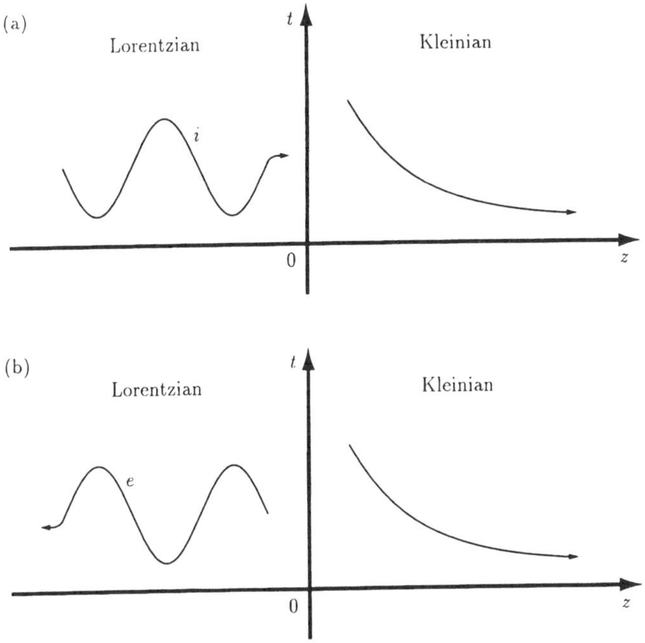

Figure 3. A second type of behaviour of wave equation solutions in a spacetime that changes signature from Lorentzian to Kleinian. In (*a*) the incident wave, labelled i, is totally absorbed by the Kleinian region, and in (*b*) a wave, labelled e, is emitted from the Kleinian region.

Case 2. $\gamma_{\mathrm{K}}^{3} = +\mathrm{i}\gamma_{\mathrm{L}}^{3}$.

This case is the same as the last except that (2.17) becomes

$$\frac{\partial \chi}{\partial z} = -\gamma_{\mathrm{L}}^{3}(\gamma_{\mathrm{L}}^{0}\omega - \boldsymbol{\gamma}_{\mathrm{L}} \cdot \boldsymbol{k} - m)\chi \tag{2.20}$$

and hence (2.18) becomes

$$\mathcal{D}V = -pV \,. \tag{2.21}$$

It follows that we may take $V = V_{-}$. Hence if χ is given by (2.19) then on applying the boundary conditions at $z = 0$, we are forced to set $b = c$ and $a = 0$. Thus the Kleinian region emits a wave which propagates through the Lorentzian region (see figure 3). There is no incoming wave solution.

We note that the Klein–Gordon equation solution looks like a superposition of the two Dirac equation solutions. This is not surprising, as the Klein–Gordon equation does not 'notice' whether $\gamma_{\mathrm{K}}^{3} = -\gamma_{\mathrm{L}}^{3}$ or $\gamma_{\mathrm{K}}^{3} = +\gamma_{\mathrm{L}}^{3}$.

2.3. Electromagnetic field

The electromagnetic field may be treated using either a set of first-order field equations for the electric and magnetic fields, or a second-order field equation for the electromagnetic potential.

First order. By introducing the dual tensor $\tilde{F}^{ab} = \frac{1}{2}\epsilon^{abcd}F_{cd}$, Maxwell's equations may be written without reference to the electromagnetic potential as

$$\partial_a F^{ab} = 0 \tag{2.22}$$

$$\partial_a \tilde{F}^{ab} = 0\,. \tag{2.23}$$

We wish to find plane-wave solutions, hence we will set

$$F^{ab} = f^{ab} e^{i(k\cdot y - \omega t)} \chi(z) \tag{2.24}$$

where $\partial_c f^{ab} = 0$. Thus from (2.22) and (2.23) we obtain

$$f^{3b}\frac{\partial \chi}{\partial z} = i(f^{0b} w - f^{\mu b} k_\mu)\chi \tag{2.25}$$

$$\epsilon^{3b}{}_{cd} f^{cd}\frac{\partial \chi}{\partial z} = i f^{cd}(\epsilon^{0b}{}_{cd} w - \epsilon^{\mu b}{}_{cd} k_\mu)\chi \tag{2.26}$$

and since f^{ab} is skew we note that at least one of the four terms in each of these equations is zero.

On the signature changing spacetime (2.1), let us write the solution as

$$F^{ab} = e^{i(k\cdot y - \omega t)} \begin{cases} a f_+^{ab} e^{ipz} + b f_-^{ab} e^{-ipz} & z < 0 \\ c f^{ab} e^{-pz} & z > 0 \end{cases} \tag{2.27}$$

where f_+^{ab}, f_-^{ab} and f^{ab} are the solutions of (2.25) and (2.26) which correspond to setting $\chi(z)$ equal to e^{ipz}, e^{-ipz} and e^{-pz}, respectively.

Noting that the equation formed by putting $b = 3$ in (2.26) is identical in both the Lorentzian and Kleinian regions, we see that one obvious solution will have

$$f_+^{\alpha\beta} = f_-^{\alpha\beta} = f^{\alpha\beta} \qquad \text{for} \quad \alpha, \beta \in \{0, 1, 2\}\,. \tag{2.28}$$

Applying the boundary conditions to these components of F^{ab}, it follows that

$$a + b = c\,. \tag{2.29}$$

Furthermore, putting $b = 0$ in (2.25) allows us to calculate f_+^{03}, f_-^{03}, and f^{03} in terms of f^{01} and f^{02}. Applying the boundary conditions to F^{03} then gives

$$a - b = -ic\,. \tag{2.30}$$

Solving (2.29) and (2.30) we find $b = ia$ and $c = (1 + i)a$. Thus the incident electromagnetic wave is completely reflected at $z = 0$ (see figure 2). Note that the remaining six field equations are satisfied by this solution if we set $\omega^2 = \boldsymbol{k}^2 + p^2$.

Second order. Choosing the Lorentz gauge $\partial_b A^b = 0$, Maxwell's equations are $\Box A^b = 0$. Thus results are as for the scalar field, we find

$$A^b = a^b e^{i(k\cdot y - \omega t)} \begin{cases} e^{ipz} - ie^{-ipz} & z < 0 \\ (1 - i)e^{-pz} & z > 0 \end{cases} \tag{2.31}$$

with $a^0\omega = \boldsymbol{a}\cdot\boldsymbol{k} + (1 + i)a^3 p$.

Once again the incoming wave experiences total reflection (see figure 2).

2.4. *A wall of Kleinian signature*

The similarity between the scalar field solution and wave behaviour in a potential well encourages us to consider whether a wave which encounters a small region of Kleinian signature could tunnel through that region.

Consider the flat signature changing spacetime given by

$$ds^2 = \begin{cases} dt^2 - dx^2 - dy^2 - dz^2 & z < 0 \\ dt^2 - dx^2 - dy^2 + dz^2 & 0 < z < l \\ dt^2 - dx^2 - dy^2 - dz^2 & z > l\,. \end{cases} \tag{2.32}$$

This spacetime consists of a wall of Kleinian signature of width l sandwiched between two Lorentzian regions. As examples of the behaviour of wave solutions on this spacetime, we will solve the Klein–Gordon and Dirac equations.

Scalar field. Solving the Klein–Gordon (2.2) on this spacetime and applying the boundary conditions at $z = 0$ and $z = l$, we find

$$\phi = \frac{e^{i(\boldsymbol{k}\cdot\boldsymbol{y}-\omega t)}}{\cosh(pl)} \begin{cases} \cosh(pl)e^{ipz} - i\sinh(pl)e^{-ipz} & z < 0 \\ \frac{1}{2}(1+i)e^{p(z-l)} + \frac{1}{2}(1-i)e^{-p(z-l)} & 0 < z < l \\ e^{ip(z-l)} & z > l\ . \end{cases} \tag{2.33}$$

Let A_i, A_r and A_t be the amplitudes of the incident, reflected and transmitted waves respectively, then

$$|A_i|^2 - |A_r|^2 - |A_t|^2 = 1 - \tanh^2(pl) - \operatorname{sech}^2(pl) = 0\,. \tag{2.34}$$

Thus the wave successfully tunnels through the Kleinian region and energy is conserved (see figure 4).

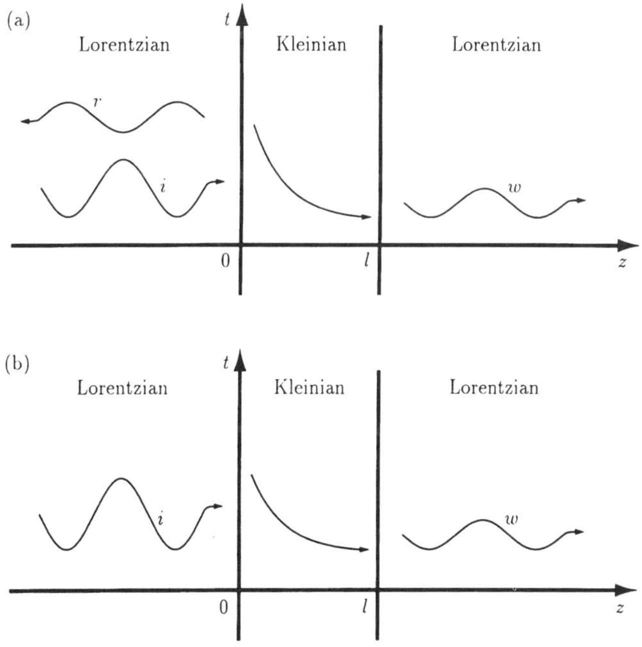

Figure 4. Two types of wave behaviour which could occur when a wave encounters a wall of Kleinian signature. The incident, reflected and transmitted waves are labelled i, r and w respectively. In (*a*) energy is conserved, the incident wave splits into a reflected wave and a transmitted wave. In (*b*) energy is lost, the incident wave tunnels through the Kleinian region without reflection, and the transmitted wave has a smaller amplitude than the incident wave.

Dirac field. In the Kleinian region we take $\gamma_{\rm K}^3 = -{\rm i}\gamma_{\rm L}^3$ then solving (2.10) and (2.11) and applying the continuous boundary conditions we find

$$\chi = V_+ \begin{cases} {\rm e}^{{\rm i}pz} & z < 0 \\ {\rm e}^{-pz} & 0 < z < l \\ {\rm e}^{-pl}{\rm e}^{{\rm i}p(z-l)} & z > l\,. \end{cases} \tag{2.35}$$

Thus the wave successfully tunnels through the Kleinian region, however, reading off the wave amplitudes we find

$$|A_{\rm i}|^2 - |A_{\rm r}|^2 - |A_{\rm t}|^2 = 1 - {\rm e}^{-2pl}\,. \tag{2.36}$$

It follows that energy is not conserved unless $l = 0$, in which case there would be no Kleinian region (see figure 4).

If instead we had used $\gamma_{\rm K}^3 = +{\rm i}\gamma_{\rm L}^3$ then the same result would have been obtained, except that the waves would be propagating from right to left rather than from left to right.

3. The Dirac equation revisited

The results of the previous section predict some strange phenomena. For example, on encountering a region of Kleinian signature a π^0-meson, described by the Klein–Gordon equation (2.2), would be totally reflected, whereas an electron, described by the Dirac equation (2.10), would be totally absorbed. Since both the first- and second-order formulations of the electromagnetic field give total reflection, an obvious conclusion is that regions of Kleinian signature distinguish between particles with integer and half-integer spin.

The purpose of this section is to determine whether it is possible, by altering our treatment of the Dirac equation, to remove this disparity between particles with integer and half-integer spin.

3.1. Partially reflecting solutions

In the previous section, our treatment of the Dirac equation used Kleinian gamma matrices given by $\gamma_{\rm K}^\alpha = \gamma_{\rm L}^\alpha$ for $\alpha \in \{0, 1, 2\}$. However, we could have chosen $\gamma_{\rm K}^\alpha = -\gamma_{\rm L}^\alpha$ for any or all of the $\gamma_{\rm K}^\alpha$. As we will show here, such choices give partially reflecting solutions of the Dirac equation, and thus give at least some degree of similarity between the solutions of the Dirac equation and other field equations. Note that the Klein–Gordon equation does not 'notice' whether we have chosen $\gamma_{\rm K}^\alpha = \gamma_{\rm L}^\alpha$ or $\gamma_{\rm K}^\alpha = -\gamma_{\rm L}^\alpha$.

We will consider solutions of (2.10) and (2.11) in the case when $\gamma_{\rm K}^\alpha = -\gamma_{\rm L}^\alpha$ for all $\alpha \in \{0, 1, 2\}$ and $\gamma_{\rm K}^3 = -{\rm i}\gamma_{\rm L}^3$.

Let Ψ have the form

$$\Psi = {\rm e}^{{\rm i}(\boldsymbol{k}\cdot\boldsymbol{y}-\omega t)} \begin{cases} aV_+{\rm e}^{{\rm i}pz} + bV_-{\rm e}^{-{\rm i}pz} & z < 0 \\ cV{\rm e}^{-pz} & z > 0 \end{cases} \tag{3.1}$$

then we find the relations

$$\mathcal{D}V_+ = pV_+ \tag{3.2}$$

$$\mathcal{D}V_- = -pV_- \tag{3.3}$$

$$\mathcal{D}V = (2m\gamma_{\rm L}^3 - p)V \tag{3.4}$$

where the operator $\mathcal{D}$ is defined by (2.14).

To proceed further, it is necessary to choose an explicit representation of the gamma matrices. We will use the 'standard representation' in which

$$\gamma_{\mathrm{L}}^0 = \begin{pmatrix} \mathbb{1} & 0 \\ 0 & -\mathbb{1} \end{pmatrix} \qquad \gamma_{\mathrm{L}}^j = \begin{pmatrix} 0 & \sigma_j \\ -\sigma_j & 0 \end{pmatrix} \qquad j = 1, 2, 3 \tag{3.5}$$

where $\mathbb{1}$ is the 2×2 identity matrix and the Pauli matrices σ_j are given by

$$\sigma_1 = \begin{pmatrix} 0 & 1 \\ 1 & 0 \end{pmatrix} \qquad \sigma_2 = \begin{pmatrix} 0 & -\mathrm{i} \\ \mathrm{i} & 0 \end{pmatrix} \qquad \sigma_3 = \begin{pmatrix} 1 & 0 \\ 0 & -1 \end{pmatrix} . \tag{3.6}$$

For simplicity, let us set $\boldsymbol{k} = 0$, then solving (3.2), (3.3) and (3.4) for V_+, V_- and V, respectively, we find

$$V_+ = \begin{pmatrix} 1 \\ 0 \\ (\omega - m)/p \\ 0 \end{pmatrix} \qquad V_- = \begin{pmatrix} 1 \\ 0 \\ (\omega - m)/(-p) \\ 0 \end{pmatrix} \qquad V = \begin{pmatrix} 1 \\ 0 \\ (\omega + m)/(-p) \\ 0 \end{pmatrix} . \tag{3.7}$$

Continuity of Ψ at $z = 0$ gives $b = -(\omega/m)\,a$ and $c = ((m - \omega)/m)\,a$, and thus Ψ has the form

$$\Psi = a\mathrm{e}^{\mathrm{i}(\boldsymbol{k}\cdot\boldsymbol{y} - \omega t)} \begin{cases} V_+\mathrm{e}^{\mathrm{i}pz} - \left(\dfrac{\omega}{m}\right) V_-\mathrm{e}^{-\mathrm{i}pz} & z < 0 \\ \left(\dfrac{m - \omega}{m}\right) V\mathrm{e}^{-pz} & z > 0 . \end{cases} \tag{3.8}$$

Defining the current $j_0 = \Psi\Psi^\dagger$ we see that

$$0 < \left(\frac{\text{incident } j_0}{\text{reflected } j_0}\right) = 1 - \frac{p^2}{w^2} < 1 \tag{3.9}$$

and thus the incoming wave is partially reflected at $z = 0$. A similar calculation with $\gamma_{\mathrm{K}}^3 = +\mathrm{i}\gamma_{\mathrm{L}}^3$ gives a solution in which the reflected wave has a greater current than the incoming wave. However, in both cases, there is no solution which totally reflects, since this would correspond to setting $p = 0$. These solutions are similar to those obtained in a Klein paradox‡ situation (see e.g. [8], but note that their interpretation of the Klein paradox solutions is incorrect [9]).

3.2. Squaring the Dirac equation

In Minkowski space, each component of a solution to the Dirac equation satisfies the standard Klein–Gordon equation. However, even in the case of a flat signature changing spacetime the analogue of this statement is not true. Here we derive the second-order equation which each component of a Dirac equation solution must satisfy if the spacetime metric is given by (2.1).

Let $\theta(z)$ be the usual step function given by

$$\theta(z) = \begin{cases} 1 & z > 0 \\ 0 & z < 0 \end{cases} \tag{3.10}$$

‡ The Klein of 'Kleinian signature' is not the same Klein as the Klein of 'Klein paradox'. The former is the leading nineteenth century mathematician Felix Klein, who is best known for his work in projective geometry. The latter is the physicist Oskar Klein, who is well known for his work in relativistic quantum mechanics and his unification of general relativity and electromagnetism.

then on defining $\tilde{\gamma}^\alpha = \gamma_L^\alpha$ for $\alpha = 0, 1, 2$, and

$$\tilde{\gamma}^3 = [\theta(-z) \pm i\theta(z)]\,\gamma_L^3 \tag{3.11}$$

the Dirac equations (2.10) and (2.11) may be concisely written as

$$(\tilde{\gamma}^a \partial_a + im)\Psi = 0\,. \tag{3.12}$$

The distributional identities

$$\delta(z)\,[\theta(z) - \theta(-z)] = 0 \qquad \delta(z)\,[\theta(z) + \theta(-z)] = \delta(z) \tag{3.13}$$

may be obtained using the definition of a product of two distributions given in [10]. This definition avoids the need to define $\theta(0)$.

Operating on (3.12) with $\tilde{\gamma}^b \partial_b$, and using the results (3.13), we obtain the second-order equation

$$\left[g^{ab}\frac{\partial}{\partial x^a}\frac{\partial}{\partial x^b} + \delta(z)\frac{\partial}{\partial z} + m^2\right]\Psi = 0\,. \tag{3.14}$$

Hence, in order to avoid distributional terms, we must restrict our solution space to functions Ψ which satisfy $(\partial\Psi/\partial z)|_{z=0} = 0$. This obviously excludes the wave solutions of the Dirac equation found in section 2. Interestingly, this condition is identical to that found by Hayward [5] when he considered the case of Euclidean signature change in which a scalar field was coupled to gravity.

3.3. Modifying the Dirac equation

We have seen that the Dirac equations (2.10) and (2.11) square up to give the Klein–Gordon equation plus an extra distributional term. It is well known that, in curved space, the square of the Dirac equation gains an extra term proportional to the scalar curvature, thus one might conclude that the signature changing spacetime has a distributional part in its curvature scalar. However, this is not the case, it is easily checked from first principles that the curvature scalar of the spacetime (2.1) is everywhere zero. Hence this additional term is a new signature effect.

The historical motivation behind the original derivation of the Dirac equation was to find an equation which squared to give the Klein–Gordon equation and which was first order. We have seen that in a discontinuously signature changing spacetime the Dirac equation (2.10) and (2.11) does not satisfy these properties. Thus one possible conclusion is that, in the case of signature change, a new 'Dirac equation' is required. Here we will investigate whether it is possible to modify the Kleinian side of the Dirac equation so that it squares to give the Klein–Gordon equation, and so that it is satisfied by solutions which have a similar form to those found in section 2.

Once again let us set

$$\Psi = e^{i(k\cdot y - \omega t)}\begin{cases} aV_+e^{ipz} + bV_-e^{-ipz} & z < 0 \\ cVe^{-pz} & z > 0 \end{cases} \tag{3.15}$$

then solving the Lorentzian Dirac equation for $z < 0$ gives

$$\mathcal{D}V_\pm = \pm pV_\pm \tag{3.16}$$

where $\mathcal{D}$ is given by (2.14). If we now require that Ψ is continuous then we must have

$$\mathcal{D}V = p\left(V - \frac{2b}{c}V_-\right). \tag{3.17}$$

Consider the vector space which is spanned by the vectors V_+ and V_-, and define the operator $\mathcal{P}$ by $\mathcal{P}V_- = V_-$ and $\mathcal{P}V_+ = 0$. Thus $\mathcal{P}$ projects onto the subspace spanned by V_-. It is easily shown that $\mathcal{P}$ commutes with $\mathcal{D}$, and furthermore, from the continuity of Ψ, that $\mathcal{P}V = (b/c)V_-$. We now see that (3.17) can be written

$$\mathcal{D}(1 - 2\mathcal{P})V = pV\,. \tag{3.18}$$

The 'Dirac equation' which gives (3.16) and (3.18) can be shown to be

$$\left[\tilde{\gamma}^a\partial_a + \mathrm{i}m - 2\theta(z)(\gamma_{\mathrm{L}}^\alpha\partial_\alpha + \mathrm{i}m)\mathcal{P}\right]\Psi = 0\,. \tag{3.19}$$

Squaring this equation by applying $\tilde{\gamma}^b\partial_b$, some miraculous cancellations conspire to give

$$\left[g^{ab}\frac{\partial}{\partial x^a}\frac{\partial}{\partial x^b} + m^2 + \delta(z)\left(\frac{\partial}{\partial z} - (1+\mathrm{i})\mathcal{D}\mathcal{P}\right)\right]\Psi = 0\,. \tag{3.20}$$

It is now easy to check, by explicit substitution of Ψ, that the distributional term does not cancel unless Ψ is everywhere zero. Hence even by modifying the Dirac equation in the Kleinian region we are unable to find an equation which squares to give the Klein–Gordon equation (2.2).

4. Continuous signature change

In this section we will examine wave solutions of the Dirac and Klein–Gordon equations in a spacetime which continuously changes signature. Thus the spacetime metric is degenerate at the surface of signature change. We will choose g^{ab} to be everywhere finite and thus g_{ab} will become infinite on the surface of signature change. This choice is more convenient since it is g^{ab} that appears in the field equations.

4.1. The scalar field

With coordinates $\{t, x, y, z\}$ we will take the metric to be

$$(g^{ab}) = \mathrm{diag}(1, -1, -1, z)\,. \tag{4.1}$$

We wish to solve

$$(\Box + m^2)\phi = 0 \tag{4.2}$$

where $\Box$ is given by

$$\Box = \frac{\partial^2}{\partial t^2} - \frac{\partial^2}{\partial x^2} - \frac{\partial^2}{\partial y^2} + z^{1/2}\frac{\partial}{\partial z}z^{1/2}\frac{\partial}{\partial z}\,. \tag{4.3}$$

We find that

$$\phi = \mathrm{e}^{\mathrm{i}(\boldsymbol{k}\cdot\boldsymbol{y} - \omega t)}\begin{cases} a\mathrm{e}^{-2pz^{1/2}} + b\mathrm{e}^{2pz^{1/2}} & z < 0 \\ c\mathrm{e}^{-2pz^{1/2}} & z > 0 \end{cases} \tag{4.4}$$

satisfies the wave equation if $\omega^2 = \boldsymbol{k}^2 + p^2 + m^2$.

Continuity of ϕ at $z = 0$ gives $a + b = c$, and continuity of $z^{1/2}\partial\phi/\partial z$ at $z = 0$ gives $a - b = c$. Hence $a = c$ and $b = 0$.

This result is remarkably similar to the Dirac equation solution in the case of discontinuous signature change (see figure 3).

4.2. The Dirac field

The Dirac equation with spacetime metric (4.1) can be shown to be

$$\left(\gamma_{\mathrm{L}}^{0}\frac{\partial}{\partial t}+\gamma_{\mathrm{L}}^{1}\frac{\partial}{\partial x}+\gamma_{\mathrm{L}}^{2}\frac{\partial}{\partial y}+\mathrm{i}\gamma_{\mathrm{L}}^{3}z^{1/2}\frac{\partial}{\partial z}+\mathrm{i}m\right)\Psi=0\,. \tag{4.5}$$

Taking

$$\Psi=V_{\pm}\mathrm{e}^{\mathrm{i}(\boldsymbol{k}\cdot\boldsymbol{y}-\omega t)}\mathrm{e}^{\pm 2pz^{1/2}} \tag{4.6}$$

we find that

$$\mathcal{D}V_{\pm}=\pm pV_{\pm} \tag{4.7}$$

where the operator $\mathcal{D}$ is given by (2.14).

Thus if we set

$$\Psi=\mathrm{e}^{\mathrm{i}(\boldsymbol{k}.\boldsymbol{y}-\omega t)}\begin{cases}aV_{+}\mathrm{e}^{2pz^{1/2}}+bV_{-}\mathrm{e}^{-2pz^{1/2}} & z<0\\ cV_{-}\mathrm{e}^{-2pz^{1/2}} & z>0\end{cases} \tag{4.8}$$

then continuity at $z=0$ gives $aV_{+}+bV_{-}=cV_{-}$. Applying $\mathcal{D}$ to this gives $aV_{+}-bV_{-}=-cV_{-}$, hence $a=0$ and $b=c$.

Thus, for the Dirac equation, we obtain qualitatively the same result that was found in the case of discontinuous signature change. There are two solutions (depending upon whether for $z<0$ one takes $z^{1/2}=-\mathrm{i}|z|^{1/2}$ or $z^{1/2}=+\mathrm{i}|z|^{1/2}$) corresponding to wave absorption by the Kleinian region and wave emission from the Kleinian region (see figure 3).

4.3. A wall of Kleinian signature

With coordinates $\{t,x,y,z\}$ we will take the metric to be

$$(g^{ab})=\mathrm{diag}(1,-1,-1,z(l-z))\,. \tag{4.9}$$

As in the case of discontinuous signature change, this spacetime consists of a wall of Kleinian signature of width l sandwiched between two Lorentzian regions. As examples of the behaviour of wave equation solutions on this spacetime, we will solve the Klein–Gordon and Dirac equations.

Scalar field. We wish to solve

$$(\Box+m^{2})\phi=0 \tag{4.10}$$

where $\Box$ is now given by

$$\Box=\frac{\partial^{2}}{\partial t^{2}}-\frac{\partial^{2}}{\partial x^{2}}-\frac{\partial^{2}}{\partial y^{2}}+z^{1/2}(l-z)^{1/2}\frac{\partial}{\partial z}z^{1/2}(l-z)^{1/2}\frac{\partial}{\partial z}\,. \tag{4.11}$$

With the requirement that both ϕ and $z^{1/2}(l-z)^{1/2}\partial\phi/\partial z$ are continuous at $z=0$ and $z=l$, we find the solution

$$\phi=a\mathrm{e}^{\mathrm{i}(\boldsymbol{k}\cdot\boldsymbol{y}-\omega t)}\begin{cases}\exp\left[-\mathrm{i}p\cosh^{-1}\left(\dfrac{l-2z}{l}\right)\right] & z<0\\ \mathrm{e}^{-p\pi}\exp\left[p\cos^{-1}\left(\dfrac{2z-l}{l}\right)\right] & 0<z<l\\ \mathrm{e}^{-p\pi}\exp\left[\mathrm{i}p\cosh^{-1}\left(\dfrac{2z-l}{l}\right)\right] & z>l\,.\end{cases} \tag{4.12}$$

Hence comparing wave amplitudes we see that

$$\left|\frac{\text{incident amplitude}}{\text{transmitted amplitude}}\right| = \left|\frac{a}{ae^{-p\pi}}\right| = e^{p\pi}. \tag{4.13}$$

Thus the wave successfully tunnels through the Kleinian region but energy is not conserved. The wave amplitude decreases as it passes through the Kleinian region (see figure 4).

Dirac field. The appropriate Dirac equation for the metric (4.9) is

$$\left(\gamma_L^0\frac{\partial}{\partial t} + \gamma_L^1\frac{\partial}{\partial x} + \gamma_L^2\frac{\partial}{\partial y} + i\gamma_L^3[z(l-z)]^{1/2}\frac{\partial}{\partial z} + im\right)\Psi = 0. \tag{4.14}$$

Solving this equation with the requirement that Ψ is continuous gives

$$\Psi = aV_{-}e^{i(\boldsymbol{k}\cdot\boldsymbol{y}-\omega t)}\begin{cases} \exp\left[-ip\cosh^{-1}\left(\frac{l-2z}{l}\right)\right] & z<0 \\ e^{-p\pi}\exp\left[p\cos^{-1}\left(\frac{2z-l}{l}\right)\right] & 0<z<l \\ e^{-p\pi}\exp\left[ip\cosh^{-1}\left(\frac{2z-l}{l}\right)\right] & z>l\,. \end{cases} \tag{4.15}$$

Thus the wave successfully tunnels through the Kleinian region, however, as in the scalar field case, the wave amplitude decreases by a factor of $e^{p\pi}$ (see figure 4).

5. Conclusion

The solutions of the wave equations in a spacetime where the metric signature changes continuously are consistent and independent of matter type, although they do predict some strange phenomena. An incoming wave can propagate through a thin wall of Kleinian signature, but in the process it would lose energy. If the wall is of infinite length then the wave is completely absorbed. For example, a beam of light could pass through a region of Kleinian signature, and although its frequency would remain unchanged, it would become less intense.

The continuous signature change results also suggest a possible resolution of the disparity between the integer and half-integer spin solutions in the case of discontinuous signature change. Although, in both approaches to signature change, the boundary conditions which we applied seemed the most natural, at closer inspection they are not equivalent. It is, however, possible to modify the boundary conditions in the discontinuous case so that they mimic those in the continuous case. Let ϕ be a scalar field on a spacetime experiencing discontinuous signature change at $z=0$, and let us take the following boundary conditions:

(i) ϕ is continuous

(ii) $\lim_{z\to 0^-}(\partial\phi/\partial z) = \pm i \lim_{z\to 0^+}(\partial\phi/\partial z)$.

Solving the wave equation with these boundary conditions gives two solutions (depending on whether + or − is taken in boundary condition (ii)). These solutions are identical in form to the Dirac equation solutions of section 2. Thus by employing these boundary conditions, the wave solutions in the presence of discontinuous signature change are consistent and independent of matter-type.

Energy is not conserved by wave equation solutions satisfying these boundary conditions, and hence it is unlikely that these solutions could be used to construct a well defined quantum field theory. Indeed, an attempt to treat the analogous situation of the Klein paradox in the context of a quantum field theory has been shown to fail [9].

An alternate view point, for those who believe that the discontinuous approach to signature change with the original boundary conditions is correct, is that regions of Kleinian signature are totally 'harmless'. On encountering a thin wall of Kleinian signature, an incoming wave splits into a reflected wave and a transmitted wave, and thus energy is conserved. If the wall is of infinite length the wave is totally reflected. The spin-$\frac{1}{2}$ results can be regarded as spurious since the components of the Dirac equation solution do not satisfy the singular Klein–Gordon equation (3.14). Furthermore, it is not possible to modify the Dirac equation (in the Kleinian region) so that its square is satisfied by solutions of the same form as those found in section 2.

In either case, it is important to note the crucial role played by boundary conditions. By choosing them carefully, it is possible to obtain qualitatively the same physical results with both the discontinuous and continuous descriptions of signature change.

Acknowledgments

I am grateful to Gary Gibbons for drawing my attention to the concept of Kleinian signature change and for many helpful discussions during the course of this work. I have also benefited from discussions with Chris Fewster. Finally, I thank Suzanne Cook for her useful comments on sections of this work.

References

[1] Hartle J B and Hawking S W 1983 Wave function of the universe *Phys. Rev.* D **28** 2960–75

[2] Dray T, Manogue C A and Tucker R W 1993 Particle production from signature change *Gen. Rel. Grav.* **23** 967–71

[3] Dray T, Manogue C A and Tucker R W 1993 Scalar field equation in the presence of signature change *Phys. Rev.* D **48** 2587–90

[4] Ellis G, Sumeruk A, Coule D and Hellaby C 1992 Change of signature in classical relativity *Class. Quantum Grav.* **9** 1535–54

[5] Hayward S A 1992 Signature change in general relativity *Class. Quantum Grav.* **9** 1851–62

[6] Gibbons G W and Hartle J B 1990 Real tunneling geometries and the large-scale topology of the universe *Phys. Rev.* D **42** 2458–68

[7] Sakharov A D 1984 Cosmological transitions with changes in the signature of the metric *Sov. Phys.–JETP* **60** 214–8

[8] Bjorken J D and Drell S D 1964 *Relativistic Quantum Mechanics* (New York: McGraw-Hill)

[9] Bongaarts P J M and Ruijsenaars S N M 1976 The Klein paradox as a many particle problem *Ann. Phys., NY* **101** 289–318

[10] Reed M and Simon B 1975 *Methods of Modern Mathematical Physics II: Fourier Analysis Self-Adjointness* (New York: Academic)

Class. Quantum Grav. **11** (1994) 2537–2543. Printed in the UK

Localized solutions of the linearized gravitational field equations in free space

A O Barut†, M G Cruz† and Y Sobouti‡

International Centre for Physics and Applied Mathematics, PK 126, Trakya University, Edirne, Turkey

Received 8 February 1994, in final form 18 July 1994

Abstract. We consider the equations of general relativity in free space in the linear approximation. Non-stationary moving solutions for these equations, which are localized and have finite energy, are explicitly constructed. The energy of the localized wave is shown to be proportional to its internal frequency and can represent a massive quantum particle 'made' of gravitational energy.

PACS numbers: 0425N, 0430N

1. Introduction

Recently, moving localized solutions of massless scalar and electromagnetic fields were constructed [1, 2]. These solutions, called wavelets, move without spreading with a dispersion relation corresponding to that of a 'massive' quantum particle with group velocity v and phase velocity u satisfying $uv = c^2$. The wavelet is characterized by an internal frequency Ω such that its central peak is within a distance c/Ω and the corresponding mass is $\hbar\Omega/c$. For a single sharp frequency Ω_0 the wavelet is stable. However, to have a finite total energy a frequency distribution $f(\Omega)$ around Ω_0 of width Δ must be taken. In this case the lifetime of the state is of the order of $1/\Delta$. The integrated energy and momentum are proportional to the frequency and wavenumber of the moving solution, respectively.

These solutions have been interpreted in a 'quantum theory of single events'§ in such a way that the standard quantum mechanics emerges after averaging over the wavelets with different initial conditions. Furthermore, for $\Omega \to \infty$, the wavelets degenerate into moving delta functions, and we obtain as a consequence a limit to classical mechanics.

The purpose of this paper is to explicitly construct the wavelet solutions for the linearized gravitational field equations.

2. Field equations

In the linear approximation to general relativity one expresses the metric $g_{\mu\nu}$ as $g_{\mu\nu} = \eta_{\mu\nu} + h_{\mu\nu}$, where $\eta_{\mu\nu}$ is the Minkowski metricf with signature $(1, -1, -1, -1, -1)$ and $h_{\mu\nu}$

† Physics Department, University of Colorado, Boulder, Colorado 80309–0390.

‡ Institute for Advanced Studies in Basic Sciences, Zanjan, PO Box 45195–159, Iran.

§ For a review see [3].

is a correction. Likewise we have that $g^{\mu\nu} = \eta^{\mu\nu} - h^{\mu\nu}$. In this approximation Einstein's equations of general relativity become [5]

$$-\tfrac{1}{2}\partial^{\gamma}\partial_{\gamma}\bar{h}_{\alpha\beta} + \partial^{\gamma}\partial_{(\beta}\bar{h}_{\alpha)\gamma} - \tfrac{1}{2}\eta_{\alpha\beta}\partial^{\gamma}\partial^{\epsilon}\bar{h}_{\gamma\epsilon} = \frac{8\pi k}{c^4}T_{\alpha\beta} \tag{1}$$

where $T_{\alpha\beta}$ is the energy–momentum tensor, k is the universal constant of gravitation, and $\bar{h}_{\alpha\beta}$ is defined as

$$\bar{h}_{\alpha\beta} = h_{\alpha\beta} - \tfrac{1}{2}\eta_{\alpha\beta} \tag{2}$$

with $h_{\alpha}{}^{\alpha} = h$. Note that we can invert this last relation as follows

$$\bar{h}_{\alpha\beta} = h_{\alpha\beta} - \tfrac{1}{2}\eta_{\alpha\beta}\bar{h} \tag{3}$$

with $\bar{h}^{\mu}_{\mu} = h$.

Let us now transform the metric as follows

$$h_{\alpha\beta} \to h_{\alpha\beta} + \partial_{\alpha}\xi_{\beta} + \partial_{\beta}\xi_{\alpha}\,. \tag{4}$$

The new metric is equivalent to the original if the vector field ξ_{α} generates an infinitesimal diffeomorphism in the spacetime [4].

Under this transformation we have

$$\bar{h}_{\alpha\beta} \to \bar{\bar{h}}_{\alpha\beta} = \bar{h}_{\alpha\beta} + \partial_{\beta}\xi_{\alpha} + \partial_{\alpha}\xi_{\beta} - \eta_{\alpha\beta}\partial^{\delta}\xi_{\delta} \tag{5}$$

and

$$\partial^{\beta}\bar{\bar{h}}_{\alpha\beta} = \partial^{\beta}\bar{h}_{\alpha\beta} + \partial^{\beta}\partial_{\beta}\xi_{\alpha}\,. \tag{6}$$

Choosing ξ_{α} such that $\partial_{\beta}\partial^{\beta}\xi_{\alpha} = -\partial^{\beta}\bar{h}_{\alpha\beta}$ one obtains $\partial_{\beta}\bar{\bar{h}}_{\alpha}\beta = 0$. Therefore assuming that $\partial^{\beta}\bar{h}_{\alpha\beta} = 0$, Einstein's equations become, with $T_{\mu\nu} = 0$, the free space wave equation $\partial^{\alpha}\partial_{\alpha}\bar{h}_{\beta\gamma}$.

3. Localized solutions

The localized solutions are first constructed in the rest frame of the lump at some point x_0 of the space. The solution in an arbitrary inertial frame will be obtained by a Lorentz transformation.

We shall assume that $\bar{h}^{1\alpha} = \bar{h}^{2\alpha} = 0$ and that $\bar{h}^{00}$, $\bar{h}^{03}$, $\bar{h}^{30}$, $\bar{h}^{33}$ are the only non-zero components. Thus, we have the metric tensors

$$(\bar{h})^{\mu}{}_{\nu} = \begin{pmatrix} \bar{h}^{0}{}_{0} & \bar{h}^{0}{}_{3} \\ \bar{h}^{3}{}_{0} & \bar{h}^{3}{}_{3} \end{pmatrix} \tag{7}$$

$$(h)^{\mu}{}_{\nu} = \begin{pmatrix} \frac{1}{2}(\bar{h}^{0}{}_{3} - \bar{h}^{3}{}_{3}) & \bar{h}^{0}{}_{3} \\ \bar{h}^{3}{}_{0} & -\frac{1}{2}(\bar{h}^{0}{}_{3} - \bar{h}^{3}{}_{3}) \end{pmatrix}. \tag{8}$$

Now we find solutions for the wave equation having azimuthal symmetry. We assume the following oscillatory form for the components of the metric

$$\bar{h}^{\alpha 0} \equiv \bar{h}^{0} = \lambda \sum_{\ell} c^{0}_{(\ell)} P_\ell(\cos\theta) \int j_\ell(\xi) f(\Omega)\, \mathrm{e}^{\mathrm{i}\Omega t} \sqrt{\frac{\Omega}{c}}\, \mathrm{d}\Omega$$
$$\bar{h}^{\alpha 3} \equiv \bar{h}^{3} = \lambda \sum_{\ell} c^{3}_{(\ell)} P_\ell(\cos\theta) \int j_\ell(\xi) f(\Omega)\, \mathrm{e}^{\mathrm{i}\Omega t} \sqrt{\frac{\Omega}{c}}\, \mathrm{d}\Omega \qquad (9)$$

where $\xi = (\Omega/c)r$ and λ is a constant with dimensions of time$\sqrt{\text{length}}$. It is implicit in the form of equation (9) that it refers to a particular inertial observer, namely, the one that observes a non-moving centre of localization. The metric $\bar{h}_{\alpha\beta}$ is dimensionaless and j_ℓ is a Bessel function

$$j_\ell(\xi) = \sqrt{\frac{\pi}{2\xi}}\, J_{\ell+1/2}(\xi). \qquad (10)$$

These solutions are required to satisfy the gauge condition

$$\partial^\alpha \bar{h}_{\alpha\beta} = \frac{1}{c}\frac{\partial}{\partial t}\bar{h}^{0} + \frac{\partial}{\partial z}\bar{h}^{3} = 0. \qquad (11)$$

Substitution into the gauge condition gives, after using the identities

$$(1-x^2)\frac{\mathrm{d}}{\mathrm{d}x}P_\ell(x) = \ell P_{\ell-1}(x) - \ell x P_\ell(x) = \frac{\ell(\ell+1)}{2\ell+1}\{P_{\ell-1} - P_{\ell+1}\}$$
$$x P_\ell = \frac{1}{2\ell+1}\{\ell P_{\ell-1} + (\ell+1)P_{\ell+1}\}$$
$$j_\ell(x) = \frac{x}{2\ell+1}\{j_{\ell-1}(x) + j_{\ell+1}(x)\} \qquad (12)$$
$$\frac{\mathrm{d}j_\ell(x)}{\mathrm{d}x} = \frac{1}{2\ell+1}\{\ell j_{\ell-1}(x) - (\ell+1) j_{\ell+1}(x)\}$$

the expressions

$$\frac{1}{c}\frac{\partial}{\partial t}\bar{h}^{0} = \lambda \sum_{\ell=0} \mathrm{i}c^{0}{}_{(\ell)} P_\ell j_{(\ell)}$$
$$\frac{\partial}{\partial z}\bar{h}^{3} = \tfrac{1}{3}\lambda c^{3}{}_{\ell+1} P_0 j_{(0)} + \lambda \sum_{\ell=1}\left\{\frac{c^{3}{}_{\ell+1}(\ell+1)}{2\ell+1} - \frac{\ell c^{3}{}_{\ell-1}}{2\ell-1}\right\} P_\ell j_{(\ell)} \qquad (13)$$

where

$$j_{(i)} \equiv \int \frac{\Omega}{c}\, j_i(\xi) f(\Omega)\, \mathrm{e}^{\mathrm{i}\Omega t} \sqrt{\frac{\Omega}{c}}\, \mathrm{d}\Omega.$$

From these derivatives one obtains the recurrence relations

$$\tfrac{1}{3}c^{3}_{(\ell)} + \mathrm{i}c^{0}_{(0)} = 0$$
$$\mathrm{i}c^{0}_{(\ell)} + \frac{\ell+1}{2\ell+3}c^{3}_{(\ell+1)} - \frac{\ell}{2\ell-1}c^{3}_{(\ell-1)} = 0 \tag{14}$$

for $\ell \geqslant 1$. The simplest solutions of these equations are

$$\bar{h}^{0} = \lambda P_1 j_{(1)} \qquad \bar{h}^{3} = \mathrm{i}j_{(0)} \tag{15}$$

and

$$\bar{h}^{0} = \tfrac{1}{3}\mathrm{i}\lambda\{j_{(0)} - 2P_2 j_{(2)}\} \qquad \bar{h}^{3} = \lambda P_1 j_{(1)} \tag{16}$$

where now

$$j_{(i)} \equiv \int j_i(\xi) f(\Omega)\, \mathrm{e}^{\mathrm{i}\Omega t}\sqrt{\frac{\Omega}{c}}\, \mathrm{d}\Omega$$

i.e. without the factor Ω/c inside the integral.

Thus our solution consists of

$$(\bar{h})^{\mu\nu} = \lambda\begin{pmatrix} \tfrac{1}{3}\mathrm{i}\{j_0 - 2P_2 j_2\} & P_1 j_1 \\ P_1 j_1 & \mathrm{i}j_0 \end{pmatrix} \qquad \bar{h} = -\tfrac{2}{3}\lambda\mathrm{i}\{j_0 + P_2 j_2\} \tag{17}$$

and

$$(h)^{\mu\nu} = \lambda\begin{pmatrix} \tfrac{1}{3}\mathrm{i}\{2j_0 - P_2 j_2\} & P_1 j_1 \\ P_2 j_1 & \tfrac{1}{3}\mathrm{i}\{2j_0 - P_2 j_2\} \end{pmatrix}. \tag{18}$$

4. Energy of the localized solution

For the purpose of obtaining the energy of the localized solution let us compute the zero–zero component of the energy–momentum pseudo-tensor for the gravitational field found by Landau and Lifshitz [5].

In the linear approximation it becomes

$$\begin{aligned} t_{00} = \frac{c^4}{16\pi k}\big\{&h^{00}{}_{,l}h^{lm}{}_{,m} - h^{0l}{}_{,l}h^{0m}{}_{,m} + \tfrac{1}{2}\eta_{lm}h^{nl}{}_{,p}h^{pm}{}_{,n} - \eta^{0l}\eta_{mn}h^{0n}{}_{,p}h^{mp}{}_{,l} \\ &- \eta^{0l}\eta_{mn}h^{0n}{}_{,p}h^{mp}{}_{,l} + \eta^{np}\eta_{lm}h^{0l}{}_{,n}h^{mp}{}_{,p} \\ &+ \tfrac{1}{8}(2\eta^{0l}\eta^{0m} - \eta^{lm})(2\eta_{pq}\eta_{qr} - \eta_{pq}\eta_{nr})h^{nr}{}_{,l}h^{pq}{}_{,m}\big\}. \end{aligned} \tag{19}$$

This expression was obtained by keeping only terms of first order in $h_{\alpha\beta}$ and using

$$g = \det(g_{\alpha\beta}) = \det(\eta_{\alpha\beta} + h_{\alpha\beta}) = -1. \tag{20}$$

Now, setting $h^{00} = h^{33}$ we obtain

$$t^{00} = \frac{c^4}{32\pi k}\{(h^{03}{}_{,1})^2 + (h^{03}{}_{,2})^2 - (h^{00}{}_{,1})^2 - (h^{00}{}_{,2})^2\}. \tag{21}$$

The following derivatives are needed to calculate the energy (with $k = 1, 2, 3$)

$$h^{00}{}_{,k} = \tfrac{1}{3}\mathrm{i}\lambda\left\{-2j_{(1)}\frac{\Omega}{c}(1+\tfrac{1}{5}P_2)\partial_k r + j_{(2)}\frac{2\cos\theta}{\sin^2\theta}(P_2-1)\partial_k\cos\theta + P_2 J_{(3)}\frac{3\Omega}{5c}\partial_k r\right\}$$

$$h^{03}{}_{,k} = \lambda\left\{j_{(0)}\frac{\Omega\cos\theta}{3c}\partial_k r + j_{(1)}\partial_k\cos\theta - j_{(2)}\frac{2\Omega}{3c}\cos\theta\,\partial_k r\right\}. \tag{22}$$

Recall that in these expressions

$$\frac{\Omega}{c}j_{(i)} \equiv \int\frac{\Omega}{c}j_i x i)f(\Omega)\,\mathrm{e}^{\mathrm{i}\Omega t}\sqrt{\frac{\Omega}{c}}\,\mathrm{d}\Omega.$$

Since for our solution

$$E = \int t^{00}\,\mathrm{d}^3x = \frac{c^4}{16\pi k}\left\{\int(h^{03}{}_{,1})^2\,\mathrm{d}x - \int(h^{00}{}_{,1})^2\,\mathrm{d}x\right\} \tag{23}$$

we just need

$$\begin{aligned}\int(h^{03}{}_{,1})^2\,\mathrm{d}^3x = \lambda^2\bigg\{&(\alpha_{00}+\alpha_{22})\int\Omega|f(\Omega)|^2\,\mathrm{d}\Omega\\ &+\alpha_{11}\int j_1 j_1'\sqrt{\frac{\Omega\Omega'}{c^2}}\,ff'\,\mathrm{e}^{\mathrm{i}(\Omega-\Omega')t}\,\mathrm{d}t\,\mathrm{d}\Omega\,\mathrm{d}\Omega'\\ &+\int\{\alpha_{01}j_0 j_1' + \alpha_{12}j_1 j_2'\}\left\{\frac{\Omega}{c}+\frac{\Omega'}{c}\right\}\sqrt{\frac{\Omega\Omega'}{c}}\\ &\times ff'\,\mathrm{e}^{\mathrm{i}(\Omega-\Omega')t}r\,\mathrm{d}r\,\mathrm{d}\Omega\,\mathrm{d}\Omega'\bigg\}\end{aligned} \tag{24}$$

and

$$\begin{aligned}-\int(h^{00}{}_{,1})^2\,\mathrm{d}^3x = \lambda^2\bigg\{&(\beta_{11}+\beta_{33})\int\Omega|f(\Omega)|^2\,\mathrm{d}\Omega\\ &+\beta_{22}\int j_2 j_2'\sqrt{\frac{\Omega\Omega'}{c}}\,ff'\,\mathrm{e}^{\mathrm{i}(\Omega-\Omega')t}\,\mathrm{d}r\,\mathrm{d}\Omega\,\mathrm{d}\Omega'\\ &+\int\{\beta_{12}j_1 j_2' + \beta_{23}j_2 j_3'\}\left\{\frac{\Omega}{c}+\frac{\Omega'}{c}\right\}\sqrt{\frac{\Omega\Omega'}{c^2}}\\ &\times ff'\,\mathrm{e}^{\mathrm{i}(\Omega-\Omega')t}r\,\mathrm{d}r\,\mathrm{d}\Omega\,\mathrm{d}\Omega'\bigg\}\end{aligned} \tag{25}$$

where

$$\alpha_{00} = \frac{4\pi}{9*15} \quad \alpha_{22} = \frac{16\pi}{9*15} \quad \alpha_{11} = \frac{4\pi}{9*15} \quad \alpha_{01} = -\frac{8\pi}{6*15} \quad \alpha_{12} = \frac{8\pi}{3*15} \tag{26}$$

and

$$\begin{aligned} \beta_{11} &= -\frac{8*287.55\pi}{9*420} \qquad \beta_{22} = -\frac{2\pi}{35} \qquad \beta_{33} = -\frac{8\pi}{21*25} \\ \beta_{12} &= -\frac{108\pi}{63*50} \qquad \beta_{23} = -\frac{12\pi}{25*63}\,. \end{aligned} \tag{27}$$

In expressions (24) and (25) we have used the notations $f = f(\Omega)$, $f' = f(\Omega')$, $j_1 = j_1(\xi)$, $j_1' = j_1(\xi')$, etc.

Hence the total energy takes the final form

$$\begin{aligned} E &= \int t^{00}\,\mathrm{d}^3 z \\ &= \frac{c^4\lambda^2}{16\pi k}\Bigg\{(\alpha_{00}+\alpha_{22}+\beta_{11}+\beta_{33})\int \Omega|f(\Omega)|^2\,\mathrm{d}\Omega \\ &\quad + \alpha_{11}\int \frac{J_{3/2}J_{3/2}'}{r} ff'\,\mathrm{e}^{\mathrm{i}(\Omega-\Omega')t}\,\mathrm{d}\Omega\,\mathrm{d}\Omega'\,\mathrm{d}r \\ &\quad + \beta_{22}\int \frac{J_{5/2}J_{5/2}'}{r} ff'\,\mathrm{e}^{\mathrm{i}(\Omega-\Omega')t}\,\mathrm{d}\Omega\,\mathrm{d}\Omega'\,\mathrm{d}r \\ &\quad + \int \{\alpha_{01}J_{1/2}J_{3/2} + (\alpha_{12}+\beta_{12})J_{3/2}J_{5/2}' + \beta_{23}J_{5/2}J_{7/2}'\}\left\{\frac{\Omega}{c}+\frac{\Omega}{c}\right\} \\ &\quad \times ff'\,\mathrm{e}^{\mathrm{i}(\Omega-\Omega')t}\,\mathrm{d}\Omega\,\mathrm{d}\Omega'\,\mathrm{d}r\Bigg\} \end{aligned} \tag{28}$$

or, after integration of r from zero to infinity,

$$\begin{aligned} E &= \frac{c^4\lambda^2}{16\pi k}\Bigg\{\gamma_1\int_0^\infty \Omega|f(\Omega)|^2\,\mathrm{d}\Omega \\ &\quad + \gamma_2\int \theta(\Omega-\Omega')\left\{\left(\frac{\Omega'}{\Omega}\right)^{3/2}+\left(\frac{\Omega'}{\Omega}\right)^{5/2}\right\} ff' 2\,\mathrm{Re}\,(\mathrm{e}^{\mathrm{i}(\Omega-\Omega')t})\,\mathrm{d}\Omega\,\mathrm{d}\Omega' \\ &\quad + \int \theta(\Omega-\Omega') ff'\,\mathrm{e}^{\mathrm{i}(\Omega-\Omega')t} \\ &\quad \times \left\{\alpha_{01}\left(\frac{\Omega}{\Omega'}\right)^{1/2}+\gamma_3\left(\frac{\Omega}{\Omega'}\right)^{3/2}+\gamma_4\left(\frac{\Omega}{\Omega'}\right)^{5/2}+\beta_{23}\left(\frac{\Omega}{\Omega'}\right)^{7/2}\right\}\mathrm{d}\Omega\,\mathrm{d}\Omega'\Bigg\} \end{aligned} \tag{29}$$

which is our final result. Here

$$\begin{aligned} \gamma_1 &= \alpha_{00}+\alpha_{22}+\beta_{11}+\beta_{33} \qquad \gamma_2 = \tfrac{1}{3}\alpha_{11}+\tfrac{1}{5}\beta_{22} \\ \gamma_3 &= \alpha_{01}+\alpha_{12}+\beta_{12} \qquad \gamma_4 = \alpha_{12}+\beta_{12}+\beta_{23}\,. \end{aligned} \tag{30}$$

5. Concluding remarks

If the frequency function $f(\Omega)$ is sufficiently localized, the second and third integrals in equation (29) could be negligible and the energy of the 'particle' becomes proportional to the average of the internal frequency. It is important to point out that without the superposition with the frequency function $f(\Omega)$, the energy of our solutions is infinite.

As in the case of electrodynamics and the massless scalar theories, when the frequency approaches infinity the localized solution behaves like a point particle. Finally, in any inertial frame the solution reads

$$h'_{\mu\nu}(r_\mu, \beta) = \Lambda_\mu^{\mu'} \Lambda_\nu^{\nu'} h_{\mu'\nu'}(r_\perp) \tag{31}$$

where

$$\begin{aligned} r_\perp^2 &= (r_\mu^\mu)^2 - r_\mu r^\mu \qquad n_\mu = \gamma(1, \beta) \\ r_\perp &= (r^2 + \beta^2\gamma^2 t^2 - 2\gamma^2(\tilde{\beta} \cdot \tilde{r}) + \gamma^2(\tilde{\beta} \cdot \tilde{r})^2)^{1/2} \\ \tilde{r} &= \tilde{x} - \tilde{a}. \end{aligned} \tag{32}$$

References

[1] Barut A O 1990 *Phys. Lett.* **143** 349

[2] Barut A O and Bracken A J 1992 *Found. Phys.* **20** 1267

[3] Barut A O 1990 *Found. Phys.* **20** 1233

[4] Wald R 1984 *General Relativity* (Chicago: University of Chicago Press)

[5] Landau L and Lifshitz E 1951 *The Classical Theory of Fields* (Reading, MA: Addison Wesley)

Class. Quantum Grav. **11** (1994) 2545–2551. Printed in the UK

On the gauge freedom of gravitational energy

Göran Bergqvist

Department of Mathematics, Mälardalen University, 721 23 Västerås, Sweden

Received 24 February 1994

Abstract. In spinorial proofs of the positive energy theorem in general relativity the total energy is expressed as the integral of a certain 3-form over a spacelike hypersurface in spacetime. The 3-form depends on a spinor field and by assuming that this spinor field satisfies the Sen–Witten equation, the 3-form becomes everywhere positive. By writing part of the 3-form as a quadratic form in first order derivatives of the spinor field, it is here shown that the Sen–Witten equation can be replaced by a large class of other differential equations, all of which can be used to prove the positive energy theorem. The class is naturally represented by eight complex functions restricted by two inequalities but otherwise arbitrary.

PACS numbers: 0420C, 0420G

1. Introduction

It is well known that it is impossible to define a pointwise energy density for the gravitational field [1]. However, in Witten's proof of the positive energy theorem [2], the total energy of an asymptotically flat spacetime is written as the integral of an everywhere positive function over a spacelike hypersurface H. Some authors (e.g. [3, p 165], [4, p 443] and [5]) have remarked that this function could be interpreted as a local energy density, although non-local in the sense that it depends on the solution of an elliptic equation on H. More reasonable though is to expect that there is a larger class of positive functions which integrated over H gives the total energy. Indeed, alternatives to Witten's original choice have been found by e.g. Parker [6] and by Horowitz and Perry [7].

It is of interest to know how large a class of functions can be used to prove the positive energy theorem. Each choice of function may be seen as a choice of gauge for the gravitational energy (or for the spinor field determining the integrating function). Furthermore, to know that one has a certain freedom is very useful if one wants to prove related results, such as mass–charge or mass–area inequalitites [8–10]. It may also affect the possibility to define quasilocal energy.

In this paper we shall begin by defining the needed quantities. We use the spinor expressions for the total energy introduced by Witten and Nester and write part of it as the integral of a quadratic form. By studying for which differential equations for the spinor field the quadratic form is positive, we find a large class of expressions which are positive and pointwise different, but which all give the total energy when integrated over H. The class contains Witten's equation and the equations given by Horowitz and Perry. Some remarks are also made for the case that the hypersurface is null.

0264-9381/94/102545+07$19.50

2. Total energy

We shall use the 2-spinor formalism which is the natural spinor formalism in general relativity [3, 4, 11]. Let λ be a spinor field on spacetime. Define the Nester–Witten 2-form $F(\lambda)$ [2, 11, 12] by

$$F_{ab} = \mathrm{i}(\bar{\lambda}_{A'}\nabla_b\lambda_A - \bar{\lambda}_{B'}\nabla_a\lambda_B). \tag{1}$$

Here capital letter indices are used for spinors and we make the usual identification $AA' = a$ with tensor indices [11]. Note that although $F(\lambda)$ is a complex 2-form, $\operatorname{Im} F(\lambda)$ is an exact 2-form. Hence $\mathrm{d}F(\lambda)$ is real as well as $F(\lambda)$ integrated over a 2-surface without boundary. Alternatively, one could replace $F(\lambda)$ by $\operatorname{Re} F(\lambda)$ in all equations below.

The ADM and Bondi momenta, which are the measures of total energy at spacelike and null infinity respectively, can be expressed or defined in terms of $F(\lambda)$. In an asymptotically flat spacetime, let $\{\Sigma_\alpha\}$ be a family of spacelike 2-surfaces approaching spacelike infinity or a cross section of future null infinity as $\alpha \to \infty$. If λ is *any* asymptotically constant spinor at spacelike or null infinity (see [4] for definitions of asymptotically constant spinors in both cases), then

$$P_a\hat{k}^a = \lim_{\alpha\to\infty}\frac{1}{8\pi}\int_{\Sigma_\alpha} F(\lambda) \tag{2}$$

where $\hat{k}^a = \lim_{\alpha\to\infty}\lambda^A\bar{\lambda}^{A'}$ is a future pointing asymptotic null translation and P_a is the ADM or Bondi momentum. To prove the positivity of energy one must prove that $P_a\hat{u}^a \geqslant 0$ for all future pointing asymptotic translations $\hat{u}^a$, but since such a $\hat{u}^a$ can be written as a linear combination of two future pointing asymptotic null translations, it is sufficient to prove $P_a\hat{k}^a \geqslant 0$ for all $\hat{k}^a$ null.

Using Stokes' theorem, and taking the limit in (2), one gets

$$P_a\hat{k}^a = \frac{1}{8\pi}\int_H \mathrm{d}F(\lambda) \tag{3}$$

where H is now a non-compact hypersurface which extends out to spacelike or null infinity. The 3-form $\mathrm{d}F(\lambda)$ is real and from (1) we see that it consists of two kinds of terms; terms containing second order derivatives of λ and terms containing products of first order derivatives of λ. As found in [2], the second order terms can be expressed in terms of the Einstein tensor G_{ab}, while the other terms can be expressed in terms of the Sparling 3-form $S(\lambda)$ which is given by $S_{abc} = Z_{[abc]}$ with $Z_{abc} = 2\mathrm{i}(\nabla_a\bar{\lambda}_{B'})(\nabla_c\lambda_B)$ (here [ ] denotes anti-symmetrization). If ${}^*S(\lambda)$ is the dual 1-form of $S(\lambda)$ and if t^a is the unit future pointing normal of H, then equation (3) can be written

$$P_a\hat{k}^a = \frac{1}{8\pi}\int_H [({}^*S(\lambda))_a - G_{ab}\lambda^B\bar{\lambda}^{B'}]t^a\,\mathrm{d}H. \tag{4}$$

By Einstein's equation $G_{ab} = -8\pi T_{ab}$ (sign convention for G_{ab} as in [11]), we obtain

$$P_a\hat{k}^a = \frac{1}{8\pi}\int_H g(\lambda)\,\mathrm{d}H \tag{5}$$

where

$$g(\lambda) = [({}^*S(\lambda))_a + 8\pi T_{ab}\lambda^B\bar{\lambda}^{B'}]t^a. \tag{6}$$

By the dominant energy condition, $T_{ab}v^a w^b \geqslant 0$ for all future pointing v^a and w^b, the second term in (6) is positive. The values of the term $({}^*S(\lambda))_a t^a$ clearly depend on how the spinor field λ varies over H. In the following section we shall study the problem: given the asymptotic values of λ, how can we propagate it over H such that $({}^*S(\lambda))_a t^a$ becomes everywhere positive?

3. Propagation equations and positivity

In the first proof of the positive energy theorem using spinors, Witten [2] found that if λ satisfies

$$D_{AA'}\lambda^A = 0 \tag{7}$$

where $D_a = (g_a^b - t_a t^b)\nabla_b$ (the projection of ∇ onto H), then $({}^*S(\lambda))_a t^a \geqslant 0$ everywhere on H. Equation (7) is called the (Sen–)Witten equation and is a first order elliptic partial differential equation. With existence of solutions to (7) for any given asymptotically constant value of λ proved in [13, 14], one has $P_a\hat{k}^a \geqslant 0$ and the positive energy theorem is established.

If (7) had been the only possible choice of equation which makes $({}^*S(\lambda))_a t^a$ positive and if uniqueness of the solution to (7) was certain, one might have thought of (6) as a kind of pointwise energy density when λ is fixed by (7). However, since no unique gravitational energy density can exist, one must expect that there are other possibilities than (7) which make (6) positive. As remarked above, Parker [6] has added a right-hand side to (7) of order zero (in derivatives) of λ containing G_{ab} (or T_{ab}) terms, and in this way found a class of equations which make the whole expression (6) positive (but not necessarily the first part $({}^*S(\lambda))_a t^a$). He also proved existence of solutions to this class of equations. His results also clearly showed that it is meaningless to consider the two parts of (6) as separated gravitational and matter energy densities. Furthermore, Horowitz and Perry [7] pointed out that if (7) is replaced by

$$D_{AA'}\lambda^A = t_{AA'}v^b D_b\lambda^A \tag{8}$$

with v^a any tangent vector to H satisfying $|v_a v^a| < 1$, then (8) is still an elliptic equation and $({}^*S(\lambda))_a t^a \geqslant 0$. Note that in (8) the principal part of the equation differs from the principal part of (7). In what follows, we shall generalize (8) and look for a general form

$$D_{AA'}\lambda^A = \delta_{AA'}{}^{BB'} D_{BB'}\lambda^A \tag{9}$$

where the spinor $\delta_{AA'}{}^{BB'}$ must be such that $({}^*S(\lambda))_a t^a \geqslant 0$ at every point of H. Every possible choice of such a spinor may be considered as a choice of gauge for the gravitational energy.

We introduce a spinor basis (o, ι) [3, 11], normalized by $o_A\iota^A = 1$. We write the unit normal of H as $t^a = 2^{-1/2}(o^A\bar{o}^{A'} + \iota^A\bar{\iota}^{A'})$. We also introduce the following notation for the eight components of the derivatives of λ:

$$\begin{aligned}
X_1 &= o^B o^A \bar{o}^{A'} \nabla_{AA'}\lambda_B & X_2 &= o^B o^A \bar{\iota}^{A'} \nabla_{AA'}\lambda_B \\
X_3 &= o^B \iota^A \bar{o}^{A'} \nabla_{AA'}\lambda_B & X_4 &= o^B \iota^A \bar{\iota}^{A'} \nabla_{AA'}\lambda_B \\
X_5 &= \iota^B o^A \bar{o}^{A'} \nabla_{AA'}\lambda_B & X_6 &= \iota^B o^A \bar{\iota}^{A'} \nabla_{AA'}\lambda_B \\
X_7 &= \iota^B \iota^A \bar{o}^{A'} \nabla_{AA'}\lambda_B & X_8 &= \iota^B \iota^A \bar{\iota}^{A'} \nabla_{AA'}\lambda_B .
\end{aligned} \tag{10}$$

The components X_2, X_3, X_6, X_7, $X_1 - X_4$ and $X_5 - X_8$ are tangential to H while $X_1 + X_4$ and $X_5 + X_8$ are normal to H. Equation (9) contains only derivatives which are tangential to H, and therefore we introduce the notation

$$X_{1,4} = X_1 - X_4 \qquad X_{5,8} = X_5 - X_8. \tag{11}$$

A straightforward calculation of $({}^*S(\lambda))_a t^a$ yields

$$
\begin{aligned}
{}^*S(\lambda))_a t^a = {} & X_2\bar{X}_2 + X_7\bar{X}_7 + X_3(-\tfrac{1}{2}\bar{X}_3 + \bar{X}_{5,8}) + \bar{X}_3(-\tfrac{1}{2}X_3 + X_{5,8}) \\
& + X_6(-\tfrac{1}{2}\bar{X}_6 - \bar{X}_{1,4}) + \bar{X}_6(-\tfrac{1}{2}X_6 - X_{1,4}).
\end{aligned} \tag{12}
$$

The question is then: for which equations (9) is the quadratic form (12) positive definite (in four variables when (9) has been substituted into it)? The two components of (9) are linear expressions in the components of $D_{AA'}\lambda_B$. Due to the form of (12), we write them as

$$
\begin{aligned}
-\tfrac{1}{2}X_3 + X_{5,8} &= a_2X_2 + a_3X_3 + a_6X_6 + a_7X_7 \\
-\tfrac{1}{2}X_6 - X_{1,4} &= b_2X_2 + b_3X_3 + b_6X_6 + b_7X_7.
\end{aligned} \tag{13}
$$

The system (13) is of the most general form we want since any system with $X_{1,4}$ or $X_{5,8}$ terms on the right-hand side can be put in the form (13) by elementary line operations if it is to have any possibility of making (12) positive definite. The coefficients a_i and b_i, $i = 2, 3, 6, 7$, are complex functions. Substituting (13) into (12) and completing the squares, we get

$$
\begin{aligned}
({}^*S(\lambda))_a t^a = {} & |X_2 + \bar{a}_2X_3 + \bar{b}_2X_6|^2 + |X_7 + \bar{a}_7X_3 + \bar{b}_7X_6|^2 + c_1|X_3|^2 \\
& + c_2|X_6|^2 + c_3X_3\bar{X}_6 + \bar{c}_3X_6\bar{X}_3
\end{aligned} \tag{14}
$$

where

$$
\begin{aligned}
& c_1 = 2\,\mathrm{Re}\,a_3 - |a_2|^2 - |a_7|^2 \qquad c_2 = 2\,\mathrm{Re}\,b_6 - |b_2|^2 - |b_7|^2 \\
& c_3 = \bar{a}_6 + b_3 - \bar{a}_2b_2 - \bar{a}_7b_7.
\end{aligned} \tag{15}
$$

Note that c_1 and c_2 are real functions and c_3 a complex function. We have that (14) is positive definite if

$$
c_1 > 0 \qquad c_1c_2 > |c_3|^2. \tag{16}
$$

Thus, any choice of equation (13) for the spinor field λ gives a positive value of $({}^*S(\lambda))_a t^a$ everywhere on H, if the eight complex functions a_i and b_i, $i = 2, 3, 6, 7$, satisfy (16).

We remark that the key ingredient in the proofs of existence of asymptotically constant solutions to (7) is in fact the positivity of (12) itself when put on the right-hand side of (4). The same positivity property of (12) holds when λ is propagated by any equation (13) which satisfies (16). The equation will again be elliptic [7], and a strict proof of existence of solutions should be done in the same way as in [6], [13] or [14]. If we allowed equalities in (16), the expression (12) still would be positive but (13) might not be elliptic in this case, and therefore the existence problem more difficult.

We may now check two special cases. First we verify that the Sen–Witten equation (7) satisfies (16). The components of (7) are

$$
\begin{aligned}
0 = \bar{o}^{A'}D_{AA'}\lambda^A &= -X_3 + \tfrac{1}{2}X_{5,8} \\
0 = \bar{\iota}^{A'}D_{AA'}\lambda^A &= X_6 + \tfrac{1}{2}X_{1,4}.
\end{aligned} \tag{17}
$$

Writing (17) in the form (13), we have

$$
-\tfrac{1}{2}X_3 + X_{5,8} = \tfrac{3}{2}X_3 \qquad -\tfrac{1}{2}X_6 - X_{1,4} = \tfrac{3}{2}X_6 \tag{18}
$$

i.e. $a_3 = b_6 = \frac{3}{2}$ and $a_2 = a_6 = a_7 = b_2 = b_3 = b_7 = 0$. Hence, for the Sen–Witten equation, $c_1 = c_2 = 3$ and $c_3 = 0$, so (16) is clearly satisfied.

The equations (8) found in [7] can be checked in the same way. The vector v^a, which is a real tangent vector to H, can be written

$$v^a = \frac{1}{\sqrt{2}}k_1(o^A\bar{o}^{A'} - \iota^A\bar{\iota}^{A'}) + \frac{1}{\sqrt{2}}k_2 o^A\bar{\iota}^{A'} + \frac{1}{\sqrt{2}}\bar{k}_2\iota^A\bar{o}^{A'} \tag{19}$$

where k_1 is a real function and k_2 a complex function. Note that (19) implies that

$$|v_a v^a| = k_1^2 + |k_2|^2. \tag{20}$$

Using (19), the components of (8) are

$$\begin{aligned} -X_3 + \tfrac{1}{2}X_{5,8} &= \tfrac{1}{2}(k_1 X_{5,8} + k_2 X_6 + \bar{k}_2 X_7) \\ X_6 + \tfrac{1}{2}X_{1,4} &= -\tfrac{1}{2}(k_1 X_{1,4} + k_2 X_2 + \bar{k}_2 X_3) \end{aligned} \tag{21}$$

which in the form (13) are

$$\begin{aligned} -\frac{1}{2}X_3 + X_{5,8} &= \left(\frac{2}{1-k_1} - \frac{1}{2}\right)X_3 + \frac{k_2}{1-k_1}X_6 + \frac{\bar{k}_2}{1-k_1}X_7 \\ -\frac{1}{2}X_6 - X_{1,4} &= \frac{k_2}{1+k_1}X_2 + \frac{\bar{k}_2}{1+k_1}X_3 + \left(\frac{2}{1+k_1} - \frac{1}{2}\right)X_6. \end{aligned} \tag{22}$$

We see that

$$\begin{aligned} &a_2 = b_7 = 0 \qquad a_6 = \bar{a}_7 = \frac{k_2}{1-k_1} \\ &b_2 = \bar{b}_3 = \frac{k_2}{1+k_1} \qquad a_3 = \frac{2}{1-k_1} - \frac{1}{2} \\ &b_6 = \frac{2}{1+k_1} - \frac{1}{2} \end{aligned} \tag{23}$$

which by (15) implies that

$$\begin{aligned} &c_1 = \frac{3 + 2k_1 - k_1^2 - |k_2|^2}{(1-k_1)^2} \qquad c_2 = \frac{3 - 2k_1 - k_1^2 - |k_2|^2}{(1+k_1)^2} \\ &c_3 = \frac{2\bar{k}_2}{(1-k_1)(1+k_1)}. \end{aligned} \tag{24}$$

We get

$$c_1 c_2 - |c_3|^2 = \frac{(k_1^2 + |k_2|^2 - 5)^2 - 16}{(1-k_1)^2(1+k_1)^2} \tag{25}$$

so $c_1 c_2 > |c_3|^2$ if

$$k_1^2 + |k_2|^2 > 9 \qquad (a)$$

or

$$k_1^2 + |k_2|^2 < 1 \qquad (b) \tag{26}$$

where (a) by (24) is inconsistent with $c_1 > 0$, and (b) always implies $c_1 > 0$. Hence, the condition that (8) implies $({}^*S(\lambda))_a t^a \geqslant 0$ is $k_1^2 + |k_2|^2 < 1$, which by (20) is $|v_a v^a| < 1$ in agreement with [7].

4. Remarks for null hypersurfaces

In the proof of the positivity of the Bondi energy at future null infinity, Ludvigsen and Vickers [15] substituted the spacelike hypersurface H by a null hypersurface N in the asymptotic region. If $l^a = o^A \bar{o}^{A'}$ is a null tangent vector to the null generators of N, then the null surface part of the total energy (3), (4) is

$$\int_N \mathrm{d}F(\lambda) = \int_N [({}^*S(\lambda))_a - G_{ab}\lambda^B\bar{\lambda}^{B'}]l^a \,\mathrm{d}N. \tag{27}$$

The tangential derivatives of λ on N are X_1, X_2, X_3, X_5, X_6 and X_7. Corresponding to (12) on H, we have on N that

$$({}^*S(\lambda))_a l^a = X_2\bar{X}_2 - X_3\bar{X}_3 + X_3\bar{X}_5 + X_5\bar{X}_3 - X_6\bar{X}_1 - X_1\bar{X}_6. \tag{28}$$

In [15] the propagation equation (7) was used on a compact spacelike part of the hypersurface while the equation

$$X_1 = 0 \qquad X_5 = 2X_3 \tag{29}$$

for λ was used on the null part N. This clearly makes the expression (28) positive on N. The necessary existence results for (29) were obtained by studying the equation explicitly in a certain coordinate system. From (28) it is clear that other propagation equations are possible on N and this freedom has been used to study generalizations of the positive energy theorem to inequalities relating the Bondi energy (or mass) to electric charge or areas of trapped surfaces [9, 10]. It has also been used to study quasilocal energy [16–18]. In [16] the equation

$$X_1 = 0 \qquad X_3 = 0 \tag{30}$$

which also makes (28) positive, was used in a definition of quasilocal momentum. Swift [19] has studied the expressions (27), (28) and (29) as the null limits of (4), (12) and (7).

5. Discussion

We have shown that there is a large class of differential equations which may be used in the proof of the positive energy theorem. The class can be represented by eight complex functions restricted by two inequalities. Our result generalizes the results of [7]. As shown in section 3 above, the class in [7] is represented by one complex and one real function restricted by one inequality. The Sen–Witten equation is only one of the possible choices in our class. Each choice of equation can be seen as a choice of gauge for the gravitational energy. This means that the existing proofs of the positive energy theorem using spinors cannot give rise to any kind of local energy density for the gravitational field. If one can obtain a unique definition of quasilocal energy (the energy contained within a spacelike 2-surface) is still an open problem.

This freedom may also be used to study generalizations of the positive energy theorem as mentioned above. A related result was given in [20] where it was shown that it is sufficient to prove existence of solutions to the Sen–Witten equation on a compact surface in order to prove the positive energy theorem. By the results given in the present paper, it is clear that also in this case the Sen–Witten equation can be replaced by any of the equations in the class given above.

Acknowledgment

Financial support by the Swedish Natural Science Research Council (NFR) is acknowledged.

References

[1] Misner C W, Thorne K S and Wheeler J A 1973 *Gravitation* (San Francisco, CA: Freeman)
[2] Witten E 1981 *Commun. Math. Phys.* **80** 381
[3] Stewart J 1990 *Advanced General Relativity* (Cambridge: Cambridge University Press)
[4] Horowitz G T and Tod K P 1982 *Commun. Math. Phys.* **85** 429
[5] Nester J M 1991 *Class. Quantum Grav.* **8** L19
[6] Parker T 1985 *Commun. Math. Phys.* **100** 471
[7] Horowitz G T and Perry M J 1982 *Phys. Rev. Lett.* **48** 371
[8] Penrose R 1982 Some unsolved problems in classical general relativity *Seminar on Differential Geometry* ed S-T Yau (Princeton, NJ: Princeton University Press)
[9] Ludvigsen M and Vickers J A G 1983 *J. Phys. A: Math. Gen.* **16** 1169
[10] Ludvigsen M and Vickers J A G 1983 *J. Phys. A: Math. Gen.* **16** 3349
[11] Penrose R and Rindler W 1986 *Spinors and Spacetime* vol I–II (Cambridge: Cambridge University Press)
[12] Nester J M 1981 *Phys. Lett.* **83A** 241
[13] Parker T and Taubes C H 1982 *Commun. Math. Phys.* **84** 223
[14] Bizón P and Malec E 1986 *Class. Quantum Grav.* **3** L123
[15] Ludvigsen M and Vickers J A G 1982 *J. Phys. A: Math. Gen.* **15** L67
[16] Ludvigsen M and Vickers J A G 1983 *J. Phys. A: Math. Gen.* **16** 1155
[17] Bergqvist G and Ludvigsen M 1991 *Class. Quantum Grav.* **8** 697
[18] Bergqvist G 1992 *Class. Quantum Grav.* **9** 1917
[19] Swift S T 1992 *Class. Quantum Grav.* **9** 1829
[20] Bergqvist G 1993 *Phys. Rev.* D **48** 628

Class. Quantum Grav. 11 (1994) 2553–2574. Printed in the UK

Spherically symmetric anisotropic fluid ICKV spacetimes

A A Coley† and B O J Tupper‡

† Department of Mathematics, Statistics and Computing Science, Dalhousie University, Halifax, NS, Canada B3H 3J5

‡ Department of Mathematics and Statistics, University of New Brunswick, Fredericton, NB, Canada E3B 5A3

Received 26 October 1993, in final form 3 June 1994

Abstract. Spherically symmetric spacetimes representing an anisotropic fluid and admitting a proper inheriting conformal Killing vector (ICKV) are studied and all such spacetimes are found. It is shown that the ICKV all lie in the (t, r)-plane except in the case of those solutions that are conformally flat, in which case there exist three angular-dependent ICKV. The physical properties of these solutions are investigated (with particular attention focused on static spacetimes), and, in the context of these solutions, examples are given which can be interpreted as stellar models (with reasonable physical properties), as models of magnetic fields in a plasma, and as models of viscous heat-conducting fluids. It is shown that global monopole solutions cannot admit a proper ICKV.

PACS numbers: 0420J, 9530L, 9530Q, 9530S

1. Introduction

In a previous paper (Coley and Tupper 1990a) the notion of an inheriting conformal Killing vector field (ICKV) in a fluid spacetime was introduced and in a second paper (Coley and Tupper 1990b) the existence of proper ICKV in perfect fluid spherically symmetric spacetimes was studied. Our motivation here is to study anisotropic fluid, spherically symmetric ICKV spacetimes. We remark that the majority of (I)CKV spacetimes known to the authors are spherically symmetric.

If a spacetime admits a conformal Killing vector (CKV) ξ^a, then

$$\mathcal{L}_\xi g_{ab} = 2\psi g_{ab} \tag{1.1}$$

where $\mathcal{L}_\xi$ signifies the Lie derivative along ξ^a and $\psi(x^a)$ is the conformal factor. If $\psi_{;ab} \neq 0$, ξ^a is called a proper CKV, else it degenerates to a special CKV (SCKV; $\psi_{;ab} = 0$, $\psi_{,a} \neq 0$), a homothetic vector (HV; $\psi_{,a} = 0$) or a Killing vector (KV; $\psi = 0$).

The spacetime symmetry, defined through the existence of a CKV (satisfying (1.1)), is said to be inherited by the physical fluid source represented by the energy–momentum tensor if (Coley and Tupper 1990a)

$$\mathcal{L}_\xi u^a = -\psi u^a \tag{1.2}$$

that is, if fluid flow lines are mapped conformally onto fluid flow lines.

We consider a spherically symmetric coordinate system in which the metric has the general form

$$ds^2 = -e^{2v}dt^2 + e^{2w}dr^2 + R^2(d\theta^2 + \sin^2\theta d\phi^2) \tag{1.3}$$

where $v = v(t, r)$, $w = w(t, r)$, $R = R(t, r)$. With the metric in this form we can always choose the fluid four-velocity to be comoving, namely

$$u^a = e^{-v}\delta_0^a \tag{1.4}$$

(i.e. a transformation to a comoving frame can always be found which leaves the form of the metric (1.3) invariant—see Kramer *et al* 1980).

We shall assume the energy–momentum tensor to be that of an anisotropic fluid, namely

$$T_{ab} = \mu u_a u_b + p_{\|} n_a n_b + p_{\perp} p_{ab} \tag{1.5}$$

where n^a is a unit spacelike vector orthogonal to u^a, p_{ab} is the projection tensor onto the subspace orthogonal to u^a and n^a, i.e. $p_{ab} = g_{ab} + u_a u_b - n_a n_b$, μ is the energy density, $p_{\|}$ is the pressure parallel to n^a, and $p_{\perp}$ is the pressure perpendicular to n^a. We note that (Maartens *et al* 1986)

$$\mathcal{L}_\xi n^a = -\psi n^a + m^a \tag{1.6}$$

where $n^a m_a = 0$ and, from equation (1.2) and $n^a u_a = 0$, $u^a m_a = 0$.

Fluids with anisotropic pressure have been studied for several reasons. First, an energy–momentum tensor can be formally of the form (1.5) (i) if the source of the gravitational field consists of two perfect fluid components with distinct four-velocities (Letelier 1980), (ii) if the source of the gravitational field is a heat-conducting viscous fluid under certain restrictions (Coley and Tupper 1991), (iii) in the region of interaction of two colliding plane impulsive gravitational waves which are followed by a null dust cloud (Taub 1991).

Second, a strong magnetic field in a plasma in which the particle collision density is low can cause the pressure along and perpendicular to the magnetic field lines to be unequal (Maartens *et al* 1986). If the source of the gravitational field can be represented by the sum of a perfect fluid and electromagnetic field so that

$$T_{ab} = (\bar{\mu} + \bar{p})u_a u_b + \bar{p} g_{ab} + T_{ab}^{\mathrm{EM}} \tag{1.7}$$

then, if H^a is the local magnetic field measured by u_a (and the local electric field E^a is zero), defining $n^a = H^a/H$, equation (1.7) can be written in the form (1.5) with (Maartens *et al* 1986) $\mu = \mu + \pi$, $p_{\|} = \bar{p} - \pi$, $p_{\perp} = \bar{p} + \pi$, where $\pi = \frac{1}{2}\lambda H^2$ and λ is the magnetic permeability.

Third, global monopoles form as a result of gauge-symmetry breaking and are predicted by grand unified theories. The gravitational field of such a monopole is static and spherically symmetric. The simplest model that gives rise to global monopoles is described by a Lagrangian $L = \frac{1}{2}\partial_a\phi^\mu\partial^a\phi^\mu - \frac{1}{4}\lambda(\phi^\mu\phi^\mu - \eta^2)^2$, where ϕ^μ, ($\mu = 1, 2, 3$) is a triplet of scalar fields (Barriola and Vilenkin 1989). The field configuration describing a monopole is $\phi^\mu = \eta\Phi(r)x^\mu/r$ (where Φ is governed by a single differential equation) and the energy–momentum tensor of the monopole is of the form of a (comoving) anisotropic fluid (see equations (5) in Barriola and Vilenkin 1989) in which the appropriate energy conditions are satisfied.

Also, it has been proposed that spontaneous symmetry breaking in grand unified theories during phase transitions in the early universe could lead to the formation of cosmic strings (Vilenkin 1981). The standard string model is described by an energy–momentum tensor of the form $T_t^t = T_z^z = -\sigma(r) = p_z$ (else zero), where σ is the energy density and p_z is the pressure along the axis of the string. Formally, therefore, the energy–momentum tensor is of the form of an anisotropic fluid with $\mu + p_{\|} = 0$, $p_{\perp} = 0$. An infinitely long string would impose cylindrical symmetry on the spacetime.

Finally, an investigation of static anisotropic spheres may have applications in relativistic astrophysics. A number of studies have been made of possible situations in which such anisotropy may be of importance. These include stellar matter in neutron stars (Ruderman 1972, Canuto 1973, Hartle 1978, Shumaker and Thorne 1983, Magli 1993), certain density ranges in superdense matter (Bowers and Liang 1974), boson stars (Ruffini and Bonazzola 1969, Gleiser 1988), galactic halos of fermionic dark matter (Ralston and Smith 1991, Madsden 1991), phase transitions (Itoh 1970, Sokolov 1980), pion condensation (Hartle et al 1975) and crystallization in white dwarfs (Kippenhahn and Weigert 1990). Attempts have been made to study the equilibrium of anisotropic materials (Magli and Kijowski 1992) and also the effect of anisotropies on dynamical instabilities in collapsing matter (Chan *et al* 1992). It should be noted that anisotropic spheres have properties that may be quite different to those of isotropic spheres (Herrera *et al* 1979, Consenza *et al* 1981, 1982, Bayin 1982, Stewart 1982).

Physical conditions. In addition to the Einstein field equations, in order to find solutions, a number of additional physical constraints can be imposed, some of which are:

(i) *energy conditions*: $\mu > 0$, $\mu \pm p_\parallel \geqslant 0$, $\mu \pm p_\perp \geqslant 0$ (dominant energy condition), $\mu + p_\parallel + 2p_\perp \geqslant 0$ (strong energy condition). In the astrophysical context it may be appropriate to impose the (more severe) positive pressures criterion $p_\parallel \geqslant 0$, $p_\perp \geqslant 0$.
(ii) *equations of state*: we may wish to impose appropriate equations of state. In the astrophysical or cosmological context it may be appropriate to demand $p_\parallel = p_\parallel(\mu)$, $p_\perp = p_\perp(\mu)$.
(iii) *conditions as* $r \to 0$: (in the astrophysical context especially) we may demand that $p_\parallel$ (and $p_\perp$) remain regular as $r \to 0$. (In addition, we may desire $p_\parallel$ and μ to be zero at $r = r_s$.)

Integrability conditions. The general integrability conditions for the existence of a CKV are $\mathcal{L}_\xi C^a{}_{bcd} = 0$, where $C^a{}_{bcd}$ is the Weyl tensor. However, we can obtain more severe conditions for the existence of an ICKV. From the CKV equations we have that

$$\mathcal{L}_\xi T_{ab} = 2\,\Box\,\psi g_{ab} - 2\psi_{;ab} \tag{1.8}$$

where $\Box\psi = \psi_{;ab}g^{ab}$. Using the form of T_{ab} given by equation (1.5) (noting equation (1.6)), taking its Lie derivative and contracting in turn with the tensors $u^a n^b$, $u^a p^{bc}$, $n^a p^{bc}$, and $p^{ac}p^{bd} - \frac{1}{2}p^{ab}p^{cd}$, when equation (1.2) holds, we obtain, respectively (Maartens *et al* 1986)

$$\psi_{;ab}u^a n^b = 0 \tag{1.9}$$
$$\psi_{;ab}u^a p^b_c = 0 \tag{1.10}$$
$$(p_\perp - p_\parallel)p^{bc}m_b = 2\psi_{;ab}n^a p^{bc} \tag{1.11}$$
$$\psi_{;ab}p^{ac}p^{bd} = \tfrac{1}{2}\psi_{;ab}p^{ab}p^{cd}. \tag{1.12}$$

Equations (1.9) and (1.10) immediately yield that u^a is an eigenvector of $\psi_{;ab}$, namely $\psi_{;ab}u^b = \lambda u_a$, whence equation (1.9) yields $\psi_{;ab}n^b = \gamma n_a + \eta p_{ab}w^b$, where w^a is any 4-vector. Since $m_a n^a = m_a u^a = 0$, we write $m_a = \pi p_{ab}w^b$, whence $p^{bc}m_b = \pi p^c_b w^b$ and equation (1.11) yields ($p^c_b w^b \neq 0$) $\eta = \frac{1}{2}(p_\perp - p_\parallel)\pi$. Using these results, equation (1.12) then yields the *integrability conditions*

$$\psi_{;ab} = \tfrac{1}{3}(\Box\psi - \lambda)g_{ab} + \tfrac{1}{3}(\Box\psi - 4\lambda)u_a u_b - \tfrac{1}{2}(\Box\psi - \lambda - 3\gamma)(n_a n_b - \tfrac{1}{3}h_{ab})$$
$$-(p_\parallel - p_\perp)k_{ab} \tag{1.13}$$

where $k_{ab} = n_{(a}m_{b)}$, $h_{ab} = g_{ab} + u_a u_b$ and we note that $\mathcal{L}_\xi p_\parallel - \mathcal{L}_\xi p_\perp + 2\psi(p_\parallel - p_\perp) = \Box\psi - \lambda - 3\gamma$.

Examples. There are a number of examples of anisotropic fluid ('spherically symmetric CKV spacetimes' in the literature). The example due to Maartens and Maharaj (1990) (MM) is a spherically symmetric spacetime in which the source, in general, is taken to be a (charged) anisotropic fluid (sphere), generalizing previous models by Herrera and Ponce de Leon (1985), in which $\xi^a = \xi^a(r)$ and the constant k in the MM metric is zero. Note that in this example the CKV is inheriting. The anisotropic fluid plus spherically symmetric charge distribution solution of the Einstein–Maxwell equations due to Rago (1989) is another example of a spherically symmetric CKV spacetime which is also a generalization of the models of Herrera and Ponce de Leon (1985). The examples due to Tello-Llanos (1988) (TL) have a general anisotropic fluid as a source; in the third (simple) model (TL3) $\mu + p_{\parallel} = 0$ and $p_{\perp} = 0$ (and the constant c in the metric is non-zero else the model is flat space). In none of these examples is the CKV inheriting.

As a result of the present investigation we find that there are, essentially, six classes of spacetimes (labelled A1, A2, B1, B2, B3, B4) satisfying the requirements, together with some special cases (A1S, A2S, B2S). The metrics of these spacetimes and the non-zero components of the corresponding ICKV (and also some non-inheriting CKV) are listed in table 1 while table 2 lists the MM and TL spacetimes with their CKV.

The organization of this paper is as follows. In section 2 we exhibit the components of equations (1.13), (1.1) and (1.2) for a spherically symmetric spacetime. In section 3 we find the metrics of all class A solutions, characterized by the condition $v_{,r} = 0$ while in section 4 we find the metrics of those class B solutions, characterized by the condition $v_{,r} \neq 0$, for which ψ is independent of the angular coordinates θ, ϕ. In section 5 we investigate those class B solutions for which ψ is dependent on the angular coordinates and find that these solutions are precisely the conformally flat members of the set of class B solutions. Section 6 is a discussion of the properties of the solutions and possible applications to problems of stellar models, global monopoles, magnetic fields and viscous fluids. The results are summarized and discussed in section 7.

2. Analysis

For simplicity we rewrite equation (1.13) in the form

$$\psi_{;ab} = Au_a u_b + Bn_a n_b + Cp_{ab} + 2Dm_{(a}n_{b)} \tag{2.1}$$

where

$$A = -\lambda = -\tfrac{1}{6}[\mathcal{L}_\xi(2\mu + p_{\parallel} + 2p_{\perp}) + 2\psi(2\mu + p_{\parallel} + 2p_{\perp})] \tag{2.2}$$

$$B = \gamma = -\tfrac{1}{6}[\mathcal{L}_\xi(\mu + 2p_{\parallel} - 2p_{\perp}) + 2\psi(\mu + 2p_{\parallel} - 2p_{\perp})] \tag{2.3}$$

$$C = \tfrac{1}{2}(\Box\psi - \lambda - \gamma) = -\tfrac{1}{6}[\mathcal{L}_\xi(\mu - p_{\parallel} + p_{\perp}) + 2\psi(\mu - p_{\parallel} + p_{\perp})] \tag{2.4}$$

$$D = \eta\pi^{-1} = \tfrac{1}{2}(p_{\perp} - p_{\parallel}). \tag{2.5}$$

From spherical symmetry it follows that μ, $p_{\parallel}$ and $p_{\perp}$ are all functions of t and r only; in particular, $D = D(t, r)$. On the other hand, the quantities A, B, C depend on the CKV ξ^a and so are, in general, functions of all four coordinates. We assume that the fluid is not a perfect fluid, so that D is non-zero.

From equations (1.3)–(1.5) the non-zero components of T_{ab} are

$$T_{00} = \mu e^{2v} \qquad T_{11} = p_{\parallel} e^{2w} \qquad T_{22} = p_{\perp} R^2 \qquad T_{33} = p_{\perp} R^2 \sin^2\theta \tag{2.6}$$

and the condition $T_{01} = 0$ implies that

$$R_{,rt} - R_{,r}w_{,t} - R_{,t}v_{,r} = 0. \tag{2.7}$$

Table 1. All spherically symmetric anisotropic fluid spacetimes admitting inheriting CKV are displayed together with the non-zero component of the ICKV and also of some non-inheriting CKV admitted by the models. In the first column the spacetime is identified by its model designation. The metric is then given in spherical polar coordinates. In the third column the non-zero components of the CKV ξ are displayed; E denotes 'exceptional' (angular dependent) case. The inheritance property is stated in the fourth column; I denotes $\mathcal{L}_\xi u^a = \psi u^a$ whereas n denotes that fluid flow lines are not mapped to fluid flow lines. The fifth column denotes the number of the equation in the text which gives the expressions for μ, $p_\parallel$ and $p_\perp$ for each model. Further comments are made in the final column.

Example	Metric	CKV	Inheritance	$\mu, p_\parallel, p_\perp$	Comments
A1	$e^{2v(t)}[-\mathrm{d}t^2 + \mathrm{d}r^2 + n^2(r)\mathrm{d}\Omega^2]$	$\xi^0 = 1$	I	(6.1)	$n(r) \neq \sin r, \sinh r, r$ (else FRW perfect fluid). $v_{tt} \neq 0$
A1S	$e^{2v(t)}[-\mathrm{d}t^2 + \mathrm{d}r^2 + \mathrm{d}\Omega^2]$	$\xi^0 = 1$	I	(6.2)	$v_{tt} \neq 0$
A1Sa	as A1S	$\xi^0 = r,\ \xi^1 = t$	n		
A2	$e^{2v(t)}[-\mathrm{d}t^2 + h^{-2}Q_u^2 e^{2Q(u)}\mathrm{d}r^2 + e^{2Q(u)}d\Omega^2]$	$\xi^0 = 1,\ \xi^1 = 1$	I	(6.3)	$u = t - r,\ v_{tt} \neq 0$
A2S	$e^{2v(t)}[-\mathrm{d}t^2 + e^{2P(u)}\mathrm{d}r^2 + \mathrm{d}\Omega^2]$	$\xi^0 = 1,\ \xi^1 = 1$	I	(6.4)	$u = t - r,\ v_{tt} \neq 0$
B1	$[S(r) + T(t)]^{-2}[-\mathrm{d}t^2 + \mathrm{d}r^2 + Q^2(r)\mathrm{d}\Omega^2]$	$\xi^0 = 1$	I	(6.5)	$T_t \neq 0$. Same as A1 if $S_r = 0$.
B1a	as B1	E	I		$Q(r) = \sinh r,\ \sin r,\ r$ *only*. Conformally flat.
B1b	as B1	$\xi^0 = t,\ \xi^1 = r$	I		$Q(r) = r$ *only*.
B2	$[S(r) + T(t)]^{-2}[-\mathrm{d}t^2 + \mathrm{d}r^2 + Q^2(t)\mathrm{d}\Omega^2]$	$\xi^1 = 1$	I	(6.6)	$S_r \neq 0$
B2S	$S^{-2}(r)[-\mathrm{d}t^2 + \mathrm{d}r^2 + \mathrm{d}\Omega^2]$	$\xi^1 = 1$	I	(6.7)	B2 with $T = 0,\ Q = 1.\ S_r^2 - SS_{rr} \neq 0$. Static
B2Sa	as B2S	$\xi^0 = r,\ \xi^1 = t$	n		
B3	$e^{2v(t,r)}[-\mathrm{d}t^2 + e^{2P(u)}\mathrm{d}r^2 + e^{2Q(u)}\mathrm{d}\Omega^2]$	$\xi^0 = 1,\ \xi^1 = 1$	I	(6.8)	$u = t - r.\ v(t,r),\ P(u),\ Q(u)$ satisfy equation (4.6). $v_r + v_t \neq$ constant.
B3a	$[S(r) + T(t)]^{-2}[-\mathrm{d}t^2 + \mathrm{d}r^2 + \mathrm{d}\Omega^2]$	$\xi^0 = 1$	I		Special case of B1 with $Q = 1$ and of B3 with $P = Q = 0$.
B3b	as B3a	$\xi^1 = 1$	I		Special case of B2 with $Q = 1$ and of B3 with $P = Q = 0$.
B4	$e^{2v(r)}[-\mathrm{d}t^2 + \mathrm{d}r^2 + \kappa^2 r^2\mathrm{d}\Omega^2]$	$\xi^0 = t,\ \xi^1 = r$	I	(6.9)	$e^v \neq r^a$. Static
B4a	as B4 with $\kappa = 1$	E	I		$v_r \neq 0$. Conformal flat.

Table 2. Examples of previous known spherically symmetric anisotropic fluid spaacetimes admitting CKV are displayed. In column 1 the spacetime is identified by the designations MM (Maartens and Maharaj 1990) and TL (Tello-Llanos 1988). The metric is then given in spherical polar coordinates. In the third column the non-zero components of the CKV ξ are displayed and the inheritance property is stated in the fourth column; I denotes $\mathcal{L}_\xi u^a = -\psi u^a$ whereas n denotes that fluid flow lines are not mapped to fluid flow lines. Further comments are made in the final column.

Example	Metric	CKV	Inheritance	Comments
MM	$-[c^2r^2\exp(-2k\int r^{-1}e^w dr)]dt^2 + e^{2w(r)}dr^2 + r^2 d\Omega^2$	$\xi^0 = \frac{1}{2}kt$ $\xi^1 = \frac{1}{2}re^{-w}$	I	Identical to B4. Transformation MM $\rightarrow$ B4 is $\exp(k\int r^{-1}e^w dr) \rightarrow \kappa r$
TL1	$-dt^2 + (ar^2+b)^{-1}dr^2 + r^2 d\Omega^2$	$\xi^0 = (ar^2+b)^{\frac{1}{2}}[c_1 e^{\sqrt{a}t} - c_2 e^{-\sqrt{a}t}]a^{-\frac{1}{2}} + c_3$ $\xi^1 = r(ar^2+b)^{\frac{1}{2}}[c_1 e^{\sqrt{a}t} + c_2 e^{-\sqrt{a}t}]$	n	
TL2	$-dt^2 + dr^2 + (be^{\sqrt{a}t} + ce^{-\sqrt{a}t})^2 d\Omega^2$	$\xi^0 = (be^{\sqrt{a}t} + ce^{-\sqrt{a}t})(c_1 e^{\sqrt{a}r} + c_2 e^{-\sqrt{a}r})$ $\xi^1 = (be^{\sqrt{a}t} - ce^{-\sqrt{a}t})(c_1 e^{\sqrt{a}r} - c_2 e^{-\sqrt{a}r})$	n	
TL3	$-dt^2 + e^{2c}dr^2 + r^2 d\Omega^2$	$\xi^0 = e^{2c}r^2 + t^2$ $\xi^1 = 2rt$	n	Special case of B4 with $v = 0$, and rescaling of r where the ICKV of B4 becomes HV.

Now n^a is an eigenvector of T_{ab} with eigenvalue $p_\parallel$ and, since $p_\parallel \neq p_\perp$, it follows from equation (2.6) that n^a has non-zero components only in the radial direction, i.e.

$$n^a = e^{-w}\delta^a_1. \tag{2.8}$$

Thus m^a lies in the 2-surface defined by p_{ab}. Choosing unit basis vectors $y^a = R^{-1}\delta^a_2$, $z^a = R^{-1}$ cosec θ δ^a_3 in this surface, so that $p_{ab} = y_a y_b + z_a z_b$, we can write m^a in the form $m^a = Yy^a + Zz^a$, so that $\mathcal{L}_\xi n^a = -\psi n^a + Yy^a + Zz^a$, where Y and Z are scalar functions of coordinates.

The components of equation (2.1) are

$$\psi_{;00} = \psi_{,tt} - v_{,r}e^{2v-2w}\psi_{,r} - v_{,t}\psi_{,t} = Ae^{2v} \tag{2.9a}$$

$$\psi_{;01} = \psi_{,tr} - v_{,r}\psi_{,t} - w_{,t}\psi_{,r} = 0 \tag{2.9b}$$

$$\psi_{;02} = \psi_{,t\theta} - R^{-1}R_{,t}\psi_{,\theta} = 0 \tag{2.9c}$$

$$\psi_{;03} = \psi_{,t\phi} - R^{-1}R_{,t}\psi_{,\phi} = 0 \tag{2.9d}$$

$$\psi_{;11} = \psi_{,rr} - w_{,t}e^{2w-2v}\psi_{,t} - w_{,r}\psi_{,r} = Be^{2w} \tag{2.9e}$$

$$\psi_{;12} = \psi_{,r\theta} - R^{-1}R_{,r}\psi_{,\theta} = DYe^{w}R \tag{2.9f}$$

$$\psi_{;13} = \psi_{,r\phi} - R^{-1}R_{,r}\psi_{,\phi} = DZe^{w}R\ \sin\,\theta \tag{2.9g}$$

$$\psi_{;22} = \psi_{,\theta\theta} - RR_{,t}e^{-2v}\psi_{,t} + RR_{,r}e^{-2w}\psi_{,r} = CR^2 \tag{2.9h}$$

$$\psi_{;23} = \psi_{,\theta\phi} - \cot\theta\,\psi_{,\phi} = 0 \tag{2.9i}$$

$$\psi_{;33} = \psi_{,\phi\phi} - RR_{,t}e^{-2v}\sin^2\theta\,\psi_{,t} + RR_{,r}e^{-2w}\sin^2\theta\,\psi_{,r} + \sin\theta\cos\theta\,\psi_{,\theta} = CR^2\sin^2\theta. \tag{2.9j}$$

The ICKV equations (1.1) and (1.2) are

$$\psi = v_{,r}\xi^1 + v_{,t}\xi^0 + \xi^0_{,t} \tag{2.10a}$$

$$= w_{,r}\xi^1 + w_{,t}\xi^0 + \xi^1_{,r} \tag{2.10b}$$

$$= R^{-1}R_{,r}\xi^1 + R^{-1}R_{,t}\xi^0 + \xi^2_{,\theta} \tag{2.10c}$$

$$= R^{-1}R_{,r}\xi^1 + R^{-1}R_{,t}\xi^0 + \cot\theta\,\xi^2 + \xi^3_{,\phi} \tag{2.10d}$$

$$e^{2w}\xi^1_{,\theta} + R^2\xi^2_{,r} = e^{2w}\xi^1_{,\phi} + R^2\sin^2\theta\,\xi^3_{,r} \tag{2.11a}$$

$$= \xi^2_{,\phi} + \xi^3_{,\theta}\sin^2\theta \tag{2.11b}$$

$$= 0 \tag{2.11c}$$

$$\xi^1_{,t} = \xi^2_{,t} = \xi^3_{,t} = 0 \tag{2.12}$$

$$\xi^0 = \xi^0(t) \equiv g(t). \tag{2.13}$$

From equation (2.11*a*) we see that either $v_{,r} = 0$ (class A) or ξ^1 is given in terms of ξ^0 and ψ (class B). We consider each class in turn.

3. Class A solutions

For this class $v_{,r} = 0$, so that equation (2.7) integrates to give

$$R_{,r} = h(r)e^{w} \tag{3.1}$$

where $h(r)$ is an arbitrary function of r and may be zero. Equations (2.10*a*) and (2.13) show that $\psi = \psi(t) = v_{,t}g + g_{,t}$ and equations (2.9*f*, *g*) imply that $Y = Z = 0$ (since $D \neq 0$) so that $\mathcal{L}_\xi n^a = -\psi n^a$, i.e. the unit spacelike vector 'inherits' the CKV symmetry. Putting

$a = 2, 3$ in this expression and using equations (2.8) and (2.11a, b) yields $\xi^2_{,r} = \xi^3_{,r} = 0$, so that, from equations (2.11a, b), $\xi^1_{,\theta} = \xi^1_{,\phi} = 0$, i.e. $\xi^1 = f(r)$. By the same argument as used in a previous paper (Coley and Tupper 1990b) we find that the non-zero ξ^2 and ξ^3 components are just those associated with the 3 KV of spherical symmetry and so may be discarded. Thus for proper ICKV we have that

$$\xi^a = (g(t), f(r), 0, 0) \qquad \psi = \psi(t). \tag{3.2}$$

Note that when ξ^a is orthogonal to u^a, i.e. $g(t) = 0$, equation (2.10a) shows that $\psi = 0$, i.e. ξ^a is a KV. Discarding this case we find that the class A solutions for proper ICKV are:

Class A1. ξ^a parallel to u^a, i.e. $f(r) = 0$. By redefining the time coordinate we can put $\xi^0 = g = 1$ and equations (2.10) become $\psi = v_{,t} = w_{,t} = R^{-1}R_{,t}$. After integration, and redefining the radial coordinate, these yield $e^w = e^v = n^{-1}(r)R$, where, comparing with equation (3.1), we have $h(r) = n_{,r}(r)$. Hence, when $h(r) \neq 0$, i.e. $n(r)$ is not constant, the *class A1* metric is of the form

$$ds^2 = e^{2v(t)}(-dt^2 + dr^2 + n^2(r)d\Omega^2), \tag{3.3}$$

where $d\Omega^2 = d\theta^2 + \sin^2\theta d\phi^2$, and the ICKV is $\xi^a = \delta^a_0$ with $\psi = v_{,t}$, i.e. $v_{,tt} \neq 0$ for a proper ICKV. The model represents a perfect fluid if and only if (iff) $nn_{,rr} - n^2_{,r} + 1 = 0$, i.e. $n = r$, $\sin r$, $\sinh r$, which are, respectively, flat, closed and open FRW models, as found by Coley and Tupper (1990b). Otherwise the model represents an anisotropic fluid. In the special case $h(r) = 0$, $n(r)$ is a constant which can be scaled to 1 and the metric becomes that of *class A1S*, namely

$$ds^2 = e^{2v(t)}(-dt^2 + dr^2 + d\Omega^2) \tag{3.4}$$

which always represents an anisotropic fluid.

Class A2. ξ^a is neither parallel to nor orthogonal to u^a, i.e. $f(r) \neq 0$, $g(t) \neq 0$. We can redefine the t and r coordinates so that $f(r) = 1$, $g(t) = 1$, i.e. $\xi^a = (1, 1, 0, 0)$. Equations (2.10) become $\psi = v_{,t} = w_{,r} + w_{,t} = R^{-1}R_{,r} + R^{-1}R_{,t}$, which imply that $R = e^{Q(u)}e^v$, $e^w = e^{P(u)}e^v$, where $u = t - r$. Equation (3.1) becomes $h(r) = Q_{,u}e^{Q-P}$, which implies that $h(r)$ is a constant. If $h \neq 0$, the *class A2* metric is of the form

$$ds^2 = e^{2v(t)}\left[-dt^2 + h^{-2}Q^2_{,u}e^{2Q(u)}dr^2 + e^{2Q(u)}d\Omega^2\right]. \tag{3.5}$$

The special case $h = 0$ leads to the *class A2S* metric, namely

$$ds^2 = e^{2v(t)}\left[-dt^2 + e^{2P(u)}dr^2 + d\Omega^2\right]. \tag{3.6}$$

4. Class B solutions

For this class $v_{,r} \neq 0$. We first assume that ψ is independent of the angular coordinates θ, ϕ, i.e. $\psi = \psi(t, r) \equiv F(t, r)$. From equations (2.9$f, g$) we again find that $\xi^a = (g(t), f(r), 0, 0)$. Equation (2.9$b$) becomes

$$F_{,tr} - v_{,r}F_{,t} - w_{,t}F_{,r} = 0. \tag{4.1}$$

The class B solutions for the proper ICKV are:

Class B1. ξ^a parallel to u^a, i.e. $f(r) = 0$. We redefine the t-coordinate to set $g = 1$ and from equations (2.7) and (2.10), after integration, we obtain, $e^{-v} = S(r) + T(t)$,

$e^w = P(r)e^v$, $R = Q(r)e^v$, and equation (4.1) is automatically satisfied. Redefining r to set $P(r) = 1$, we obtain the *class B1* metric

$$ds^2 = \frac{1}{[S(r)+T(t)]^2}(-dt^2 + dr^2 + Q^2(r)d\Omega^2). \tag{4.2}$$

The ICKV is given by $\xi^a = \delta^a_0$ with $\psi = -T_{,t}(S+T)^{-1}$.

Class B2. ξ^a is orthogonal to u^a, i.e. $g(t) = 0$. We redefine the r-coordinate to set $f = 1$ and equations (2.10) become $\psi = v_{,r} = w_{,r} = R^{-1}R_{,r}$ which, together with equation (2.7), integrate to yield $e^{-v} = K(t)[S(r)+T(t)]$, $e^w = K(t)e^v$, $R = Q(t)e^v$ so that equation (4.1) is satisfied. Redefining t to set $K(t) = 1$ we obtain the *class B2* metric, namely

$$ds^2 = \frac{1}{[S(r)+T(t)]^2}(-dt^2 + dr^2 + Q^2(t)d\Omega^2). \tag{4.3}$$

The ICKV is given by $\xi^a = \delta^a_1$ with $\psi = -S_{,r}(S+T)^{-1}$. This analysis holds true in the special case when the spacetime is static, i.e. $T = 0$, $Q = 1$, which gives the *class B2S* metric, namely

$$ds^2 = S^{-2}(r)(-dt^2 + dr^2 + d\Omega^2) \tag{4.4}$$

with $\xi^a = \delta^a_1$, $\psi = -S^{-1}S_{,r}$ $(S^2_{,r} - SS_{,rr} \neq 0)$.

Class B3. ξ^a is neither parallel to nor orthogonal to u^a, i.e. $f(r) \neq 0$, $g(t) \neq 0$. We redefine the t and r coordinates so that $f = g = 1$. However, we cannot do this if the spacetime is static since the redefinition of t will re-introduce time dependence into the metric tensor. Assuming that the spacetime is not static, equations (2.10) become $\psi = v_{,r} + v_{,t} = w_{,r} + w_{,t} = R^{-1}R_{,r} + R^{-1}R_{,t}$, which can be integrated to obtain the *class B3* metric

$$ds^2 = e^{2v}(-dt^2 + e^{2P(u)}dr^2 + e^{2Q(u)}d\Omega^2) \tag{4.5}$$

where $u = t - r$. Equation (2.7) then leads to the differential equation

$$v_{,rt} - v_{,r}v_{,t} - Q_{,uu} - P_{,u}v_{,r} + P_{,u}Q_{,u} - Q^2_{,u} = 0. \tag{4.6}$$

The ICKV is $\xi^a = \delta^a_0 + \delta^a_1$.

Note that only the class B1 spacetime (4.2) with $Q(r) = r$ appears to admit more than one ICKV of the form $\xi^a = (\xi^0, \xi^1, 0, 0)$. However, consider the spacetime with metric

$$ds^2 = [S(r) + T(t)]^{-2}(-dt^2 + dr^2 + d\Omega^2) \tag{4.7}$$

which can be considered as a special case of the class B1 spacetime (4.2) with $Q(r) = 1$, or of the class B2 spacetime (4.3) with $Q(t) = 1$, or of the class B3 spacetime (4.5) with $P(u) = Q(u) = 0$ (in which case equation (4.6) leads to $e^{-v} = S(r)+T(t)$). The spacetime (4.7) admits two independent ICKV, namely $\xi^a_{(1)} = \delta^a_0$ and $\xi^a_{(2)} = \delta^a_1$, i.e. the ICKV associated with the class B3 spacetime (4.5) is the sum of two proper ICKV in this case.

Class B4. ξ^a is neither parallel to nor orthogonal to u^a and the spacetime is static. Equations (2.10a, b) show that $\psi = \psi(r)$ and g_t = constant. A rescaling of the time coordinate will set $g = t$ without affecting the time-independence of the metric tensor. Redefining the radial coordinate to put $R = r$ and using equation (2.10) yields $\psi = v_{,r}f + 1 = w_{,r}f + f_{,r} = r^{-1}f$, so that $\psi = \psi(r)$, $f = -r(rv_{,r} - 1)^{-1}$ (provided that $rv_{,r} \neq 1$), and $e^w = \kappa^{-1}(rv_{,r} - 1)$, where κ is a constant. The transformation $\kappa^{-1}re^{-v} \to r$ then changes the *class B4* metric to the form

$$ds^2 = e^{2v(r)}(-dt^2 + dr^2 + \kappa^2 r^2 d\Omega^2) \tag{4.8}$$

with $\xi^a = (t, r, 0, 0)$ and $\psi = rv_{,r} + 1$. The spacetime is conformally flat iff $\kappa^2 = 1$ and when $v_{,r} = 0$, ξ^a is a HV. When $rv_{,r} = 1$, this case is a special subcase of the B2S metric (4.4) with $e^v = r$.

5. Solutions with $\psi = \psi(t, r, \theta, \phi)$.

When $\psi = \psi(t, r, \theta, \phi)$, the solutions must be of class B. Equations (2.9c, d, h, i, j) yield $\psi = R\sigma + F(t, r)$, where

$$\sigma(r, \theta, \phi) = a(r)\cos\phi\sin\theta + b(r)\sin\phi\sin\theta + c(r)\cos\theta. \tag{5.1}$$

Differentiating equation (2.10a) with respect to (wrt) t, eliminating ξ^1 and integrating, we obtain

$$v_{,r} = m(r)R \tag{5.2}$$

where m is an arbitrary function. Differentiating equation (2.10a) wrt θ and wrt ϕ and using equation (5.2) we obtain $\xi^1_{,\theta} = m^{-1}\sigma_{,\theta}$, $\xi^1_{,\phi} = m^{-1}\sigma_{,\phi}$. Since $\xi^1_{,\theta} \neq 0$, it follows from equation (2.12) that $R = q(r)e^w$ and by the transformation $\int q^{-1}(r)\mathrm{d}r \to r$ we can put $q(r) = 1$, i.e.

$$R = e^w. \tag{5.3}$$

Setting $a = 2, 3$, respectively, in the expression $\mathcal{L}_\xi n^a = -\psi n^a + Yy^a + Zz^a$, we obtain $\xi^2_{,r} = -Y$, $\xi^3_{,r} = -Z \operatorname{cosec}\theta$. Hence, from equations (2.11) $\xi^1_{,\theta} = Y = m^{-1}\sigma_{,\theta}$, $\xi^1_{,\phi} = Z\sin\theta = m^{-1}\sigma_{,\phi}$. These expressions and equations (2.9f, g) lead to $\sigma_{,r\theta} = Dm^{-1}R\sigma_{,\theta}$, $\sigma_{,r\phi} = Dm^{-1}R\sigma_{,\phi}$, and the expression (5.1) yields $a_{,r}/a = b_{,r}/b = c_{,r}/c = DR/m = h_{,r}/h$, where $h = h(r)$. Thus we can put $a = \alpha h$, $b = \beta h$, $c = \gamma h$, where α, β, γ are arbitrary constants, and so obtain $\sigma = h\Sigma$, where

$$\Sigma = \alpha\cos\phi\sin\theta + \beta\sin\phi\sin\theta + \gamma\cos\theta \tag{5.4}$$

so that we have

$$\psi = Rh\Sigma + F, \quad Y = m^{-1}h\Sigma_{,\theta} \qquad Z = m^{-1}h\Sigma_{,\phi}. \tag{5.5}$$

Equations (5.2), (5.3) and (2.11a, b) then lead to

$$\begin{aligned} &\xi^1 = m^{-1}h\Sigma + m^{-1}R^{-1}(F - v_{,t}g - g_{,t}) \\ &\xi^2 = \Sigma_{,\theta}\int m^{-1}h\mathrm{d}r \qquad \xi^3 = -\Sigma_{,\phi}\int m^{-1}h\mathrm{d}r \operatorname{cosec}^2\theta \end{aligned} \tag{5.6}$$

where we have factored out the components of the three KV of spherical symmetry. Equations (5.3) and (2.10b, c) then lead to $(m^{-1}h)_{,r}\Sigma = -\int m^{-1}h\mathrm{d}r\,\Sigma_{,\theta\theta} = \int m^{-1}h\,\mathrm{d}r\,\Sigma$ so that, putting $k(r) = \int m^{-1}h\mathrm{d}r$, we have $k_{,rr} = k$ which, with suitable translations of the r coordinate and a rescaling of the t coordinate, yields the following solutions:

$$(a)\quad k = \sinh r \qquad (b)\quad k = \cosh r \qquad (c)\quad k = e^r. \tag{5.7}$$

From equations (2.11b), (5.3) and (5.6), picking out coefficients of Σ and integrating, we obtain $e^v = Rk_{,r}$ which, using equation (5.3), integrating equation (2.7) first wrt r and then wrt t, leads to

$$R = e^w = \left[M(r) + T(t)k_{,r}\right]^{-1} \tag{5.8}$$

$$e^v = \left[M(r) + T(t)k_{,r}\right]^{-1} k_{,r} \tag{5.9}$$

where $M(r)$, $T(t)$, are arbitrary functions. The coordinate transformation $\int k_{,r}^{-1}\mathrm{d}r \to r$ and the relabelling of $k_{,r}^{-1}M(r)$ as $S(r)$ results in the metric (4.2), i.e. the metric of class B1 models, with $Q(r)$ given by one of the three functions

$$(a)\quad Q(r) = \sin r \qquad (b)\quad Q(r) = \sinh r \qquad (c)\quad Q(r) = r \tag{5.10}$$

corresponding to the cases (*a*), (*b*), (*c*), respectively, of equation (5.7). These spacetimes are precisely the conformally flat members of the class of B1 spacetimes.

Note that the function $F(t, r)$ in the expression (5.5) for ψ is the same as the solution for ψ in the case when $\psi = \psi(t, r)$ so that the ICKV ξ^a and the corresponding conformal scalar are given by

$$\xi^a = (\delta t + \varepsilon,\ \delta r + \Sigma,\ Q^{-1}Q_{,r}\Sigma_{,\theta},\ Q^{-1}Q_{,r}\operatorname{cosec}^2\theta\,\Sigma_{,\phi}) \tag{5.11}$$

$$\psi = (S+T)^{-1}\left[-S_{,r}\Sigma + \delta(S + T - rS_{,r} - tT_{,t}) - \varepsilon T_{,t}\right]. \tag{5.12}$$

where δ, ε are constants and $\delta = 0$ when Q is given by (5.10*a*, *b*). Thus in case (5.10*c*) there are five proper ICKV while in cases (*a*), (*b*) there are four proper ICKV.

If the spacetime is static, i.e. if v, w, R are independent of t, the equations (2.7) and (5.2) do not exist, so it is not clear that the metric (4.2) with $T(t) =$ constant will be the only solution. Hence we consider the case in which $v_{,t} = w_{,t} = R_{,t} = 0$, $v_{,r} \neq 0$, $\psi = \psi(t, r, \theta, \phi)$. We fix the radial coordinate by choosing $R = r$, and equations (2.1) become

$$\psi = v_{,r}\xi^1 + g_{,t} = w_{,r}\xi^1 + \xi^1_{,r} = r^{-1}\xi^1 + \xi^2_{,\theta} = r^{-1}\xi^1 + \cot\theta\quad \xi^2 + \xi^3_{,\phi} \tag{5.13}$$

which show that $\psi_{,t} = 0$ and $\xi^0 = g(t) = \delta t + \varepsilon$, where δ, ε are arbitrary constants. From equations (2.9*c*, *d*, *h*, *i*, *j*) we obtain $\psi = \sigma + F(r)$ where σ is given by equation (5.1).

From these expressions and equations (2.11*a*, *b*) it follows that $\xi^1_{,\theta} = Yre^{-w} = v_{,r}^{-1}\sigma_{,\theta}$, $\xi^1_{,\phi} = Zre^{-w}\sin\theta = v_{,r}^{-1}\sigma_{,\phi}$, and equations (2.9*f*, *g*) yield $\sigma_{,r\theta} = r^{-1}\sigma_{,\theta} + Dv_{,r}^{-1}\sigma_{,\theta}e^{2w}$, $\sigma_{,r\phi} = r^{-1}\sigma_{,\phi} + Dv_{,r}^{-1}\sigma_{,\phi}e^{2w}$, from which it follows that $\sigma_{,\theta}^{-1}\sigma_{,r\theta}$ and $\sigma_{,\phi}^{-1}\sigma_{,r\phi}$ are equal functions of r only so that, as before, σ has the form $\sigma = h\Sigma$ where $h = h(r)$ is arbitrary and Σ is given by equation (5.4). Using equations (2.11*a*, *b*) and integrating we obtain $\xi^2 = -k\Sigma_{,\theta}$, $\xi^3 = -k\operatorname{cosec}^2\theta\,\Sigma_{,\phi}$, where the KV of spherical symmetry have been factored out and k is defined by $k_{,r} = hr^{-2}v_{,r}^{-1}e^{2w}$.

Discarding the constant ε, since this corresponds to the timelike KV of the static spacetime, the ICKV is of the form $\xi^a = \left[\delta t,\ (h\Sigma + F - \delta)v_{,r}^{-1},\ -k\Sigma_{,\theta},\ -k\operatorname{cosec}^2\theta\,\Sigma_{,\phi}\right]$, $\psi = h\Sigma + F(r)$. Substituting these expressions into equation (5.13) equating coefficients of like terms, and integrating we obtain $F = \delta(1 - rv_{,r})^{-1}$, $h = e^{v-w}v_{,r}$ and, provided that $rv_{,r} \neq 1$, $e^{2w} = (rv_{,r} - 1)^2$. The transformation $re^{-v} \to r$ changes the metric to the form

$$ds^2 = e^{2v(r)}(-dt^2 + dr^2 + r^2 d\Omega^2) \tag{5.14}$$

which is the conformally flat special case of the class B4 metric (4.7). The spacetime thus admits four proper ICKV given by

$$\xi^a = (\delta t,\ \Sigma + \delta r,\ r^{-1}\Sigma_{,\theta},\ r^{-1}\operatorname{cosec}^2\theta\,\Sigma_{,\phi}) \tag{5.15}$$

$$\psi = v_{,r}\Sigma + \delta(rv_{,r} + 1). \tag{5.16}$$

Hence, while the class B4 spacetime with metric (4.7) in general admits only the ICKV $\xi^a = (t, r, 0, 0)$, the conformally flat special case with $\kappa = 1$ also admits three angular dependent proper ICKV. As before the case $v_{,r} = r^{-1}$ is a special subcase of the class B2S metric (4.4) admitting only the ICKV $\xi^a = \delta^a_1$.

When $e^v = r^a (a \neq 1)$ in the metric (5.14), the ICKV $\xi^a = (t, r, 0, 0)$ becomes an HV, so that the spacetime admits only angular dependent proper ICKV. We note that such ICKV occur only when the spacetime is conformally flat in both static and non-static cases.

6. The models

The expressions for μ, $p_{\|}$, $p_{\perp}$ for each of the models is found in section 3 and 4 are as follows:

Model A1.
$$\begin{aligned} \mu &= e^{-2v}(-2n^{-1}n_{,rr} - n^{-2}n_{,r}^2 + n^{-2} + 3v_{,t}^2) \\ p_{\|} &= e^{-2v}(n^{-2}n_{,r}^2 - n^{-2} - 2v_{,tt} - v_{,t}^2) \\ p_{\perp} &= e^{-2v}(n^{-1}n_{,rr} - 2v_{,tt} - v_{,t}^2) \end{aligned} \tag{6.1}$$

Model A1S.
$$\begin{aligned} \mu &= e^{-2v}(1 + 3v_{,t}^2) \\ p_{\|} &= -e^{-2v}(1 + v_{,t}^2 + 2v_{,tt}) \\ p_{\perp} &= -e^{-2v}(v_{,t}^2 + 2v_{,tt}) \end{aligned} \tag{6.2}$$

Model A2.
$$\begin{aligned} \mu &= (1-h^2)e^{-2v-2Q} + e^{-2v}(2Q_{,u}^{-1}Q_{,uu} + 3Q_{,u} + 3v_{,t})(Q_{,u} + v_{,t}) \\ p_{\|} &= -(1-h^2)e^{-2v-2Q} - e^{-2v}(2Q_{,uu} + 3Q_{,u}^2 + 4Q_{,u}v_{,t} + 2v_{,tt} + v_{,t}^2) \\ p_{\perp} &= -e^{-2v}(Q_{,u}^{-1}Q_{,uuu} + 5Q_{,uu} + 4v_{,t}Q_{,u} + 2v_{,t}Q_{,u}^{-1}Q_{,uu} + 2v_{,tt} + v_{,t}^2) \end{aligned} \tag{6.3}$$

Model A2S.
$$\begin{aligned} \mu &= e^{-2v}(1 + 3v_{,t}^2 + 2v_{,t}P_{,u}) \\ p_{\|} &= -e^{-2v}(1 + v_{,t}^2 + 2v_{,tt}) \\ p_{\perp} &= -e^{-2v}(P_{,uu} + P_{,u}^2 + 2v_{,t}P_{,u} + 2v_{,tt} + v_{,t}^2) \end{aligned} \tag{6.4}$$

Model B1.
$$\begin{aligned} \mu &= (S+T)^2(Q^{-2} - Q^{-1}Q_{,rr} - Q^{-2}Q_{,r}^2) \\ &\quad + 2(S+T)(S_{,rr} + 2S_{,r}Q^{-1}Q_{,r}) - 3S_{,r}^2 + 3T_{,t}^2 \\ p_{\|} &= (S+T)^2(-Q^{-2} + Q^{-2}Q_{,r}^2) - 4(S+T)S_{,r}Q^{-1}Q_{,r} \\ &\quad + 2T_{,tt}(S+T) + 3S_{,r}^2 - 3T_{,t}^2 \\ p_{\perp} &= (S+T)^2Q^{-1}Q_{,rr} - 2(S+T)(S_{,rr} + S_{,r}Q^{-1}Q_{,r}) \\ &\quad + 2T_{,tt}(S+T) + 3S_{,r}^2 - 3T_{,t}^2 \end{aligned} \tag{6.5}$$

Model B2.
$$\begin{aligned} \mu &= (S+T)^2(Q^{-2} + Q^{-2}Q_{,t}^2) + 2(S+T)(S_{,rr} - 2T_{,t}Q^{-1}Q_{,t}) \\ &\quad - 3S_{,r}^2 + 3T_{,t}^2 \\ p_{\|} &= -(S+T)^2(Q^{-2} + Q^{-2}Q_{,t}^2 + 2Q^{-1}Q_{,tt}) \\ &\quad + 2(S+T)(T_{,tt} + 2T_{,t}Q^{-1}Q_{,t}) + 3S_{,r}^2 - 3T_{,t}^2 \\ p_{\perp} &= -(S+T)^2Q^{-1}Q_{,tt} + 2(S+T)(-S_{,rr} + T_{,tt} + T_{,t}Q^{-1}Q_{,t}) \\ &\quad + 3S_{,r}^2 - 3T_{,t}^2 \end{aligned} \tag{6.6}$$

Model B2S.
$$\begin{aligned} \mu &= S^2 + 2SS_{,rr} - 3S_{,r}^2 \\ p_{\|} &= -S^2 + 3S_{,r}^2 \\ p_{\perp} &= -2SS_{,rr} + 3S_{,r}^2 \end{aligned} \tag{6.7}$$

Model B3.
$$\begin{aligned} \mu &= -e^{-2v-2P}(2v_{,rr} + v_{,r}^2 + 2v_{,r}P_{,u} - 4v_{,r}Q_{,u} + 2Q_{,uu} + 3Q_{,u}^2 - 2P_{,u}Q_{,u}) \\ &\quad + e^{-2v}(3v_{,t}^2 + 2v_{,t}P_{,u} + 4v_{,t}Q_{,u} + 2P_{,u}Q_{,u} + Q_{,u}^2) + e^{-2v-2Q} \\ p_{\|} &= -e^{-2v-2Q} + e^{-2v-2P}(3v_{,r}^2 - 4v_{,r}Q_{,u} + Q_{,u}^2) \\ &\quad - e^{-2v}(2v_{,tt} + v_{,t}^2 + 4v_{,t}Q_{,u} + 2Q_{,uu} + 3Q_{,u}^2) \\ p_{\perp} &= e^{-2v-2P}(2v_{,rr} + v_{,r}^2 + 2v_{,r}P_{,u} - 2v_{,r}Q_{,u} + Q_{,uu} + Q_{,u}^2 - P_{,u}Q_{,u}) \end{aligned} \tag{6.8}$$

$$- e^{-2v}(2v_{,tt} + v_{,t}^2 + 2v_{,t}P_{,u} + 2v_{,t}Q_{,u} + P_{,uu} + Q_{,uu} + P_{,u}^2 + P_{,u}Q_{,u} + Q_{,u}^2)$$

Model B4.
$$\mu = e^{-2v}[(\kappa^{-2} - 1)r^{-2} - 2v_{,rr} - 4v_{,r}r^{-1} - v_{,r}^2]$$
$$p_{\|} = e^{-2v}[-(\kappa^{-2} - 1)r^{-2} + 3v_{,r}^2 + 4v_{,r}r^{-1}] \tag{6.9}$$
$$p_{\perp} = e^{-2v}[2v_{,rr} + v_{,r}^2 + 2v_{,r}r^{-1}]$$

We note that, in general, it is difficult to obtain any information from the various energy conditions unless other restrictions are first applied.

There are a number of special cases of interest which we now consider.

(i) *Equations of state.* Suppose we assume that there exist equations of state of the form

$$p_{\|} = p_{\|}(\mu) \qquad p_{\perp} = p_{\perp}(\mu), \tag{6.10}$$

whence it follows that

$$\frac{\partial p_{\|}}{\partial t}\frac{\partial \mu}{\partial r} = \frac{\partial p_{\|}}{\partial r}\frac{\partial \mu}{\partial t} \qquad \frac{\partial p_{\perp}}{\partial t}\frac{\partial \mu}{\partial r} = \frac{\partial p_{\perp}}{\partial r}\frac{\partial \mu}{\partial t}. \tag{6.11}$$

For models A1S, B2S, and B4, μ, $p_{\|}$ and $p_{\perp}$ depend on one variable only and so immediately satisfy equation (6.10). For model A2S, equation (6.11) implies that $P_{,u} = 0$, so this model satisfies the equations of state iff it reduces to model A1S. For model A1, equations (6.11) imply either that $n_{,r} = 0$, so that the model reduces to A1S, or that $nn_{,rr} - n_{,r}^2 + 1 = 0$, which implies that the model represents a perfect fluid.

For the remaining models, i.e. models A2, B1, B2, and B3, equations (6.11) are highly non-linear expressions (quadratic in third derivatives of various functions) containing of the order of 100 terms. In an effort to utilize these equations (in each case) we used the REDUCE algebraic manipulation system to analyse the equations $A^i_{,t}A^j_{,r} - A^i_{,r}A^j_{,t} = 0$ for $A^1 = \mu$, $A^2 = p_{\|}$, $A^3 = p_{\perp}$, $A^4 = \mu + p_{\|}$, $A^5 = \mu + p_{\perp}$, $A^6 = p_{\|} - p_{\perp} A^7 = \mu - p_{\|} + 2p_{\perp}$ (etc) and sums and differences thereof.

In the case of model B1 (equations (6.5)) these equations can be REDUCED to two independent expressions with 36 and 35 terms, namely

$$\begin{aligned}
&(S+T)^3\{\dot{T}[Q'''Q'^2Q^{-3} - Q'''Q^{-1} - 2Q''^2Q'Q^{-3} + Q''Q'^3Q^{-4} - Q''Q'Q^{-4}] \\
&\quad + \dddot{T}\,[Q'''Q^{-1} + 3Q''Q'Q^{-2} - 4Q'^3Q^{-3} + 4Q'QQ^{-3}]\} \\
&\quad +2(S+T)^2\{\dot{T}[S'(-Q'''Q'Q^{-2} + 2Q''^2Q^{-2} - Q''Q'^2Q^{-3} - Q''Q^{-3}) \\
&\quad +S''(2Q''Q'Q^{-2} + Q'^3Q^{-3} - Q'Q^{-3})] + \dot{T}\ddot{T}[-Q'''Q^{-1} + Q'^3Q^{-3} \\
&\quad -Q'Q^{-3}] + \dddot{T}\,[S'(-2Q''Q^{-1} + 5Q'^2Q^{-2} - 2Q^{-2}) - 3S''Q'Q^{-1}]\} \\
&\quad +6(S+T)\{\dot{T}(\ddot{T} + S'')S'[-Q''Q^{-1} - Q'^2Q^{-2} + Q^{-2}] + \dddot{T}\,S'[-S'Q'Q^{-1} \\
&\quad +S'']\} + 6\dot{T}S'(\ddot{T} + S'')(\ddot{T} + S'Q'Q^{-1}) = 0
\end{aligned} \tag{6.12}$$

and

$$\begin{aligned}
&[\dot{T}][3(S+T)^3\{Q'''Q'Q^{-2} - 2Q''^2Q'Q^{-3} + Q''Q'^3Q^{-2} - Q''Q'Q^{-4} - Q'''Q^{-3}\} \\
&\quad +(S+T)^2\{3\ddot{T}(-Q'''Q^{-1} + 3Q''Q'Q^{-2} - 2Q'^3Q^{-3} + 2Q'Q^{-3}) \\
&\quad +2S'(-2Q'''Q'Q^{-2} + 4Q''^2Q^{-2} - 2Q''Q'^2Q^{-3} + Q''Q^{-3}) \\
&\quad +S''(Q'''Q^{-1} + 11Q''Q'Q^{-2} - 6Q'^3Q^{-3} + 6Q'Q^{-3}) \\
&\quad +2S'''(-Q''Q^{-1} - 2Q'^2Q^{-2} + 2Q^{-2})\} \\
&\quad +6(S+T)\{(\ddot{T} + S'')S'(-2Q''Q^{-1} + 2Q'^2Q^{-2} - Q^{-2}) \\
&\quad +\ddot{T}(S''' - S''Q'Q^{-1}) \\
&\quad +(S'S''' - S''^2)Q'Q^{-1}\} + 6S'(\ddot{T} + S'')(-S'Q'Q^{-1} + S'')] = 0
\end{aligned} \tag{6.13}$$

where, for convenience, we have used an overdot and a prime to denote differentiation wrt t and r, respectively. In this second expression $\dot{T}$ is a factor, leading to the trivial solution $\dot{T}=0$ in which the metric components and μ, $p_{\parallel}$ and $p_{\perp}$ (from (6.5)) are all functions of r alone. We also note that in expression (6.13) there are no terms involving $\dddot{T}$. Even in the special cases $Q = \sinh r$, $\sin r$, or r, for which equations (6.12), (6.13) appear in a somewhat more tractable form, it has proved too difficult to find solutions to the equations, other than quite simple solutions that correspond to perfect fluid spacetimes. For example, when $Q=r$, equations (6.11) are satisfied by $S'=-r$, $\ddot{T}=1$, but this yields the de Sitter spacetime satisfying $p_{\parallel}=p_{\perp}=-\mu$. It is not clear if equations (6.11) can be satisfied by any functional values of S and T which correspond to anisotropic fluids.

The case of model B2 is similar, although the corresponding expressions are a little more complicated. The expressions can be *reduced* to two independent equations, one of which has 40 terms and the second of which has somewhat more terms but does not contain a term of the form $S'''\,\dddot{T}$ (although other $\dddot{T}$ terms do appear).

In the case of model B3 (equations (6.8)) even in the special case $P(u) = Q(u)$, equations (6.11) have 421 and 391 terms, respectively. Using (4.6) (and its derivatives), these equations can be *reduced* to around 300 terms, and their difference (after simplification) to 199 terms. No further progress is possible in this case.

In the case of model A2, the equations are, again, very complicated.

(ii) *Shear-free spacetimes.* We note that most known spherically symmetric solutions have vanishing shear (Kramer *et al* 1980). The condition for the spacetimes under consideration to be shear-free is

$$w_{,t} = R_{,t}R^{-1}. \tag{6.14}$$

Applying this to the models listed in table 1, we see that all models are shear-free with the following exceptions:

(a) model A2 with $Q_{,uu} \neq 0$,
(b) model A2S with $P_{,u} \neq 0$ (i.e. the model does not reduce to model A1S),
(c) model B2 with $Q_{,t} \neq 0$ (i.e. the model does not reduce to model B3a),
(d) model B3 with $P_{,u} \neq Q_{,u}$.

Note that the special case of model B3 with $P(u) = Q(u)$ and $v(t,r) = -P(u)+m(r)$, where $m(r)$ is arbitrary, has zero shear, expansion and twist. The metric is

$$ds^2 = e^{2m(r)}(-e^{-2P(u)}dt^2 + dr^2 + d\Omega^2) \tag{6.15}$$

and equation (4.6) is satisfied.

(iii) *Static spacetimes.* The two (classes of) static spacetimes admitting proper ICKV are models B2S and B4. There are various special solutions that are of interest from astrophysical (e.g. stellar models) and cosmological (e.g. cosmic strings) considerations. These are given by the equations of state (i) $p_{\perp}=0$, (ii) $\mu+p_{\parallel}=0$ or (iii) $p_{\parallel}=p_{\perp}$ (i.e. perfect fluid). The implications of these equations of state for the B2S and B4 models are listed in table 3.

The two classes of perfect fluid solutions listed in table 3 contain the static perfect fluid solutions found by Coley and Tupper (1990b) except for the case of the general Schwarzschild interior solution. The absence of this solution is due to the fact that here we have assumed that D, defined by equation (2.5), is non-zero. However, from equations (2.9f, g), this assumption does not affect any of our arguments as long as ψ is independent of θ and ϕ, so any perfect fluid solution which admits an ICKV that is not angular dependent (even if it admits angular dependent ICKV in addition) will appear as a special case of an anisotropic fluid solution. However, the general Schwarzschild interior solution

Table 3. Static spherically symmetric anisotropic fluid spacetimes admitting ICKV and satisfying the equations of state (i) $p_\perp = 0$, (ii) $\mu + p_\parallel = 0$, and (iii) $p_\parallel = p_\perp$ (perfect fluid) are listed together with the corresponding expressions for μ, $p_\parallel$ and $p_\perp$.

	Model B2S	Model B4
(i) $p_\perp = 0$	$S = (ar+b)^{-2}$	$e^v = (a \ln r + b)^2$
	$\mu = S^2$	$\mu = \left[(\kappa^{-2}-1)e^{-2v} - 4ae^{-5v/2}\right]r^{-2}$
	$p_\parallel = 12a^2S^3 - S^2$	$p_\parallel = \left[-(\kappa^{-2}-2)e^{-2v} + 8ae^{-5v/2} + 12a^2e^{-3v}\right]r^{-2}$
(ii) $\mu + p_\parallel = 0$	$S = ar + b$	$e^v = (ar+b)^{-1}$
	$\mu = -p_\parallel = S^2 - 3a^2$	$\mu = -p_\parallel = (\kappa^{-2}-1)r^{-2}e^{-2v} + 4abr^{-1} + a^2$
	$p_\perp = 3a^2$	$p_\perp = a^2 - 2abr^{-1}$
(iii) $p_\parallel = p_\perp$	$S_{,rr} = \frac{1}{2}S$	$e^{-v} = ar^{1+n} + br^{1-n}$
		$(2n^2 \equiv \kappa^{-2} + 1)$.

admits the three angular dependent ICKV and no others, and so is lost by the assumption $D \neq 0$.

(c1) *Stellar models.* When modelling a star (of radius r_s) there are a number of other physical conditions (in addition to the energy conditions) that we may also wish to impose for $r \leqslant r_s$ (cf Maharaj and Maartens, 1989):

(a) $p_\parallel \geqslant 0$, $p_\perp \geqslant 0$ (positive pressures criterion).

(b) $\mu(0)$, $p_\parallel(0)$, $p_\perp(0)$ finite (regularity conditions) $[p_\perp(0) = p_\parallel(0)]$.

(c) $p_\parallel(r_s) = 0$ $[\mu(r_s) = 0;\ \ p_\perp(r_s)$ finite].

(d) smooth matching (at $r = r_s$) to the Schwarzschild exterior solution (and stellar radius is beyond the horizon, i.e. $r_s > r_h = 2M$).

(e) $\mathrm{d}p_\parallel/\mathrm{d}r < 0$.

(f) $\mathrm{d}\mu/\mathrm{d}r < 0$ (stability condition).

(g) $0 \leqslant \mathrm{d}p_\parallel/\mathrm{d}\mu \leqslant 1$, $0 \leqslant \mathrm{d}p_\perp/\mathrm{d}\mu \leqslant 1$ [(speed of sound)$^2 \leqslant$ (speed of light)^2].

There are a number of ways of proceeding, one of which (cf Maharaj and Maartens, 1989) is to specify an appropriate ansatz for $\mu(r)$, $p_\parallel(r)$ or $p_\perp(r)$ and then check that is gives rise to a suitable solution.

For example, using the model B2S, suppose we assume that (for $0 \leqslant r \leqslant r_s$)

$$p_\parallel = S^2 \left\{ \frac{C^2(2 - r^2/r_s^2)^2}{[C + 1 - r^2/r_s^2]^2} - 1 \right\} , \tag{6.16}$$

where C is a constant, so that $p_\parallel(0) = S^2(0)(1+3C)(C-1)(1+C)^{-2}$ (> 0 if $C > 1$), $p_\parallel(r_s) = 0$, and (for $C > 1$) $p'_\parallel \leqslant 0$ for all r, as desired. From equation (6.7) we have

$$\frac{S'}{S} = -\frac{C}{\sqrt{3}} \frac{(2 - r^2/r_s^2)}{[C + 1 - r^2/r_s^2]} , \tag{6.17}$$

which integrates to yield

$$S = S(0) \exp \left\{ -\frac{C}{\sqrt{3}} \left[r + \frac{r_s(1-C)}{2\sqrt{1+C}} \ln \left| \frac{\sqrt{C+1} + r/r_s}{\sqrt{C+1} - r/r_s} \right| \right] \right\} . \tag{6.18}$$

For small r, this becomes

$$S = S(0) \left\{ 1 - \frac{2}{\sqrt{3}} \frac{C}{(1+C)} r + \frac{2C^2}{3(1+C)^2} r^2 + O(r^3) \right\} . \tag{6.19}$$

From equation (6.7) we find that

$$\mu(r) = \frac{S^2}{[C+1-r^2/r_s^2]^2} F(r);$$

$$F(r) \equiv \left\{ \frac{1}{3}(-C^2+6C+3) + \frac{4C(C-1)}{\sqrt{3}r_s}\frac{r}{r_s} - \frac{2}{3}(-2C^2+3C+3)\frac{r^2}{r_s^2} + \frac{1}{3}(3-C^2)\frac{r^4}{r_s^4} \right\}. \tag{6.20}$$

We note that $\mu(0) = \frac{1}{3}S^2(0)(2+6C-C^2)(1+C)^{-2}$, which is positive for $3-2\sqrt{3} < C < 3+2\sqrt{3}$, and

$$\frac{p_{\parallel}(0)}{\mu(0)} = \frac{3(1+3C)(C-1)}{6C+3-C^2},$$

whence $p_{\parallel}(0) \leqslant \mu(0)$ for $\frac{3-2\sqrt{6}}{5} < C < \frac{3+2\sqrt{6}}{5}$. For $1 < C < 3+2\sqrt{3}$, $F(0) > 0$, $F(r_s) > 0$, and by Descarte's rule of sign (DRS) for $C \geqslant \sqrt{3}$, $F(r)$ has at most one real positive root, and hence $F(r) \geqslant 0$ for $0 \leqslant r \leqslant r_s$, and for $C < \sqrt{3}$, $F(r)$ has at most two real positive roots. Now, for $1 < C < \sqrt{3}$, $r_sF'(r) = \frac{4C(C-1)}{\sqrt{3}r_s} - \frac{4}{3}(-2C^2+3C+3)\frac{r}{r_s} + \frac{4}{3}(3-C^3)\frac{r^3}{r_s^3}$ has at most two real positive roots (DRS), and since $r_sF'(0) > 0$ and $r_sF'(r_s) < 0$ $(r_s > \sqrt{3}(C-1)(3-C)^{-1})$, it follows that there exists a root $r_c \in (0, r_s)$ such that $F'(r) > 0$ for $r \in (0, r_c)$ and $F'(r) \leqslant 0$ for $r \in (r_c, r_s)$, and hence again $F(r) \geqslant 0$ for $0 \leqslant r \leqslant r_s$. Therefore, $\mu(r) \geqslant 0$ for all $0 \leqslant r \leqslant r_s$ for these values of the parameters. Also,

$$\mu'(r) = \frac{S^2}{\left[C+1-\frac{r^2}{r_s^2}\right]^3} \left\{ \frac{4C}{3\sqrt{3}}\left[(C^2-6C-3) + \frac{3(C^2-1)}{r_s^2}\right] + \frac{8}{3r_s}(-C^3+C^2)\frac{r}{r_s} + \frac{2C}{\sqrt{3}}\left[(-3C^2+6C+5) + \frac{6(C-1)}{r_s^2}\right]\frac{r^2}{r_s^2} + \frac{4}{3r_s}(C^3-C^2)\frac{r^3}{r_s^3} + \frac{4C}{\sqrt{3}}(C^2-C-2)\frac{r^4}{r_s^4} + \frac{2C(3-C^2)}{3\sqrt{3}}\frac{r^6}{r_s^6} \right\} \tag{6.21}$$

which can be shown to be negative (for all $r \leqslant r_s$) for an appropriate range of values for C. Finally, calculating $p'_{\parallel}(0)$ and $\mu'(0)$, we find that

$$\left.\frac{\mathrm{d}p_{\parallel}}{\mathrm{d}\mu}\right|_0 = 3(3C+1)(C-1)[3+6C-C^2-3(C^2-1)r_s^{-2}]^{-1}$$

which is positive and less than unity if $C < 3+2\sqrt{6}$ (and suitable r_s). Also, from (6.7), $p_{\perp}(0) = \frac{4}{3}S^2(0)C^2(1+C)^{-2}$, whence

$$\frac{p_{\perp}(0)}{p_{\parallel}(0)} = \frac{4C^2}{3(1+3C)(C-1)} \qquad \frac{p_{\perp}(0)}{\mu(0)} = \frac{4C^2}{3+6C-C^2}$$

so that $p_{\perp}(0) \geqslant p_{\parallel}(0)$ and $p_{\perp}(0) \leqslant \mu(0)$ for $C \leqslant (3+2\sqrt{6}/5)$.

However, this model is artificial and simply serves to illustrate the method and that the ICKV spacetimes found here may be useful in describing stellar models. In particular, from a physical point of view, the model can at best be valid for $r > r_c$ (some finite constant) because $r = 0$ does not mark a symmetry centre of spherical symmetry (i.e. a worldline along which the orbits of $SO(3)$ vanish) in this model since, from equation (6.18) or (6.19), the proper radius of the orbits of $SO(3)$ is given by $R = 1/S(r) \to 1/S(0) \neq 0$ as $r \to 0$. We have stated the various regularity conditions as $r \to 0$ implicitly assuming this locus to be a symmetry centre of spherical symmetry (i.e. $R \to 0$ as $r \to 0$); formally this is true only in the limit $S(0) \to \infty$ here.

(c2) *Global monopoles.* For a global monopole the energy–momentum tensor is of the form of a static spherically symmetric comoving anisotropic fluid satisfying $\mu + p_{\|} = 0$, $p_{\perp} = 0$. If a proper ICKV exists then the metric is of the form of either model B2S or model B4. However, from table 3, it follows that the equations of state are satisfied iff $S =$ constant in B2S, which implies that the ICKV is a KV, and $v =$ constant in B4, which implies that the ICKV is a HV.

Using the simplest model described earlier with the field configuration $\phi^a = \eta\Phi(r)x^a/r$, it follows that the solution outside the monopole core is given by (Barriola and Vilenkin, 1989):

$$e^{2v} = e^{-2w} = 1 - \eta^2 - \frac{M}{4\pi r}, \quad R = r, \tag{6.22}$$

where $M \sim M_{\text{core}}$ is a constant of integration. It is easily checked that these metric tensor components do not satisfy the ICKV equations. For reasonable values (of η and λ) M is totally negligible on the astrophysical scale leading to (after rescaling the r and t coordinates) the well known monopole metric

$$\mathrm{d}s^2 = -\mathrm{d}t^2 + \mathrm{d}r^2 + (1 - \eta^2)r^2(\mathrm{d}\theta^2 + \sin^2\theta\mathrm{d}\phi^2), \tag{6.23}$$

which describes a space with a deficit solid angle. This metric is a special case of B4 with $v = 0$ and $\kappa^2 \neq 1$; it is also a rescaled version of the TL3 metric. It admits a HV and also a non-inheriting CKV, but no proper ICKV. Thus we have shown that global monopole solutions do not admit proper ICKV.

(iv) *Magnetic fields.* Suppose that we have a perfect fluid with a magnetic field as given by the energy–momentum tensor (1.7). We note that the subsequent set of equations implies that

$$\bar{p} = \tfrac{1}{2}(p_{\|} + p_{\perp}) \qquad \bar{\mu} = \tfrac{1}{2}(2\mu - p_{\perp} + p_{\|}) \qquad \pi = \tfrac{1}{2}(p_{\perp} - p_{\|}) > 0 \tag{6.24}$$

and the only model that always satisfies the last inequality is model A1S.

Assuming that the perfect fluid satisfies an equation of state of the form $\bar{p} = (\gamma - 1)\bar{\mu}$, the quantities μ, $p_{\|}$, $p_{\perp}$ are related by the equation

$$2(\gamma - 1)\mu + (\gamma - 2)p_{\|} - \gamma p_{\perp} = 0 \tag{6.25a}$$

and in particular cases this becomes

$$\text{(i) dust } (\gamma = 1) \qquad p_{\|} + p_{\perp} = 0 \tag{6.25b}$$

$$\text{(ii) radiation } (\gamma = \tfrac{4}{3}) \qquad \mu = p_{\|} + 2p_{\perp} \tag{6.25c}$$

$$\text{(iii) stiff matter } (\gamma = 2) \qquad \mu = p_{\perp}. \tag{6.25d}$$

We now present three examples, each representing a perfect fluid with magnetic field. In the first two examples the fluid is dust, the model is time-dependent in one case and static in the other, while in the third example the fluid is static incoherent radiation.

Example 1. Consider model A1S representing dust with a magnetic field. Equations (6.2) and (6.25b) yield $4v_{tt} + 2v_t^2 + 1 = 0$, which integrates to give $e^v = a\cos^2\left[2^{-\frac{3}{2}}(t - t_0)\right]$, where a and t_0 are arbitrary constants. The density and magnetic field are given by

$$\bar{\mu} = \tfrac{1}{2}a^{-2}\left\{\sec^2\left[2^{-\frac{3}{2}}(t - t_0)\right] + 3\sin^2\left[2^{-\frac{3}{2}}(t - t_0)\right]\right\}$$

$$\lambda H^2 = a^{-2}\sec^2\left[2^{-\frac{3}{2}}(t - t_0)\right]$$

and all energy conditions are satisfied for all t.

Example 2. Consider model B2S representing a static solution of dust with a magnetic field. Equations (6.7) and (6.25b) yield $2SS_{,rr} - 6S_{,r}^2 + S^2 = 0$. Choosing the particular solution $S = e^{\frac{1}{2}r}$, we find that $\bar{\mu} = \frac{1}{2}e^r$, $\lambda H^2 = \frac{1}{2}e^r$, and all energy conditions are satisfied for all r.

Example 3. Consider model B4 representing a static solution of a radiating fluid with a magnetic field. Equations (6.9) and (6.25c) yield $3v_{,rr} + 3v_{,r}^2 + 6v_{,r}r^{-1} = (\kappa^{-2} - 1)r^{-2}$, which has the general solution $e^v = C_1 r^{-\frac{1}{2}+m} + C_2 r^{-\frac{1}{2}-m}$, where $m = \frac{1}{6}(12\kappa^{-2} - 3)^{\frac{1}{2}}$ and C_1 and C_2 are arbitrary constants. Putting $C_1 = 1, C_2 = 0$ we obtain the particular solution

$$ds^2 = r^{2m-1}[-dt^2 + dr^2 + (3m^2 + \tfrac{1}{4})^{-1}d\Omega^2]$$

$$\bar{\mu} = 3\bar{p} = \tfrac{3}{8}(4m^2 - 1)r^{-1-2m} \qquad \lambda H^2 = \tfrac{1}{8}(2m - 1)(2m - 3)r^{-1-2m}$$

and all energy conditions are satisfied for all r provided that $m > \frac{3}{2}$.

(v) *Lie derivative of n^a*. The radially-directed spacelike unit vector n^a will be mapped onto itself by the action of an ICKV ξ^a, i.e. n^a will satisfy $\mathcal{L}_\xi n^a = -\psi n^a$, if the scalar functions Y and Z in equation (2.9) are both zero. From equations (2.9f, g) it follows that an anisotropic fluid (i.e. $D \neq 0$) will admit an ICKV satisfying this condition in addition to equation (1.2) provided that ψ satisfies the equations

$$\psi_{,r\theta} = R^{-1}R_{,r}\psi_{,\theta} \qquad \psi_{,r\phi} = R^{-1}R_{,r}\psi_{,\phi}\,. \tag{6.26}$$

These equations are always satisfied when ψ is independent of θ and ϕ which is true for all models and all ICKV except the angular-dependent ICKV of models B1a and B4a. For model B1a, equations (5.12) and (6.26) yield $S_{,r}^{-1}S_{,rr} = Q^{-1}Q_{,r}$ which, from equations (5.10) and (6.5), is precisely the condition for the model to represent a perfect fluid. Similarly, for model B4a, equations (5.16) and (6.26) yield $v_{,rr} - v_{,r}^2 - v_{,r}r^{-1} = 0$ which, from equation (6.9) (with $\kappa = 1$) is the condition for a perfect fluid. Thus we have the following theorem.

Theorem. For a spherically symmetric anisotropic fluid, the radially-directed spacelike unit vector n^a will be mapped onto itself by the action of an ICKV ξ^a if and only if the corresponding conformal scalar is independent of the spherical angular coordinates.

(vi) *Viscous fluid solutions.* A viscous heat-conducting fluid (VHCF) has energy–momentum tensor of the form

$$T_{ab} = (\mu^* + p^*)v_a v_b + p^* g_{ab} - 2\eta\sigma_{ab} + q_a v_b + q_b v_a \tag{6.27}$$

where μ^*, p^* are the density and isotropic pressure, q^a is the heat flux vector relative to the four-velocity v^a satisfying $q^a v_a = 0$, $\eta (\geqslant 0)$ is the shear viscosity coefficient, and σ_{ab} is the shear tensor. In order to satisfy the energy conditions T_{ab} must be either of Segré type {1,111} or of type {2,11} or any of their degeneracies (Hall 1984). We consider the particular case when T_{ab} is of Segré type {1,1(11)}, i.e. when the VHCF energy–momentum tensor is formally of the form (1.5). Writing $q_a = qe_a$, where e_a is a spacelike unit vector satisfying $e_a v^a = 0$, and assuming that e^a is an eigenvector of σ_{ab} (Kolassis *et al* 1988) and that q^a is not zero, we see that v^a is not the timelike eigenvector of T_{ab}, i.e. unlike the situation for perfect fluids and anisotropic fluids, the timelike eigenvector for a VHCF does not coincide with the four-velocity. Now, an ICKV is defined by

$$\mathcal{L}_\xi v^a = -\psi v^a \tag{6.28}$$

i.e. fluid flow lines are mapped onto themselves. Thus inheritance is a physical property. It may be of interest to study spacetimes in which a type of 'geometric inheritance' is defined (independent of the properties of the source) in which the congruence defined by the timelike eigenvector of T_{ab} is mapped onto itself, i.e. equation (1.2) is satisfied. For example, we note that FRW perfect fluid models, which all possess an (I)CKV parallel to the comoving four-velocity, may be reinterpreted as VHCF models with a non-comoving four-velocity (Coley and Tupper 1984) which does not necessarily satisfy equation (6.28), but the timelike eigenvector (i.e. the four-velocity of the perfect fluid interpretation) does inherit the CKV symmetry.

If the models found in section 3, 4 and 5 are interpreted as VHCF models with energy–momentum tensor of Segré type {1,1(11)}, the condition (1.2) will always hold for the timelike eigenvector u^a but, in general, the condition (6.28) will not hold, i.e. ξ^a is not an ICKV in the VHCF interpretation. However, under certain restrictions, it may be possible for both equations (1.2) and (6.28) to be satisfied (i.e. ξ^a is an ICKV that also maps the timelike eigenvector of T_{ab} onto itself), as shown in the following example.

Example. Consider model B1, i.e. metric (4.5) with equation (6.6). Let v^a, e^a be defined by $v^a = u^a \cosh\chi + n^a \sinh\chi$, $e^a = u^a \sinh\chi + n^a \cosh\chi$ where $\chi = \chi(t,r)$ is an arbitrary function. From equations (6.6) and (6.27) we obtain $\mu^* = \mu\cosh^2\chi + p_\parallel \sinh^2\chi$, $p^* = \frac{1}{3}(\mu\sinh^2\chi + p_\parallel\cosh^2\chi + 2p_\perp)$, $q = -(\mu + p_\parallel)\cosh\chi\sinh\chi$, $\eta = \frac{1}{2}(p_\perp - \mu\sinh^2\chi - p_\parallel\cosh^2\chi)(\chi_t\sinh\chi + \chi_r\cosh\chi - Q^{-1}Q_r\sinh\chi)(S+T)^{-1}$, where μ, $p_\parallel$ and $p_\perp$ are given by equations (6.6). The energy conditions will be satisfied if they are satisfied for the corresponding anisotropic fluid solution and S, T, Q and χ must be such that $\eta \geqslant 0$ is satisfied.

Now u^a and n^a each inherit the symmetry corresponding to the CKV $\xi^a = \delta^a_0$ and it is easily seen that v^a and e^a will also satisfy equation (6.28) and the similar equation for e^a iff $\mathcal{L}_\xi\chi = 0$, i.e. $\chi_t = 0$. Hence, if the 'tilt angle' is independent of the time coordinate, each of v^a, e^a, u^a and n^a will be mapped onto itself by the action of ξ^a.

If we specify the model to be B1b (i.e. B1 with $Q(r) = r$), and consider the CKV $\xi^a = (t, r, 0, 0)$, the condition $\mathcal{L}_\xi\chi = 0$ now yields $t\chi_t + r\chi_r = 0$, i.e. $\chi = \chi(tr^{-1})$ is the requirement for each of v^a, e^a, u^a and n^a to be mapped onto itself by the action of ξ^a.

7. Conclusion

We have discussed spherically symmetric anisotropic fluid spacetimes which admit inheriting CKV and we have found all such spacetimes (up to the integration of equation (4.6) in the case of model B3). Not surprisingly, the spacetimes found here are more general than the spherically symmetric perfect fluid spacetimes admitting ICKV. Some of these spacetimes (B1a, B4a) are conformally flat and these are the only models that admit the three angular-dependent ICKV characterized by the quantity Σ defined by equation (5.4). One of the models (B3a,b) is not conformally flat and admits two ICKV, a situation which does not arise in the perfect fluid case.

All of the spacetimes that have been obtained are presented in table 1 (in terms of their metrics), along with their corresponding ICKV. There are essentially six classes of solutions (and their special subcases). Solutions have been classified according to whether $v_r = 0$ (class A) or $v_r \neq 0$ (class B), and according to whether the ICKV ξ^a is (i) parallel to the four-velocity u^a (models A1 and B1 with metrics (3.3) and (4.2), respectively), (ii) orthogonal to u^a (model B2 with metric (4.3), and the special static subcase with metric (4.4)), or (iii) neither parallel nor perpendicular to u^a (model A2 with metric (3.5), the special case model A2S with metric (3.6) corresponding to $h(r) = 0$, the non-static model B3 with metric (4.5) and the static model B4 with metric (4.7)). In general, in the models above the conformal factor is of the form $\psi = \psi(t,r)$. However, models with $\psi = \psi(t,r,\theta,\phi)$, in which there exist 'exceptional' angular dependent ICKV (denoted by E in table 1), occur as special (conformally flat) cases of class B1 with metric (4.2) where $Q(r)$ is given by (5.10) and the static class B4 with metric (4.7) (the ICKV are given explicitly by (5.11) and (5.15), respectively, in each case). Previously known spherically symmetric anisotropic fluid ICKV spacetimes (see table 1 in Coley and Tupper 1990b), some of which occur as special cases of the models appearing in table 1, are listed in table 2; the model by Maartens and Maharaj (1990) (denoted by MM) is identical to model B4 (under the transformation $\exp(k \int r^{-1} e^{w} dr) \to \kappa r$) and the model by Tello-Llanos (1988) (denoted by TL3) is a special case of model B4 in which the ICKV degenerates to a HV.

The models, and particularly their physical properties, were studied in detail in section 6. For example, μ, $p_{\|}$ and $p_{\perp}$ were calculated for each model and the conditions under which an equation of state is admitted was discussed. In addition, the questions of whether the models are shear-free (all models that are *not* shear-free were identified) and whether the radically-directed spacelike unit vector satisfies the condition $\mathcal{L}_{\xi} n^a = -\psi n^a$ (it does iff the corresponding conformal scalar is independent of the spherical angular coordinates) were addressed.

Static spacetimes were studied in detail since they are of particular physical interest; for example, such solutions may be interpreted as stellar models or global monopoles. Static spacetimes admitting a proper ICKV belong to models of types B2S or B4. Such spacetimes necessarily admit an equation of state; additional constraints (on each of models B2S and B4) are listed in table 3 for various (particular) equations of state of physical interest ($p_{\perp} = 0$, $\mu + p_{\perp} = 0$, and $p_{\|} = p_{\perp}$ (the perfect fluid case)).

Spacetimes that can be interpreted as stellar models were then considered. Such models must satisfy a number of additional physical conditions. A class of models, of type B2S and satisfying (6.16), were then obtained (equations (6.18) and (6.20)). While this model is artificial and not entirely satisfactory, it does satisfy some reasonable physical conditions and illustrates the method of obtaining such models. Spacetimes representing global monopoles were also considered, but it was shown that there are no such spacetimes that admit a proper ICKV.

Spacetimes which can be interpreted as models with a perfect fluid plus magnetic field and also those which can be interpreted as viscous heat-conducting fluid models were also studied. Three models with a magnetic field and a perfect fluid with equation of state $\bar{p} = (\gamma - 1)\bar{\mu}$ were obtained; (1) a model of type A1S with $\gamma = 1$ (dust), (2) a static model of type B1S with $\gamma = 1$ (dust), and (3) a static model of type B4 with $\gamma = \frac{4}{3}$ (radiation). The possibility of a viscous heat-conducting fluid model admitting a CKV which not only maps the physical fluid flow lines onto themselves but also maps the geometric congruence of timelike eigenvectors of the energy–momentum tensor onto itself (cf equations (1.2) and (6.28)) was discussed and a particular example, a model of type B1, was given.

Finally we note that any spherically symmetric spacetime must have a corresponding T_{ab} which is either of Segré type {1,1(11)} or of type {2,(11)} (assuming that the usual energy conditions are satisfied). In the former case T_{ab} has the form (1.5), with u^a as the unit timelike eigenvector, which is not necessarily a fluid 4-velocity, while in the latter case the spacetime must be the Vaidya solution (Kramer *et al* 1980), which possesses no CKV. Furthermore, the usual argument that a spherically symmetric perfect fluid spacetime metric can be written in the form (1.3) depends only on the fact that T_{ab} possesses a timelike eigenvector (see, for example, problem 16.26 in Lightman *et al* (1975)) and a similar argument can be used to show that the metric of any spherically symmetric spacetime for which T_{ab} is of Segré type {1,1(11)} can be written in the form (1.3). It then follows that if, as in section 6(f), we define ICKV by requiring that the unit timelike eigenvector,

u^a, of T_{ab} inherits the CKV symmetry, as in equation (1.2), then the spacetime solutions presented here, together with those perfect fluid spacetimes found previously (Coley and Tupper 1990b) which are not special cases of the present anisotropic solutions, are the only spherically symmetric spacetimes, irrespective of the material distribution considered, admitting one or more CKV that map the unit timelike eigenvector of T_{ab} onto itself. Only in the case of perfect fluids or anisotropic fluids with T_{ab} of the form (1.5) is the unit timelike eigenvector identical to the fluid 4-velocity, thus corresponding to our physical definition of fluid flow lines mapped onto fluid flow lines.

Acknowledgments

We would like to thank José Castejón-Amenedo for his help at the beginning of this research. This work was supported, in part, by the Natural Sciences and Engineering Research Council of Canada through an operating grant to each of the authors.

References

Barriola M and Vilenkin A 1989 *Phys. Rev. Lett.* **63** 341
Bayin S 1982 *Phys. Rev.* D **26** 1262
Bowers R L and Liang E P T 1974 *Astrophys. J.* **188** 657
Canuto V 1973 *Neutron Stars: General Review* presented at the *Solvay Conf. on Astrophysics and Gravitation (Brussels)*
Chan R, Herrera L and Santos N O 1992 *Class. Quantum Grav.* **9** L133
Coley A A and Tupper B O J 1984 *Astrophys. J.* **280** 26
——1990a *Class. Quantum Grav.* **7** 1961
——1990b *Class. Quantum Grav.* **7** 2195
——1991 *Gen. Rel. Grav.* **23** 1113
Consenza M, Herrera L, Esculpi M and Witten L 1981 *J. Math. Phys.* **22** 118
——1982 *J. Math. Phys.* **23** 2527
Gleiser M 1988 *Phys. Rev.* D **38** 2376
Hall G S 1984 *Arabian J. Sci. Eng.* **9** 88
Hartle J B 1978 *Phys. Rep.* **46** 202
Hartle J B, Sawyer R F and Scalapino D J 1975 *Astrophys. J.* **99** 471
Herrara L, Ruggeri G J and Witten L 1979 *Astrophys. J.* **234** 1094
Herrera L and Ponce de Léon 1985 *J. Math. Phys.* **27** 2987
Itoh N 1970 *Prog. Theor. Phys.* **44** 291
Kippenhahn R and Weigert A 1990 *Stellar Structure and Evolution* (Berlin: Springer)
Kolassis C A, Santos N O and Tsoubelis D 1988 *Class. Quantum Grav.* **5** 1329
Kramer D, Stephani H, MacCallum M and Herlt E 1980 *Exact Solutions of Einstein's Field Equations* (Cambridge: Cambridge University Press)
Kuchowicz B 1968 *Acta Phys. Polon.* **33** 541
Letelier P S 1980 *Phys. Rev.* D **22** 807
Lightman A P, Press W H, Price R H and Teukolsky S A 1975 *Problem Book in Relativity and Gravitation* (Princeton, NJ: Princeton University Press)
Maartens R and Maharaj M S 1990 *J. Math. Phys.* **31** 151
Maartens R, Mason D P and Tsamparlis M 1986 *J. Math. Phys.* **27** 2987
Magli G and Kijowski J 1992 *Gen. Rel. Grav.* **24** 139
Magli G 1993 *Gen. Rel. Grav.* **25** 441
Maharaj S D and Maartens R 1989 *Gen. Rel. Grav.* **21** 899
Rago M 1989 *J. Math. Phys.* **30** 2110
Ruderman M 1972 *Ann. Rev. Astron.* A **10** 427
Ruffini R and Bonazzola S 1969 *Phys. Rev.* **187** 1767
Shumaker B L and Thorne K S 1983 *Mon. Not. R. Astron. Soc.* **203** 457
Sokolov A I 1980 *Sov. Phys.–JETP* **52** 575
Stewart B W 1982 *J. Phys. A: Math. Gen.* **15** 2419

Taub A H 1991 *J. Math. Phys.* **32** 1322
Tello-Llanos R A 1988 *Gen. Rel. Grav.* **20** 765
Tolman R C 1939 *Phys. Rev.* **55** 364
Vilenkin A 1981 *Phys. Rev.* D **23** 851

Class. Quantum Grav. **11** (1994) 2575–2583. Printed in the UK

Non-metricity induced by dilaton gravity in two dimensions

T Dereli†§ and Robin W Tucker‡||

† Department of Mathematics, Middle East Technical University, 06531 Ankara, Turkey

‡ School of Physics and Materials, University of Lancaster, Bailrigg, Lancs LA1 4YB, UK

Received 25 April 1994, in final form 26 August 1994

Abstract. We discuss how theories of dilaton gravity can be constructed in terms of a torsion-free but non-metric-compatible connection and explore the nature of connection-induced dilaton couplings to gravity when the theory is reformulated in terms of the Levi-Civita connection. Exact classical solutions are presented in the context of a string-inspired dilaton model in two dimensions.

PACS numbers: 0420C, 0450

1. Introduction

We recall that the linear connection ∇ of a (pseudo-) Riemannian geometry is torsion-free and metric-compatible. Thus for all vector fields X, Y on the spacetime manifold:

$$T(X, Y) \equiv \nabla_X Y - \nabla_Y X - [X, Y] = 0 \tag{1}$$

and

$$Q \equiv \nabla g = 0. \tag{2}$$

Here g denotes the metric tensor, T the torsion tensor of ∇ and Q the gradient tensor of g with respect to ∇. Einsteinian gravitational interactions with matter fields are conventionally formulated in terms of this unique connection and the notion of geodesic curves with respect to g coincides with the notion of autoparallel curves of ∇. This Levi-Civita connection provides a useful reference connection since it depends entirely on the metric structure of the manifold. However, in the presence of matter, gravitational interactions can sometimes be derived from more general connections. Torsional connections have been shown to accommodate gravitational interactions between spinors [1] as well as scalar fields [2]. In these approaches it is of interest to see if the theory can be rewritten in terms of the Levi-Civita connection so that the torsion and non-metricity tensors can then be interpreted as matter induced couplings for Einsteinian gravity. Alternatively theories with new local symmetries (such as supergravity) often necessitate complicated matter couplings with respect to the Levi-Civita connection and these can be greatly simplified by reformulating the theory in terms of a connection with a non-vanishing torsion. Surprisingly little attention has been devoted to a similar strategy of including also connections that are not metric-compatible [3]–[11].

§ Email address: dereli@trmetu.bitnet

|| Email address: rwt@lavu.physics.lancaster.ac.uk

0264-9381/94/102575+09$19.50

In this paper we demonstrate how theories of dilaton gravity can be constructed in terms of a torsion-free non-metric connection and deduce the corresponding renormalization of the connection induced couplings when the theory is reformulated in terms of the Levi-Civita connection. In view of the intense interest of gravities in two dimensions we shall illustrate our results in terms of an action adapted to a string-inspired dilaton model [12]–[15].

2. The non-metric-compatible connection

The torsion tensor can be written in terms of a set of 2-forms $\{T^a\}$ with respect to a coframe $\{e^a\}$:

$$T^a(X,Y) = \tfrac{1}{2}e^a(\boldsymbol{T}(X,Y)). \tag{3}$$

In the following it will prove convenient to choose a g-orthonormal coframe in which g takes the form $g = \eta_{ab}e^a \otimes e^b$ with $\eta_{ab} = \pm 1$. In this coframe we define the connection 1-forms Λ^a_b by

$$\nabla_{X_a} e^c = -\Lambda^c{}_b(X_a)\, e^b.$$

It follows that the torsion and curvature forms of ∇ are given by

$$T^a = De^a \equiv de^a + \Lambda^a{}_b \wedge e^b \tag{4}$$

$$R^a{}_b = d\Lambda^a{}_b + \Lambda^a{}_c \wedge \Lambda^c{}_b. \tag{5}$$

We shall often use the exterior covariant derivative D whose general definition can be found in [16].

We may decompose the connection 1-forms as follows;

$$\Lambda^a{}_b = \Omega^a{}_b + Q^a{}_b \tag{6}$$

where $\Omega_{ab} = -\Omega_{ba}$ and $Q_{ab} = Q_{ba}$. Thus

$$Q_{ab} = \tfrac{1}{2}(\Lambda^c{}_a\eta_{cb} + \Lambda^c{}_b\eta_{ac}) = \tfrac{1}{2}(\Lambda_{ba} + \Lambda_{ab}) \tag{7}$$

$$Q^a{}_b = \tfrac{1}{2}(\Lambda^a{}_b + \eta^{ac}\eta_{bd}\Lambda^d{}_c). \tag{8}$$

The symmetric non-metricity forms Q_{ab} are tensorial and it follows from (7) that

$$D\eta_{ab} = -2Q_{ab}.$$

We may solve (1) and (2) for the connection 1-forms :

$$\Lambda_{ab} = \tfrac{1}{2}i_{X_a}(\mathcal{T}_b) - \tfrac{1}{2}i_{X_b}(\mathcal{T}_a) - \tfrac{1}{2}i_{X_a}i_{X_b}(\mathcal{T}_c)\, e^c + Q_{ab} \tag{9}$$

where

$$\mathcal{T}_a = T_a - Q_{ac} \wedge e^c - d\, e_a. \tag{10}$$

From (9) it follows that the general connection forms may be expressed in terms of the Levi-Civita forms:

$$\omega_{ab} = \tfrac{1}{2} i_{X_a}(d\, e_b) - \tfrac{1}{2} i_{X_b}(d\, e_a) - \tfrac{1}{2} i_{X_a} i_{X_b}(d\, e_c)\, e^c, \tag{11}$$

the tensorial contorsion forms:

$$K_{ab} = \tfrac{1}{2} i_{X_a}(T_b) - \tfrac{1}{2} i_{X_b}(T_a) - \tfrac{1}{2} i_{X_a} i_{X_b}(T_c)\, e^c, \tag{12}$$

and the non-metricity forms Q_{ab}:

$$\Lambda_{ab} = \omega_{ab} + K_{ab} - i_{X_a} Q_{bc}\, e^c + i_{X_b} Q_{ac}\, e^c + Q_{ab}. \tag{13}$$

From these formulae and the defining relations $d\, e^a + \omega^a{}_b \wedge e^b = 0$ and $K^a{}_b \wedge e^b = T^a$ we may decompose the general curvature 2-forms in (5) into Levi-Civita curvature forms, contorsion and non-metricity tensor contributions.

It is clear from (9) that for any 1-forms $B^a{}_b$ the forms $\mathcal{T}_a$ remain invariant under the substitutions:

$$e^a \mapsto e^a \tag{14}$$

$$T^a \mapsto T^a + B^a{}_b \wedge e^b \tag{15}$$

$$Q^a{}_b \mapsto Q^a{}_b + B^a{}_b \tag{16}$$

and hence

$$\Omega^a{}_b \mapsto \Omega^a{}_b. \tag{17}$$

For certain choices of $B^a{}_b$ it follows that the generalized Einstein forms $R^a{}_b \wedge \star(e_a \wedge e^b)$ are also invariant. This is the origin of the invariances discussed in [17] where the above transformations have been combined with the Weyl transformations [2] to give

$$e^a \mapsto e^\sigma\, e^a \tag{18}$$

$$T^a \mapsto e^\sigma\,(T^a + B^a{}_b \wedge e^b + d\,\sigma \wedge e^a) \tag{19}$$

$$Q^a{}_b \mapsto Q^a{}_b + B^a{}_b \tag{20}$$

for $B^a{}_b = (C-1)\delta^a{}_b\, d\,\sigma$ where σ is any 0-form and C an arbitrary constant.

3. Dilaton gravity with a non-metric-compatible connection

The simplest model of two-dimensional gravity invokes the presence of a dilaton scalar ϕ with a coupling to the curvature scalar in order to define a non-trivial action. We shall also demand that the gradient of the metric with respect to the connection is determined by the gradient of the dilaton scalar. In two (and only two) dimensions the Hodge dual of the gradient can also compete so we shall develop the theory with a non-metric compatible connection determined by the non-metricity 1-forms

$$Q^b{}_a = k\delta^b{}_a d\phi + l\delta^b{}_a \star d\phi \tag{21}$$

where k and l are fundamental coupling constants. Our theory is based on the action 2-form

$$\begin{aligned}\mathcal{L}(\phi, e^a, \Lambda^a{}_b, \lambda_a, \rho^a{}_b) \\ &= \tfrac{1}{2}\phi^p R^a{}_b \wedge \star(e_a \wedge e^b) + \tfrac{1}{2}\alpha d\phi \wedge \star d\phi + \tfrac{1}{2}\beta\phi^2 \star 1 + \tfrac{1}{4}\mu Q^a{}_b \wedge \star Q^b{}_a \\ &+ \lambda_a(de^a + \Lambda^a{}_b \wedge e^b) + \rho^a{}_b \wedge (Q^b{}_a - k\delta^b{}_a d\phi - l\delta^b{}_a \star d\phi).\end{aligned} \tag{22}$$

The field equations are derived using the techniques of constrained variations developed in [18], [19]. Variations with respect to the set of 0-form multipliers $\{\lambda_a\}$ ensure that the connection is torsion-free while variation with respect to the 1-forms $\{\rho^a{}_b\}$ where $\rho_{ab} = \rho_{ba}$, constrain the gradient of the metric in accordance with (21). In the following $\rho = \rho^a{}_a$. Variations with respect to ϕ, e^a and $\Lambda^a{}_b$ give the remaining field equations:

$$\tfrac{1}{2}p\phi^{p-1} R^a{}_b \star (e_a \wedge e^b) - \alpha\, d \star d\phi + \beta\phi \star 1 - k\, d\rho + l\, d \star \rho = 0 \tag{23}$$

$$\tfrac{1}{2} - \alpha\tau_a[\phi] + \tfrac{1}{2}\beta\phi^2 \star e_a - \tfrac{1}{4}\mu\tau_a[Q] + D\lambda_a + l(i_a d\phi \star \rho + \rho i_a \star d\phi) = 0 \tag{24}$$

$$D(\tfrac{1}{2}\phi^p \star e_a \wedge e^b) + \tfrac{1}{2}\mu \star Q^b{}_a + \lambda_a e^b - \rho^b{}_a = 0. \tag{25}$$

In these expressions α, β, and μ are real coupling constants and p is a positive definite integer. The stress forms for the dilaton are

$$\tau_a[\phi] = i_a d\phi \star d\phi + d\phi i_a \star d\phi \tag{26}$$

while

$$\tau_a[Q] = i_a Q^b{}_c \star Q^c{}_b + Q^b{}_c i_a \star Q^c{}_b \tag{27}$$

where $i_b \equiv i_{X_b}$. A simple equation follows from the trace of (24) and the torsion-free condition:

$$\beta\phi^2 \star 1 = d(e^a\lambda_a). \tag{28}$$

It is possible to solve (25) for the multipliers. Writing it as

$$\mathcal{M}^b{}_a + \lambda_a e^b - \rho^b{}_a = 0 \tag{29}$$

where

$$\mathcal{M}^b{}_a = D(\tfrac{1}{2}\phi^p \star (e_a \wedge e^b)) + \tfrac{1}{2}\mu \star Q^b{}_a. \tag{30}$$

Contracting (29) with i_b gives

$$i_b\mathcal{M}^b{}_a + 2\lambda_a - i_b\rho^b{}_a = 0. \tag{31}$$

while contracting (29) with i_c and taking care not to lower indices under D gives:

$$\eta^{ac}(i_c\mathcal{M}^b{}_a) + \lambda^b - \eta^{ac}\, i_c\rho^b{}_a = 0 \tag{32}$$

or

$$\eta_{ab}\eta^{dc}(i_c\mathcal{M}^b{}_d) + \lambda_a - i_c\rho^c_a = 0. \tag{33}$$

Subtracting (32) from (29) yields

$$\lambda_a = -i_b\mathcal{M}^b{}_a + \eta_{ab}\eta^{cd}\,(i_c\mathcal{M}^b{}_d). \tag{34}$$

Inserting this in (29) then gives

$$\rho^b{}_a = \mathcal{M}^b{}_a - (i_c\mathcal{M}^c{}_a)\, e^b + \eta_{ak}\eta^{cn}\,(i_c\mathcal{M}^k{}_n)\, e^b \tag{35}$$

and

$$\rho = \mathcal{M}^a{}_a - (i_c\mathcal{M}^c{}_a)\, e^a + \eta^{bc}\,(i_b\mathcal{M}^a{}_c)\, e_a \tag{36}$$

$$d\rho = d\mathcal{M}^a{}_a + D(i_c\mathcal{M}^c{}_a) \wedge e^a - 2Q^{bc} \wedge (i_b\mathcal{M}^a{}_c)\, e_a - \eta^{bc}\, D(i_b\mathcal{M}^a{}_c) \wedge e_a \tag{37}$$

where

$$\mathcal{M}^b{}_a = D(\tfrac{1}{2}\phi^p \star (e_a \wedge e^b)) + \tfrac{1}{2}\mu \star \delta^b{}_a(k\, d\phi + l \star d\phi). \tag{38}$$

Inserting these expressions for the multipliers into (24) and (23) gives the field equations for the metric and dilaton following from the action (22). In the following we shall take $p = 2$.

4. Reduction to a theory with the Levi-Civita connection

Using the equations (11)–(13) the connection forms $\Lambda^a{}_b$ can be expressed in terms of the Levi-Civita connection forms $\omega^a{}_b$:

$$\Lambda^a{}_b = \omega^a{}_b + \epsilon^a{}_b(k \star d\phi + l d\phi) + Q^a{}_b. \tag{39}$$

In a similar way the curvature forms $R^a{}_b$ can be expressed in terms of the Levi-Civita curvature forms $R^a{}_b(\omega)$:

$$R^a{}_b = R^a{}_b(\omega) + (k\epsilon^a{}_b + l\delta^a{}_b)\, d \star d\phi. \tag{40}$$

Thus the Einstein forms become

$$R^a{}_b \star (e_a \wedge e^b) = R^a{}_b(\omega) \star (e_a \wedge e^b) - 2k\, d \star d\phi \tag{41}$$

while

$$Q^a{}_b \wedge \star Q^b{}_a = 2(k^2 - l^2)\, d\phi \wedge \star d\phi \tag{42}$$

and

$$D\lambda_a = D(\omega)\lambda_a - \{2k\phi + \mu(k^2 + l^2)\}\, \tau_a[\phi] - (2\,l\phi + 2\,\mu kl) \star \tau_a[\phi] \tag{43}$$

where

$$\lambda_a = i_a \star d\phi^2 + \mu(k\, i_a \star d\phi + l\, i_a d\phi) \tag{44}$$

and $D(\omega)$ denotes the Levi-Civita exterior covariant derivative. Inserting these expressions into (23), (24) one finds that the field equations for ϕ and the metric may be written solely in terms of the standard metric-compatible torsion-free connection as:

$$R^a{}_b(\omega) \star (e_a \wedge e^b) - 4k\phi d \star d\phi - 2kd\phi \wedge \star d\phi - \{\alpha + \mu(k^2 - l^2)\}\, d \star d\phi + \beta\phi \star 1 = 0 \tag{45}$$

$$-\, 2k\phi\tau_a[\phi] - \left\{\frac{\alpha + \mu(k^2 - l^2)}{2}\right\} \tau_a[\phi] + \frac{\beta}{2}\phi^2 \star e_a + D(\omega)\lambda_a = 0. \tag{46}$$

We have also verified that these are the variational equations that arise from rewriting the action in terms of the standard connection:

$$\begin{aligned}\mathcal{L}(\phi, e^a, \Lambda^a{}_b, \lambda_a, \rho^a{}_b) &= \tfrac{1}{2}\phi^2 R^a{}_b(\omega) \wedge \star(e_a \wedge e^b)\\ &\quad + (2k\phi + \tfrac{1}{2}\{\alpha + \mu\,(k^2 - l^2)\}\,) d\phi \wedge \star d\phi\\ &\quad + \tfrac{1}{2}\beta\phi^2 \star 1 + \lambda_a(d\, e^a + \Lambda^a{}_b \wedge e^b) + \rho^a{}_b \wedge Q^b{}_a \quad \text{mod } (d).\end{aligned} \tag{47}$$

It is interesting to observe how the non-metricity forms give rise to standard kinetic terms for the scalar field, rescale the scalar stress forms and for $k \neq 0$ introduce a novel derivative scalar interaction in the above action.

5. Classical solutions

The theory depends on the five real couplings $\{\alpha, \beta, \mu, k, l\}$. To make contact with existing models we note that for $\mu = 0$, $k = l = 0$ the theory reduces to the action of [13] which for $\alpha = 4$ possesses the so called *dilaton black hole*:

$$e^1 = \frac{dx}{F(x)} \tag{48}$$

$$e^C = F(x)\, dt \tag{49}$$

$$\phi(x) = e^{c_1 x} \tag{50}$$

$$F(x)^2 = \frac{\beta + 4c_3 c_1^2 \phi(x)^{-2}}{4c_1^2} \tag{51}$$

for constants c_1, c_2. This is also a solution to the above theory when $l \neq 0$. Furthermore it follows from an analysis of the coupling shifts in the section above that the dilaton black hole is a solution to the theory with arbitrary α and β provided $\mu = (\alpha - 4)/l^2$.

A completely different static solution exists in the sector of the theory $\alpha = \beta = \mu = l = 0$.

$$F(x) = ck\sqrt{\frac{x}{e\,W(4xk^2/e)}} \tag{52}$$

$$\phi(x) = \frac{1 + W(4xk^2/e)}{2k}. \tag{53}$$

Here $W(z)$ is the function defined implicitly by the relation

$$W(z)e^{W(z)} = z. \tag{54}$$

It is real-valued for $-\frac{1}{e} < z < \infty$. [20]. We have chosen inessential constants of integration to simplify the form of the solution. Again by rewriting the theory in Levi-Civita form it follows that the same solution can be obtained for the couplings $\alpha \neq 0$, $\beta = 0$, $k \neq 0$, $l \neq 0$, but $\mu = \alpha/(l^2 - k^2)$. A more complex static solution with a non-trivial dependence of the constants $\{c_1, c_2, c_3\}$ follows if we study the sector of the theory with $\alpha = 2$ and $k = l$:

$$\phi(x) = c_2 - \frac{L(x)}{2k} \tag{55}$$

$$\begin{aligned} F(x)^2 = \{&8(\beta x c_2^2 - c_3)k^3 + 8\beta x c_2(1 - L(x))k^2 \\ &+ 2\beta k(xL(x)^2 - 2xL(x) + 2x - 2c_1c_2 + 2c_1c_2L(x)) \\ &- 2\beta c_1 - \beta c_1 L(x)^2 + 2c_1\beta L(x)\}(c_1 - 2kx)/(8k^2(2c_2k - L(x))) \end{aligned} \tag{56}$$

where $L(x) = Ln(2k\,x - c_1)$.

Finally, we have examined cosmological type solutions where

$$e^1 = F(t)\,dx \tag{57}$$

$$e^0 = dt. \tag{58}$$

If we take $\alpha = 6$, $\mu = 0$, then

$$\phi(t) = \frac{\beta t^2}{16k} \tag{59}$$

$$F(t) = \frac{ce^{\frac{1}{8}\beta t^2}}{t^3}. \tag{60}$$

Alternatively we may leave α arbitrary and choose a value of μ so that $\alpha = 6+\mu(l^2-k^2)$. Each of the solutions above arises from imposing at least one relation among the couplings in the theory. Finally, for $k = 0$ and $\beta/(\alpha - \mu l^2 - 8) > 0$ we have the inflationary type solutions:

$$F(t) = e^{c_1 t}, \qquad \phi(t) = e^{c_2 t} \tag{61}$$

where

$$c_2 = \pm\sqrt{\frac{\beta}{\alpha - \mu l^2 - 8}}$$

and

$$c_1 = c_2\{\tfrac{1}{2}\mu l^2 - \tfrac{1}{2}\alpha + 2\}.$$

6. Conclusions

We have constructed a consistent dilaton theory of gravity in two dimensions in terms of a torsion-free non-metric-compatible connection determined by a two parameter non-metricity tensor. The theory is a generalization of that discussed in [13]. The structure of the field equations has enabled us to solve for the constraint forms algebraically and a number of exact classical solutions have been made explicit. The theory has been recast in terms of the standard Levi-Civita connection, thereby exposing the nature of the dilaton interactions induced by the non-metric-compatible formulation. Remarkably most of the induced interactions in the action can be absorbed into redefinitions of the couplings. However a particular derivative dilaton coupling remains and is responsible for the detailed properties of the solutions with $k \neq 0$. The approach developed for this theory may be contrasted with the formulation of scalar–tensor theories of gravity formulated in terms of a metric-compatible connection with torsion [2]. A similar redefinition of the couplings is possible when such theories are recast in terms of the Levi-Civita connection.

Weyl was perhaps the first to consider a theory of gravity in terms of a non-metric-compatible connection. Einstein's criticism of Weyl's original theory was based on the assumption that the world lines of atoms with a constant mass move on autoparallels of the Weyl connection rather than the autoparallels of the Levi-Civita connection associated with the spacetime metric g, (i.e. geodesics). If this assumption is true then the criticism is valid for Weyl's choice of non-metricity tensor $S = A \otimes g$ where A is the Maxwell electromagnetic 1-form. However, one no longer identifies the electromagnetic interaction via the Weyl connection. Thus the physical implications of the departure of worldlines of emitting atoms from geodesics depend on the magnitude of the departure induced by the non-metricity. One of our motivations here is to give a geometrical description of interactions in string-inspired two-dimensional models in terms of a non-metric-compatible connection. We have explicitly demonstrated how such a description can be re-expressed in terms of the standard Levi-Civita connection. Such a formulation is entirely consistent with the notion of emitting atoms on geodesic worldlines rather than autoparallels of a non-metric-compatible connection. One reason for searching for a non-metric formulation is to make easier the search for interesting symmetries in the action for the theory. This philosophy is in the spirit of the elegant formulation of supergravity in terms of a connection with torsion.

We conclude that in the search for solutions and symmetries of more general theories accommodating gravitational interactions with fields, it may prove worthwhile to investigate formulations in terms of a metric and a connection that is neither torsion-free nor metric compatible. We shall report elsewhere on the implications of this approach for theories in other dimensions and involving spinors.

Acknowledgments

TD thanks the School of Physics, Lancaster University, Lancaster for hospitality and the Scientific and Technical Research Council of Turkey (TUBITAK) for partial support under project TBAG-CG1. RWT thanks the Middle East Technical University, Ankara for hospitality, the British Council and Human Capital and Mobility Programme of the European Union for partial support.

References

[1] Deser D and Zumino B 1976 *Phys. Lett.* **62B** 335
[2] Dereli T and Tucker R W 1982 *Phys. Lett.* **110B** 206
[3] Hehl F W, Lord and Smalley L L 1981 *Gen. Rel. Grav.* **13** 1037
[4] Ponomariev V N and Obukhov Y 1982 *Gen. Rel. Grav.* **14** 309
[5] McCrea J D 1992 *Class. Quantum Grav.* **9** 553
[6] Coley A A 1983 *Phys. Rev.* D **27** 728
[7] Coley A A 1983 *Phys. Rev.* D **28** 1829, 1844
[8] Coley A A 1982 *Nuov. Cimento* B **69** 89
[9] Gasperini M 1988 *Class. Quantum Grav.* **5** 521
[10] Stelmach J 1991 *Class. Quantum Grav.* **8** 897
[11] Aringazin A K and Mikhailov A L 1991 *Class. Quantum Grav.* **8** 1685
[12] Mandal G, Sengupta A M and Wadia S R 1991 *Mod. Phys. Lett.* A **6** 1685
[13] Witten E 1991 *Phys. Rev.* D **44** 314
[14] Callan C G, Jr, Giddings S B, Harvey J A and Strominger A 1992 *Phys. Rev.* D **45** R1005
[15] Mann R B, Morris M S and Ross S F 1993 *Class. Quantum Grav.* **10** 1477
[16] Benn I M and Tucker R W 1987 *An Introduction to Spinors and Geometry with Applications in Physics* (Bristol: Adam Hilger)
[17] Smalley L L 1986 *Phys. Rev.* D **33** 3590
[18] Dereli T and Tucker R W 1982 *J. Phys. A: Math. Gen.* **15** L7
[19] Dereli T and Tucker R W 1987 *Class. Quantum Grav.* **4** 791
[20] Fritsch F N, Schafer R E and Crowley W P 1973 Solution of the transcendental equation $We^W = x''$ *Commun. ACM* **16**(2)

Class. Quantum Grav. **11** (1994) 2585–2587. Printed in the UK

Perfect fluid sources in 2+1 dimensions*

Metin Gürses†

Department of Mathematics, Faculty of Science, Bilkent University, 06533 Ankara, Turkey

Received 20 December 1993, in final form 17 August 1994

Abstract. We show that a class of metrics of Einstein theory with perfect fluid sources are also the solutions of the Deser–Jackiw–Templeton (DJT) theory with a cosmological constant. Because of this analogy we interpret a recently found black hole solution of the DJT equations as a rotating fluid solution of the Einstein theory in three dimensions.

PACS numbers: 0420J, 0450, 0460K, 1125

Recently [1] it has been shown that topologically massive gravity (TMG) [2] can be obtained from three-dimensional string theory with constant dilaton and vanishing gauge fields. Hence the black hole solutions of TMG might bring some new insights to three-dimensional gravitational theories [3] and also to the string theory in three dimensions [4].

In this paper we first connect the solutions of the Einstein equations with perfect fluid source to the solutions of TMG and then interpret the recently found black hole solution [5] as a rotating fluid in three dimensions.

TMG equations with a cosmological constant are given as follows.

$$G^{\mu}_{\nu} + \frac{1}{\mu}\, \eta^{\mu\beta\alpha}\, \nabla_{\alpha}\, (R_{\nu\beta} - \tfrac{1}{4}\, R\, g_{\nu\beta}) = \lambda\, \delta^{\mu}_{\nu}. \tag{1}$$

Here $G_{\mu\nu}$ and $R_{\mu\nu}$ are the Einstein and Ricci tensors respectively, $\eta_{\mu\alpha\beta} = \sqrt{-g}\, \epsilon_{\mu\alpha\beta}$ is the Levi-Civita tensor in three dimensions, μ and λ are respectively the DJT parameter and the cosmological constant.

Einstein's equations with a perfect fluid source are given as follows

$$G_{\mu\,\nu} = T_{\mu\,\nu}, \tag{2}$$

where

$$T_{\mu\,\nu} = (p + \rho)\, u_{\mu}\, u_{\nu} + p\, g_{\mu\,\nu}, \tag{3}$$

and where the fluid equations are obtained through the conservation equation $\nabla_{\mu}\, T^{\mu\nu} = 0$, p and ρ are respectively the pressure and mass density of the fluid and u_{μ} is the fluid's timelike unit 4-velocity, i.e, $u^{\mu}\, u_{\mu} = -1$.

We have the following:

Lemma. If p, ρ are constants and

$$\nabla_{\mu}\, u_{\nu} = \frac{\mu}{3}\, \eta_{\mu\nu\alpha}\, u^{\alpha} \tag{4}$$

* This work was partially supported by the Scientific and Technical Research Council of Turkey (TUBITAK).

† E-mail address: gurses@fen.bilkent.edu.tr

then any solution of the Einstein equations $G_{\mu\nu} = T_{\mu\nu}$ is also a solution of the TMG (1) with a cosmological constant $\lambda = (2p - \rho)/3$.

Proof of the lemma is straightforward. The Einstein and Ricci tensors in (1) are first expressed in terms of the perfect fluid energy–momentum tensor through (2) and (3). Then DJT equations (1) reduce to coupled equations for p, ρ and u^{μ}. These equations are identically satisfied for the conditions given by the lemma. Here we assumed that the 4-velocity vector is timelike. One has a similar lemma for the null case.

As an application of the above lemma let us consider the following spacetime metric

$$ds^2 = -\phi\, dt^2 + 2q\, dt\, d\theta + \frac{-q^2 + h^2 \psi}{\phi}\, d\theta^2 + \frac{1}{\psi}\, dr^2, \tag{5}$$

where

$$\phi = a_0 \qquad \psi = b_0 + \frac{b_1}{r^2} + \frac{3\lambda_0}{4} r^2$$

$$q = c_0 + \frac{e_0 \mu}{3} r^2 \qquad h = e_0 r \qquad \lambda_0 = \lambda + \frac{\mu^2}{27}$$

and where a_0, b_0, b_1, c_0 and e_0 are arbitrary constants. This metric solves the Einstein equations (2) with the following fluid parameters

$$p = \frac{\mu^2}{9} \qquad \rho = \frac{\mu^2 - 9\lambda_0}{3}$$

$$u_\mu = \sqrt{a_0}(1, 0, \Omega) \qquad \Omega = -\frac{\mu e_0 r^2 + 3c_0}{3a_0}.$$

To have a physically reasonable solution we must have $\mu^2 > 9\lambda_0$ or $\lambda > \frac{2}{27}\mu^2$. When $p + \rho = 0$ the solution reduces to the Banados–Teitelboim–Zanelli (BTZ) metric. This limit is $\mu^2 = 9\lambda$. After performing the transformation $\theta = -\theta' + \omega t$,the parameters a_0, b_0, b_1, c_0, e_0 and ω can be arranged in such a way that the above metric (5) becomes

$$\begin{aligned} ds^2 = &-\left[-M + \frac{\mu^2}{9} r^2 + \frac{\mu^2}{18} \frac{\mu^2 - 9\lambda}{(2J\mu - 6M)} r^4\right] dt^2 \\ &+ \left(J - \frac{\mu}{3} \frac{\mu^2 - 9\lambda}{2J\mu - 6M} r^4\right) dt\, d\theta + r^2 \left[1 - \frac{\mu^2 - 9\lambda}{2(2J\mu - 6M)} r^2\right] d\theta^2 \\ &+ \frac{dr^2}{-M + (J^2/4r^2) + [(\mu^2 + 27\lambda) r^2/36]} \end{aligned} \tag{6}$$

with $-\infty < t < \infty$, $0 < r < \infty$, and $0 \leqslant \theta \leqslant 2\pi$. Here the constants are chosen as

$$\begin{aligned} &a_0 = (J\mu - 3M)/3, \qquad b_0 = -M, \qquad c_0 = -J/2, \\ &e_0 = 1, \qquad b_1 = J^2/4, \qquad \omega = -\mu/3. \end{aligned} \tag{7}$$

The metric given in equation (6) describes a black hole with horizons r_+ and r_- defined through the following relations

$$M = \frac{r_+^2 + r_-^2}{l^2}, \qquad J = \frac{2r_+ r_-}{l} \tag{8}$$

where

$$\frac{1}{l^2} = \frac{\mu^2 + 27\lambda}{36}$$

and $r_\pm$ are given by

$$r_\pm^2 = \tfrac{1}{2} l\,(M\,l \pm \sqrt{M^2\,l^2 - J^2}). \tag{9}$$

Here we assume $\lambda_0 \neq 0$ ($l \neq \infty$) and $M^2\,l^2 - J^2 \geqslant 0$. Two horizons coincide when $J^2 = M^2\,l^2$, which is the extremal limit. There is also another extremal limit which corresponds to $l \to \infty$ or $\lambda_0 = 0$. In this case the horizon is located at $r^2 = J^2/4M$. The curvature invariants $R = -6\,\lambda$ and $R_{\mu\nu}\,R^{\mu\nu} = \frac{2}{27}\,(\mu^4 - 18\,\mu^2\lambda + 243\lambda^2)$ are regular everywhere. Further properties of the solution (6) will be discussed elsewhere.

Fluid parameters p and ρ remain the same but the velocity 4-vector becomes

$$u_\mu = \sqrt{a}_0(1 + \tfrac{1}{3}\mu\Omega', 0, \Omega') \tag{10}$$

where Ω' becomes

$$\Omega' = \frac{2\,\mu\,r^2 - 3\,J}{2\mu\,J - 6\,M}$$

and a_0 is given in (7). In the limit $\mu^2 = 9\,\lambda$ the above metric reduces exactly to the form given by BTZ. In that form M and J are respectively interpreted as the mass and angular momentum of the solution.

References

[1] Kaloper N 1993 Topological mass generation in three dimensional string theory *Preprint* Alberta-Thy-46-93, September 1993 (to appear in *Phys. Lett.* **B**)

[2] Deser S, Jackiw R and Templeton S 1982 *Phys. Rev. Lett.* **48** 975; 1982 *Ann.Phys., NY* **140** 372

[3] Banados M, Teitelboim C and Zanelli J 1992 *Phys. Rev. Lett.* **69** 1849

[4] Horowitz G and Welch D L 1993 *Phys. Rev. Lett.* **71** 328

[5] Gürses M 1993 Black hole solutions of DJT equations with cosmological constant (unpublished). The solution given in equation (5) has been utilized by Y Nutku in his recent publication 1993 *Class. Quantum. Grav.* **10** 2657

Class. Quantum Grav. **11** (1994) 2589–2595. Printed in the UK

Primordial gravitational waves and multidimensional theories

Marek Biesiada
Copernicus Astronomical Centre, Bartycka 18, 00-716 Warsaw, Poland

Received 10 February 1994, in final form 16 August 1994

Abstract. The idea that our world has more than four dimensions remains attractive from the point of view of theories unifying gravity with other fundamental interactions. However, it is extremely difficult to verify multidimensional theories experimentally. One of the potential possibilities of such verification comes from investigating cosmological consequences of higher dimensions.

In this paper we discuss the creation of primordial gravitons (in physical space) induced by the dynamics of a multidimensional world. The graviton spectrum is discussed for several cosmological models in which dynamical dimensional reduction takes place. It turns out that perspectives for observing gravitons dynamically created by extra dimensions are poor (at least for the classes of models discussed).

PACS numbers: 0440, 0450

1. Introduction.

Recent attempts to unify all elementary interactions often invoke the concept of a multidimensional world. There are actually three classes of such theories. One of them contains extensions of Kaluza–Klein theory which was originally designed to unify gravity and electromagnetic interactions. The renewed interest in Kaluza–Klein theories stems from the fact that multidimensional general relativity can generate non-Abelian gauge theories out of the symmetries of compact internal space (for an introduction see for example [1]). The heterotic superstring theories describing bosonic and fermionic fields in ten-dimensional spacetime (see [2] for details) and supersymmetric theories are examples of the remaining two approaches to constructing a unified theory of elementary interactions.

Up to now there is no experimental evidence of the multidimensionality of our world. However, we can speculate about dynamical effects of higher dimensions in the early universe, such as inflation induced by higher dimensions, or contemplate the evolution of the Ω parameter in multidimensional cosmologies [3].

Accepting the point of view that higher dimensions are physical reality, one faces the problem of dimensional reduction. The essence of this problem is to provide the mechanism that leads to compactification of the microspace and emergence of physical four-dimensional spacetime. Several mechanisms of dimensional reduction, including the Freund–Rubin mechanism [4], thc Casimir effect [5], or higher-derivative terms in gravitational action [6] have been proposed, but one of the most elegant was that of Chodos and Detweiler [7] seeking a natural mechanism for compactification in the dynamics of the multidimensional world itself.

While considering small perturbations about a homogeneous background metric one can decompose these perturbations into scalar, vector and tensor harmonics. The gravitons

0264-9381/94/102589+07$19.50

described by (transverse and traceless) tensor harmonics are decoupled from the scalar and vector harmonics, and as such they depend only on the behaviour of the background. This property points towards gravitons as good candidates for investigating dynamical properties of the early universe in general, and in the context of multidimensional theories in particular. It is well known that the transition from the inflationary phase to the decelerated one is capable of producing the background of relic gravitons [8] which can be used as a diagnostic tool to reconstruct the early history of the Hubble parameter [9]. The idea that the process of compactification can analogously contribute to the stochastic background of primordial gravitons was formulated in [10]. The toy model employed there was a kind of artificial construction formulated in the hope of capturing the essentials of dimensional reduction. A much more realistic situation was considered in [11], where the authors calculated the spectrum of cosmic gravitons in a higher-dimensional Brans–Dicke model in which the (power law) compactification phase is followed by subsequent transitions to radiation and matter dominated regimes. They found that the resulting spectra may be decreasing, flat or increasing, depending on actual values of power law exponents. However, their parameter space was artificially expanded, since they had not considered constraints imposed upon the power law exponents by multidimensional field equations. This was actually the price they paid for being model independent.

An alternative approach to the problem of primordial gravitons has been presented by Demiański, Polnarev and Naselsky [12]. They considered the (phenomenological) multidimensional cosmological model described in [13] and investigated graviton production in this particular model.

The main objective of the present paper is to examine the contribution of dynamical dimensional reduction to the production of background cosmic gravitons in several models known from the literature, in which the dimensional reduction itself is a dynamical effect of multidimensional Einstein equations.

2. Brief theory of primordial graviton production

Throughout the paper we shall assume the product structure of higher-dimensional spacetime $\mathcal{M}^{n+4} = \mathcal{M}^4 \times \mathcal{B}^n$ equipped with the metric:

$$g_{AB} = \mathrm{diag}\left(1, -R(t)^2, -R(t)^2, -R(t)^2, -r(t)^2, \ldots, -r(t)^2\right) \tag{2.1}$$

where $\mathcal{M}^4$ is our familiar physical space and $\mathcal{B}^n$ is compact microspace.

The problem of graviton production is usually treated by starting from the linearized wave equation

$$\frac{1}{\sqrt{|g|}} \partial_C \left(\sqrt{|g|}\, g^{CD} \partial_D\right) h_{AB} = 0 \tag{2.2}$$

for a gravitational perturbation h_{AB} propagating freely in the background field (2.1). In this paper, as well as is implicitly done in [10–12] we adopt the point of view that the behaviour of macrospace components $h_{\mu\nu}$ is exactly what we are about. This assumption is non-trivial in the light of conceptual problems with multidimensional objects moving along null geodesics discussed in [14]. Regardless of this, once we focus on $h_{\mu\nu}$ it is a standard procedure to switch into conformal time η such that $\mathrm{d}t = R(\eta)\,\mathrm{d}\eta$. Each polarization component $h(\eta)$ can be Fourier analysed $h(\eta, \boldsymbol{x}) = \sum_k h_k(\eta)\, e^{i\boldsymbol{k}\boldsymbol{x}}$ and individual modes can be parametrized in the following way:

$$h_k(\eta) = \frac{\mu_k(\eta)}{R(\eta)\, r(\eta)^{n/2}} \tag{2.3}$$

i.e. physical degrees of freedom are disentangled from scale factors. The time dependence of $h_k(\eta)$ is therefore encoded in $\mu_k(\eta)$ which satisfy the equation:

$$\mu_k'' + \left[k^2 - V(\eta)\right]\mu_k = 0. \tag{2.4}$$

This can be recognized as an equation of a parametrically excited oscillator. In our case

$$V(\eta) = \frac{R''}{R} + n\,\frac{R'r'}{Rr} + \tfrac{1}{2}n\,\frac{r''}{r} + \tfrac{1}{4}n(n-2)\left(\frac{r'}{r}\right)^2 \tag{2.5}$$

where, as well as in (2.4), the prime denotes differentiation with respect to η.

All the model solutions which we shall refer to later exhibit the power law behaviour:

$$R(t) \sim t^{\alpha} \qquad r(t) \sim t^{\beta} \tag{2.6}$$

and we shall assume that the compactification process is followed by transition (at $\eta = -\eta_0$) to the standard radiation dominated regime where the internal scale factors are frozen. This is required by consistency with the known limits on the time variation of the fundamental couplings which imply that the internal space must have been constant, or very nearly so, since nucleosynthesis [15]. The effective potential (2.5) for the power law behaviour of the scale factors (2.6) simplifies to

$$V_{\mathrm{pl}}(\eta) = \frac{1}{4\eta^2}\frac{1}{(1-\alpha)^2}\left[4\alpha(2\alpha-1) + 2n\beta(\alpha-1) + n^2\beta^2 + 4n\alpha\beta\right]. \tag{2.7}$$

Equation (2.4) has the general 'out' solution

$$\mu_k = \frac{1}{\sqrt{k}}\left(A_k\, e^{-ik\eta} + B_k\, e^{ik\eta}\right) \tag{2.8}$$

which should match continously the 'in' solution given (in the case of potential (2.7)) by second order Hankel functions†

$$\mu_k = \sqrt{\eta}\, H_\nu^{(2)}(k|\eta|)C \tag{2.9}$$

where C is the normalizing constant and

$$\nu = \frac{|3\alpha + n\beta - 1|}{2|1-\alpha|}. \tag{2.10}$$

It is known that in the process of tunnelling through the barrier (2.5) the initial vacuum state of gravitons goes over to the final multiparticle state. The Bogoliubov transformation connecting 'in' and 'out' states is determined by $c_\pm$ coefficients which in turn can be

† The general solution is a linear combination of $H_\nu^{(1)}$ and $H_\nu^{(2)}$. Our choice is dictated by the limit behaviour $\sqrt{\eta}\, H_\nu^{(2)}(k|\eta|) \to e^{-ik\eta}/\sqrt{k}$ when $\eta \to -\infty$.

calculated from junction conditions for the solutions (2.9) and (2.8). In our case (see also [10, 11]): $|c_-(k)| \sim (k\eta_0)^{-|\nu|-1/2}$. The energy density of gravitons produced during the transition from the compactification to the radiation dominated epoch with frequencies between ω and $\omega + \mathrm{d}\omega$ can be estimated in the following way: two polarization states of gravitons each of energy $\hbar\omega$ times the density of states times the expectation number of gravitons in mode k (equal to $|c_-|^2$):

$$\mathrm{d}\varepsilon(\omega) = 2\hbar\omega \frac{\omega^2}{2\pi^2 c^3} |c_-|^2 \,\mathrm{d}\omega = \frac{\hbar}{\pi^2 c^3} \omega^3 |c_-|^2 \,\mathrm{d}\omega.$$

Thus the spectral density $\varepsilon = \omega(\mathrm{d}\varepsilon/\mathrm{d}\omega)$ may be written in the form:

$$\varepsilon(\omega) = \frac{\hbar}{\pi^2 c^3} \omega^{3-2|\nu|} (R(\eta)\eta_0)^{-1-2|\nu|}. \tag{2.11}$$

3. Models of dynamical dimensional reduction

We shall now concentrate on some multidimensional model solutions which exhibit the power law behaviour of the scale factors. Our toy models comprise:

(i) dynamical dimensional reduction of vacuum multidimensional cosmologies (in the spirit of Chodos and Detweiler [7]);

and power law solutions arising from the bosonic sector of $N = 1$, $D = 10$ supergravity [16] with

(ii) zero Kalb–Ramond fields F_{ABC} and non-zero dilaton field ϕ;

(iii) both ϕ and F_{ABC} being functions of time.

Power law solutions (2.6) of the case (i) must satisfy multidimensional Einstein equations

$$R_{AB} - \tfrac{1}{2} R g_{AB} = 0$$

hence they are necessarily of Kasner type, i.e. α and β satisfy the following conditions:

$$3\alpha + n\beta = 1 \qquad 3\alpha^2 + n\beta^2 = 1. \tag{3.1}$$

Therefore we obtain that (cf e.g. [17])

$$\alpha = \frac{1}{n+3}\left[1 + \sqrt{\tfrac{1}{3}n(n+2)}\right] \qquad \beta = \frac{1}{n+3}\left[1 - \sqrt{\frac{3(n+2)}{n}}\right]. \tag{3.2}$$

The plus and minus signs in α and β have been chosen so as to guarantee the dimensional reduction (expanding macrospace and shrinking microspace).

In the cases (ii) and (iii) discussed by Gleiser and Stein-Schabes [16] in the context of $N = 1$, $D = 10$ supergravity, the field equations are much more complicated:

$$R_{AB} - \tfrac{1}{2} R g_{AB} = -\, T_{AB}$$

where

$$T_{AB} = -\,\partial_A\phi\partial_B\phi + \tfrac{1}{2}(\partial\phi)^2 g_{AB} - \mathrm{e}^{\phi}(F_{ACD}F_B^{CD} - \tfrac{1}{6}F_{CDE}F^{CDE} g_{AB})$$

$$\partial_A \left[\sqrt{g^{(10)}}\, \mathrm{e}^{\phi} F^{ABC}\right] = 0 \tag{3.3}$$

$$\frac{1}{\sqrt{g^{(10)}}} \partial_A\phi \left(\sqrt{g^{(10)}}\, \partial^A\right) = \tfrac{1}{6}\, \mathrm{e}^{\phi} F_{ABC}F^{ABC}.$$

The Einstein equations can be solved only after imposing a restrictive ansatz on the F_{ABC} and ϕ fields.

In case (ii) it is assumed that $F_{ABC} = 0$ and $\phi = \phi(t)$. Then the solution compatible with the dimensional reduction is [16]

$$R \sim t^{2/3} \qquad r \sim t^{-1/6} \qquad \phi \sim \ln t. \tag{3.4}$$

Case (iii) is based on the assumption that $F_{ijk} = \epsilon_{ijk}\sqrt{g^{(3)}}\, F(t)$, $(i, j, k) \in \{1, 2, 3\}$, $F_{mnp} = F_{\mu\nu p} = F_{\mu np} = 0$, $(m, n, p) \in \{4, ..., 10\}$, $\mu \in \{0, 1, 2, 3\}$. Then the behaviour of the scale factors is

$$R \sim t^{3/7} \qquad r \sim t^{-1/7} \qquad \phi \sim \tfrac{4}{7} \ln t. \tag{3.5}$$

In the presence of a dilaton field ϕ the effective potential (2.5) is modified to (see also [11])

$$\widehat{V}(\eta) = V(\eta) - \tfrac{1}{2}\phi'' + \tfrac{1}{4}\phi'^2 - \frac{R'}{R}\phi' - \tfrac{1}{2}n\frac{r'}{r}\phi'$$

and consequently

$$\widehat{V}_{\rm pl}(\eta) = V_{\rm pl}(\eta) + \frac{\gamma}{4\eta^2(1-\alpha)^2}[\gamma - 2(n\beta + 3\alpha) + 2].$$

Hence

$$\nu = \frac{|3\alpha + n\beta + \gamma - 1|}{2|1-\alpha|}.$$

Substituting the actual values of α, β and γ we see that $\nu = 0$ for case (i) and (iii) and $\nu = \frac{3}{2}$ for case (ii). Note that in the case (i) $\nu = 0$ is a consequence of one of the Kasner constraints (3.1).

Finally the spectrum of primordial gravitons is

$$\varepsilon(\omega) \sim \omega^3 \tag{3.6}$$

for cases (i) and (iii) and $\varepsilon(\omega) \sim \omega^0$ i.e. flat Zeldovich–Harrison spectrum for case (ii).

In order to contemplate the observability of primordial gravitons from dimensional reduction it is convenient to parametrize the spectrum with an $\Omega_{\rm grav} = \varepsilon(\omega)/\varepsilon_{\rm cr}$ parameter which is the fraction of critical energy $\varepsilon_{\rm cr} = 3H^2/8\pi G$ needed to close the Universe contained in gravitons. Then the following formula is valid [11]:

$$\Omega_{\rm grav} \simeq GH_0^2\Omega_\gamma \left(\frac{\omega}{\omega_0}\right)^{3-2|\nu|} = \left(\frac{H_0}{M_{\rm pl}}\right)^2 \Omega_\gamma \left(\frac{\omega}{\omega_0}\right)^{3-2|\nu|}$$

where $\Omega_\gamma \sim 10^{-4}$ is the present fraction of closure energy density contained in the background radiation, H_0 and ω_0 denote the Hubble constant and the (present) frequency of modes entering the barrier at the moment of transition from the compactification to the radiation dominated epoch, $M_{\rm pl}$ is the Planck mass. The frequency ω_0 can be estimated (see [11]) as $\omega_0 \sim 10^{29}(H_0/M_{\rm pl})^{1/2}\omega_{\rm H} \sim 10^{11}$ Hz where $\omega_{\rm H} \sim 10^{-18}$ Hz is the frequency of modes currently entering the horizon. Hence we can estimate for the Kasner epoch

$$\Omega_{\rm grav} \sim 10^{-37}\left(\frac{H_0}{M_{\rm pl}}\right)^{1/2}\left(\frac{\omega}{1Hz}\right)^3 .$$

This is far beyond the possibilities of any gravity wave experiment planned in the foreseeable future. Indeed even under a secure assumption that $H_0 < 10^2 M_{\rm pl}$ we have $\Omega_{\rm grav} < 10^{-36}\omega^3$ and it is extremely small compared with the LIGO peak sensitivity of $\Omega \sim 10^{-11}$ at 10 Hz [22] or the peak sensitivity of $\Omega \sim 10^{-16}$ at 10^{-4} Hz expected for beam-in-space experiments [23] planned for the future. This demonstrates that the prospects for observational tests of multidimensional theories by means of primordial gravitons are poor, at least, in some simple toy models employed here to guide our intuition. A similar conclusion has been reached by Demiański *et al* [12] for their model.

4. Conclusions

We have investigated primordial gravitons in the physical sector of three types of multidimensional cosmological models. Our results are rather discouraging for seeking the multidimensionality of the Universe imprinted on gravitons. However, it should be stressed that the power law behaviour of scale factors considered in the present paper is by no means a unique scenario for dynamical dimensional reduction. One of alternative models in which the isotropic internal sector is not reached monotonically, but exhibits never-ending small oscillations around the isotropic state, is given in [18]. Power law solutions (being simple and illustrative) are, however, very good to start with when one tackles the question of today's observational consequences of the early multidimensional universe.

The original idea of Maeda [19] of viewing the dimensional reduction as an attractor in multidimensional phase space initiated a series of papers [3, 20] in which the methods of dynamical systems were applied. This approach made it possible to discuss the dynamical properties and asymptotic states of multidimensional universes on the respective phase portraits without actually knowing explicit solutions of the field equations. The extensive studies of higher-dimensional cosmological models performed in the last decade revealed the importance (and universality) of Kasner solutions for the description of various asymptotic states of a multidimensional universe [20, 21]. Therefore the growing spectrum $\varepsilon(\omega) \sim \omega^3$ of primordial gravitons is also, in a sense, 'universal', and should leave an imprint on the overall spectrum at least in some frequency range.

The (potential) possibility that observational determination of the spectrum of primordial gravitational waves could reveal information about the epoch of dimensional reduction is extremely fascinating. Therefore despite the discouraging results of the simplest estimates it is very important to broaden our experience with this aspect of multidimensional theories, which still remain serious candidates for the ultimate dream of theoretical physicists—the unification of all interactions.

Acknowledgments

This project was supported by the Foundation For Polish Science and by the KBN Grant 2 P304 022 06.

References

[1] Emel'yanov V M, Nikitin Yu P, Rozental I L and Berkov A V 1986 *Phys.Rep.* **143** 1–68

[2] Green M B, Schwarz J H and Witten E 1987 *Superstring Theory* (Cambridge: Cambridge University Press)

[3] Szydłowski M and Biesiada M 1990 *Phys. Rev.* D **41** 2487
Biesiada M and Szydłowski M 1990 *Phys. Lett.* **145A** 401
Szydłowski M, Heller M and Łapeta B 1993 *Phys. Rev.* D **47** 460–7

[4] Freund P G O and Rubin M A 1980 *Phys. Lett.* **97B** 233

[5] Candelas P and Weinberg S 1984 *Nucl.Phys.* B **237** 397

[6] Shafi Q and Wetterich C 1983 *Phys. Lett.* **129B** 387

[7] Chodos A. and Detweiler S 1980 *Phys. Rev.* D **21** 2167

[8] Grishchuk L P 1988 *Usp. Fiz. Nauk* **156** 297; 1977 *Ann. NY Acad. Sci.* **302** 439; 1975 *Sov. Phys.–JETP* **40** 409

[9] Grishchuk L P and Solokhin M 1991 *Phys. Rev.* D **43** 2566

[10] Gasperini M and Giovannini M 1992 *Class. Quantum Grav.* **9** L137

[11] Gasperini M and Giovannini M 1993 *Phys. Rev.* D **47** 1519

[12] Demiański M, Polnarev A G and Naselsky P 1993 *Phys. Rev.* D **47** 5275–9

[13] Belobrodov A, Demiański M, Ivanov P and Polnarev A G 1993 *Phys. Rev.* D **48** 503–12

[14] Biesiada M, Szydłowski M and Szczęsny J 1988 *Phys. Lett.* **211B** 42

[15] Kolb E W, Perry M J and Walker T P 1986 *Phys. Rev.* D **33** 869

[16] Gleiser M and Stein-Schabes J A 1986 *Phys. Rev.* D **34** 1739

[17] Lorenz-Petzold D 1986 *Phys. Lett.* **167B** 157
Biesiada M, Szydłowski M and Szczęsny J 1988 *Acta Phys. Pol.* B **19** 3

[18] Stawicki T, Szydłowski M and Heller M 1993 *Phys. Lett.* **173A** 247

[19] Maeda K 1986 *Class. Quantum Grav.* **3** 233; *Phys. Lett.* **166B** 59

[20] Wiltshire D L 1987 *Phys. Rev.* D **36** 1634
Szydłowski M 1988 *Gen. Rel. Grav.* **20** 221
Szydłowski M and Piekut-Łapeta B 1989 *Gen. Rel. Grav.* **21** 349, 383
Demiański M and Polnarev A 1990 *Phys. Rev.* D **41** 3003

[21] Halpern P 1984 *Phys. Rev.* D **33** 354
Demaret J, Henneaux M and Spindel P 1985 *Phys. Lett.* **164B** 27
Ishihara H 1985 *Prog. Theor. Phys.* **74** 490
Szydłowski M, Biesiada M and Szczęsny J 1987 *Phys. Lett.* **195B** 27
Szydłowski M, Szczęsny J and Biesiada M 1987 *Class. Quantum Grav.* **4** 1731

[22] Christensen N 1992 *Phys. Rev.* D **46** 5250

[23] Thorne K S 1988 *300 Years of Gravitation* ed S W Hawking and W Israel (Cambridge: Cambridge University Press)

Class. Quantum Grav. **11** (1994) 2597–2613. Printed in the UK

Conformal transformations of the Wigner function and solutions of the quantum corrected Vlasov equation

Oleg A Fonarev
Racah Institute of Physics, The Hebrew University, Jerusalem 91904, Israel

Received 18 February 1994, in final form 29 July 1994

Abstract. We study conformal properties of the quantum kinetic equations in curved spacetime. A transformation law for the covariant Wigner function under conformal transformations of a spacetime is derived by using the formalism of tangent bundles. The conformal invariance of the quantum corrected Vlasov equation is proved. This provides a basis for generating new solutions of the quantum kinetic equations in the presence of gravitational and other external fields. We use our method to find explicit quantum corrections to the class of locally isotropic distributions, to which equilibrium distributions belong. We show that the quantum corrected stress–energy tensor for such distributions has, in general, a non-equilibrium structure. Local thermal equilibrium is possible in quantum systems only if an underlying spacetime is conformally static (not stationary). Possible applications of our results are discussed.

PACS numbers: 0530, 0520D, 0490

1. Introduction

In a regime where the curvature of a spacetime is less than the Planck scale, a semiclassical approximation to the quantum gravity is considered to be sufficient [1, 2]. This approximation is valid for 'nearly' classical systems, for which quantum fluctuations in the stress–energy tensor are small enough [3]. In such cases the method of Wigner functions [4–6] is particularly convenient. It allows one to derive, from quantum field equations, quantum kinetic equations for Wigner functions. Iterative solutions of the kinetic equations give quantum corrections to classical distribution functions, which are used to compute expectation values of local observables such as the stress–energy tensor, the number-flux vector, etc. The standard definition of the flat-space Wigner function exploits the Wigner–Weyl transformation [5] which fails in curved spacetime. Different approaches have been proposed to extend the notion of the Wigner–Weyl transformation to curved spacetime. In [8] a Wigner–Weyl type transformation is defined by considering the bundle of geodesic lines on a manifold. [9] utilizes Riemann normal coordinate systems and adiabatic expansions similar to one used by [12] for studying ultraviolet properties of Green functions in curved spacetime. In [10], a covariant Wigner function is defined by using the formalism of tangent bundles. All three approaches mentioned, being intrinsically different, are locally equivalent to each other [11], in the sense of adiabatic expansions, and yield the same quantum corrected Vlasov type of kinetic equations. The mathematical structure of the equations is rather complicated, even at lowest adiabatic order. This makes difficult the study of properties of generic solutions of the quantum kinetic equations in an arbitrary background spacetime. No doubt any explicit computation of the Wigner function may be useful for better understanding of quantum kinetic processes in gravitating systems.

0264-9381/94/102597+17$19.50

If a spacetime manifold is endowed with some symmetries then both classical and quantum kinetic equations can often be solved explicitly. Up to now, only in a very few cases have explicit solutions of the quantum corrected Vlasov equation in curved spacetime been obtained. They include the Friedmann–Robertson–Walker cosmology [13, 14], stationary spacetimes [15] and some others [16]. A common property of those solutions is that the quantum corrections are local in phase space. Once a solution of the Vlasov equation (a classical distribution function) is known, one can compute, to a given adiabatic order, the Wigner function which corresponds to this solution, by applying a certain finite order differential operator to the classical distribution function. This property implies that one can express the expectation value of the stress–energy tensor or of any other local observable in terms of moments of the classical distribution function multiplied by some local geometrical quantities. This facilitates to a great extent the analysis of the back-reaction problem in those cases.

The natural question arises of how symmetries of a spacetime manifest themselves in Wigner functions. In the present paper, we consider a particular, but important, type of geometrical symmetry, the conformal symmetry. For simplicity, attention will be restricted to the case of a scalar field; spin-$\frac{1}{2}$ and spin-1 fields can be treated similarly [11, 16]. The conformal symmetry is usually understood as the invariance of the generalized massless Klein–Gordon equation (with the term $\frac{1}{6}R\varphi$ involved [1]) under the conformal rescaling of the metric and of the scalar field: $g_{\alpha\beta} \to a^2 g_{\alpha\beta}$, $\varphi \to \varphi/a$. If the scalar field is coupled to an external potential, the latter must also be transformed in such a way that the scalar field equation remains invariant under this extended group of conformal transformations.

Two different views at the conformal symmetry are possible. When the external potential has a direct physical meaning, the conformal symmetry is a *dynamical* symmetry of the physical system under consideration. This is the case, for example, in the φ^4 theory where the expectation value $\langle\varphi^3\rangle$ plays the role of an external potential for quantum perturbations. The second point of view is formal. One can treat, for example, the mass of the scalar field as an external potential and use the conformal symmetry as a convenient tool for generating new solutions of the field equation [17].

In this paper we study transformation properties of the covariant Wigner function under the conformal transformation. As we have mentioned, different curved space extensions of the flat space Wigner function are possible, all of them giving similar adiabatic expansions. The use of the tangent bundle technique is particularly convenient for exploring mathematical properties of the Wigner function. By using the formalism of tangent bundles we derive, to second adiabatic order, the transformation law for the Wigner function and prove the conformal invariance of the quantum-corrected Vlasov equation. We then use our result to find explicit quantum corrections to the class of distribution functions which are locally isotropic in momentum space, i.e. have the form $F_{\rm cl}(x, p) = F(x, u^\alpha(x)p_\alpha)$. As was shown by [18, 19] (see also [20–22]), in the case where external fields other than gravitation are absent, such a form of the distribution function imposes the following restriction on the underlying spacetime: it must be either stationary or conformally static (with a special form of the conformal factor). In the presence of an external potential this condition has to be modified. We do this by using the conformal properties of the Vlasov equation. We also analyse the structure of the quantum corrected stress–energy tensor and number-flux vector, and show that in general the quantum corrections do not preserve perfect fluid structure, even if the classical distribution implies local thermal equilibrium.

The paper is organized as follows. In section 2 we outline the formalism of the Wigner function in curved spacetime. Section 3 derives the transformation law for the Wigner function. The lowest order quantum corrections to the isotropic distributions are evaluated in section 4, and the quantum corrected stress–energy tensor and number-flux vector are calculated in section 5. Section 6 contains some remarks concerning possible applications of our results. In appendix A technical details used for the evaluation of the transformation law are presented, and in appendix B, the conformal invariance of the quantum corrected Vlasov equation is proved. Appendix C recalls some properties of the stress–energy tensor.

Conventions used throughout are the following. Greek indices run from 0 to 3, Latin indices from 1 to 3. The signature for the metric tensor is $(+\,-\,-\,-)$. The Riemann tensor $R^{\mu}_{\ \nu\alpha\beta}$ is defined by

$$\left[\nabla_\alpha, \nabla_\beta\right] X^\mu = R^{\mu}_{\ \nu\alpha\beta}\, X^\nu .$$

The Ricci tensor is $R_{\alpha\beta} = R^{\nu}_{\ \alpha\nu\beta}$, the scalar curvature $R = R^{\alpha}_{\ \alpha}$.

2. The covariant Wigner function

To keep the paper self contained, we shall outline our definition and some properties of the covariant Wigner function in curved spacetime [10, 11]. Let $\varphi(x)$ be a scalar field on a spacetime manifold $\mathcal{M}$ with the metric $g_{\alpha\beta}(x)$, and $\mathcal{T}_x(\mathcal{M})$ be the tangent space at a point x. The horizontal lift of the covariant derivative operator ∇_α to the tangent bundle $\mathcal{T}(\mathcal{M})$ is defined by [23]

$$\widehat{\nabla}_\alpha = \nabla_\alpha - \Gamma^{\beta}_{\alpha\nu}\, y^\nu \frac{\partial}{\partial y^\beta} \tag{2.1}$$

where y^α are components of a tangent vector pertaining to $\mathcal{T}_x(\mathcal{M})$, and $\Gamma^{\beta}_{\alpha\nu}(x)$ are the Christoffel symbols. Let us introduce the horizontal lift of the scalar field:

$$\Phi(x, y) = \exp(y^\alpha \widehat{\nabla}_\alpha)\, \varphi(x). \tag{2.2}$$

Then, the covariant Wigner function is defined as follows:

$$f(x, p) = (\pi\hbar)^{-4} \sqrt{-g(x)} \int_{\mathcal{T}_x(\mathcal{M})} \mathrm{d}^4 y\, \mathrm{e}^{-2\mathrm{i} y^\alpha p_\alpha/\hbar} \langle \Phi(x, -y)\Phi^\dagger(x, y)\rangle \tag{2.3}$$

where the bracket stands for an ensemble averaging with some density matrix defined on a Cauchy hypersurface [7], the dagger indicates the Hermitian conjugation.

If the field $\varphi(x)$ is a solution of the generalized Klein–Gordon equation:

$$\left(\hbar^2\, \nabla_\alpha \nabla^\alpha - \tfrac{1}{6}\hbar^2\, R(x) + V(x)\right) \varphi(x) = 0\,, \tag{2.4}$$

then the Wigner function obeys the quantum corrected Liouville–Vlasov equation supplemented by the quantum corrected mass shell constraint [8, 9, 11]. The explicit form of the equations to second adiabatic order is given in appendix B. We note here that a Wigner function satisfying these equations (to second adiabatic order) is represented as the series in derivatives of the Dirac δ-function [9, 11]:

$$\begin{aligned} f(x, p) = {} & \left(F_{\mathrm{cl}}(x, p) + \hbar^2 F_{\mathrm{qu}}(x, p)\right) \delta(\Omega) + \hbar^2 F_1(x, p)\, \delta'(\Omega) + \hbar^2 F_2(x, p)\, \delta''(\Omega) \\ & + \hbar^2 F_3(x, p)\, \delta'''(\Omega) \end{aligned} \tag{2.5}$$

with the argument of the δ-function being

$$\Omega = g^{\alpha\beta}(x) p_\alpha p_\beta - V(x). \tag{2.6}$$

The functions $F_n(x, p)$, $n = 1, 2, 3$, are expressed in terms of the classical distribution function $F_{\rm cl}(x, p)$, in such a way that the quantum corrected mass shell constraint is satisfied (see appendix B), assuming that $F_{\rm cl}(x, p)$ obeys the Vlasov equation:

$$\delta(\Omega)\left(\widehat{\mathcal{L}} F_{\rm cl}(x, p)\right) = 0. \tag{2.7}$$

Here $\widehat{\mathcal{L}}$ is the Liouville–Vlasov operator [24, 25, 27]:

$$\widehat{\mathcal{L}} = p^\alpha D_\alpha + \tfrac{1}{2} V_{,\alpha} \frac{\partial}{\partial p_\alpha} \tag{2.8}$$

D_α being the horizontal lift of the covariant derivative operator to the cotangent bundle [23]:

$$D_\alpha = \nabla_\alpha + \Gamma^\mu_{\alpha\beta}\, p_\mu \frac{\partial}{\partial p_\beta}\,. \tag{2.9}$$

The function $F_{\rm qu}(x, p)$ is found by integrating the quantum transport equation (see appendix B) along classical trajectories [9, 11].

Equation (2.7) defines the classical distribution function only on the mass shell $\Omega = 0$. This is sufficient so far as one is interested in classical observables defined on the mass shell [25]. In quantum kinetic theory the situation is different [11]. Though the function $F_{\rm cl}(x, p)$ is tied up with the δ-function in the expansion (2.5), its properties out of the mass shell influence upon both the off-shell quantum corrections, the F_n, and the on-shell quantum corrections, $F_{\rm qu}(x, p)$. The latter becomes evident if we get rid of the δ-function in equation (2.7). The function $F_{\rm cl}(x, p)$ must satisfy the generalized Vlasov equation:

$$\widehat{\mathcal{L}}\, F_{\rm cl}(x, p) = \Omega\, \Delta_F(x, p) \tag{2.10}$$

Δ_F being an arbitrary function which is assumed to be non-singular on the mass shell. Then the equation governing the evolution of $F_{\rm qu}(x, p)$ explicitly involves the function Δ_F (see appendix B). In the next section we shall see that, in general, one can not get rid of Δ_F. Even if the RHS side of equation (2.10) happens to vanish for some system, for systems conformally related to the one under consideration, the Liouville–Vlasov operator annihilates $F_{\rm cl}$ on the mass shell only.

3. The Wigner function and conformal transformations of a spacetime

Let us now consider the conformal transformation on a spacetime manifold (we shall use throughout an overline for quantities in the conformally related spacetime):

$$g_{\alpha\beta}(x) = a(x)^2 \overline{g}_{\alpha\beta}(x) \qquad \varphi(x) = \frac{\overline{\varphi}(x)}{a(x)} \qquad V(x) = \frac{\overline{V}(x)}{a(x)^2} \tag{3.1}$$

$a(x)$ being an arbitrary smooth function. Equation (2.4) is known to be invariant under such a transformation [1]. If $\varphi(x)$ is a solution of equation (2.4) on the manifold $\mathcal{M}$ with the metric $g_{\alpha\beta}(x)$ and in the presence of the external potential $V(x)$, then $\overline{\varphi}(x)$ is a solution of the conformally transformed equation on the manifold $\overline{\mathcal{M}}$ with the metric $\overline{g}_{\alpha\beta}(x)$ and in the presence of the external potential $\overline{V}(x)$, and vice versa.

One can expect that the Wigner function $f(x, p)$ associated with the field φ must be somehow connected with the Wigner function $\overline{f}(x, p)$ associated with the field $\overline{\varphi}$. In the classical limit ($\hbar \to 0$) the connection is very simple. The mass shell (2.6) is only scaled under the conformal transformation (we treat covariant components of the momentum vector, p_α, and contravariant components of the vector y^α as parameters which are not changed under conformal transformations, thus making a distinction between the tangent and cotangent spaces). Thus,

$$\Omega = \overline{\Omega}\, a^{-2}. \tag{3.2}$$

Next, under the conformal transformation (3.1), the Liouville–Vlasov operator is transformed as follows:

$$\widehat{\mathcal{L}} = \widehat{\overline{\mathcal{L}}}\, a^{-2} + a_\alpha \frac{\partial}{\partial p_\alpha} \Omega \tag{3.3}$$

where

$$a_\alpha = \frac{\partial}{\partial x^\alpha} \ln a. \tag{3.4}$$

By comparing equations (3.2), (3.3) with equations (B.2) and (B.1), one can conclude that the transformation law for the Wigner function should be

$$f(x, p) = a(x)^2\, \overline{f}(x, p) + \text{quantum corrections}. \tag{3.5}$$

We could also arrive at equation (3.5) by writing the transformation law for the Liouville–Vlasov operator in the form

$$\widehat{\mathcal{L}} = a^{-2}\, \widehat{\overline{\mathcal{L}}} + \Omega\, a_\alpha \frac{\partial}{\partial p_\alpha}. \tag{3.6}$$

From equations (2.7) and (3.6) it follows that the classical distribution function remains invariant under the conformal transformation. Then equations (2.5) and (3.2) imply (3.5).

One can also see from equation (3.6) that, unlike equation (2.7), the off-shell Vlasov equation (2.10) is not invariant under the conformal transformation. Namely, the function Δ_F is changed as follows:

$$\Delta_F(x, p) = \overline{\Delta_F}(x, p) + a_\alpha \frac{\partial}{\partial p_\alpha} F_{\rm cl}(x, p). \tag{3.7}$$

In order to find the quantum corrections in equation (3.5), we must turn to the definition of the covariant Wigner function (2.3). First, we write the transformation law for the operator $\widehat{\nabla}_\alpha$, equation (2.1), when acting as a scalar function in the tangent bundle. Under the conformal transformation of the metric, equation (3.1), it is transformed as follows (cf equation (B.14)):

$$\widehat{\nabla}_\alpha = \widehat{\overline{\nabla}}_\alpha - a_\nu y^\nu \frac{\partial}{\partial y^\alpha} - a_\alpha y^\nu \frac{\partial}{\partial y^\nu} + y_\alpha a^\nu \frac{\partial}{\partial y^\nu}. \tag{3.8}$$

Let us now write the field (2.2) in the following way (compare equation (3.1)):

$$\Phi(x,y)=\left(e^{(y^\alpha\widehat{\nabla}_\alpha)}a^{-1}\,e^{(-y^\alpha\widehat{\nabla}_\alpha)}\right)\left(e^{(y^\alpha\widehat{\nabla}_\alpha)}\,e^{(-y^\alpha\widehat{\overline{\nabla}}_\alpha)}\right)\left(e^{(y^\alpha\widehat{\overline{\nabla}}_\alpha)}\overline{\varphi}(x)\right). \tag{3.9}$$

The last factor on the RHS of equation (3.9) is, by definition, $\overline{\Phi}(x, y)$. The second factor, which is an operator acting in the tangent bundle, can be expanded as follows (see appendix A):

$$\begin{aligned}\widehat{\overline{Z}}(x,y) &:= e^{(y^\alpha\widehat{\nabla}_\alpha)}\,e^{(-y^\alpha\widehat{\overline{\nabla}}_\alpha)}\\ &= 1+A^\alpha\left(\tfrac{1}{2}\widehat{\nabla}_\alpha-\frac{\partial}{\partial y^\alpha}\right)+B^\alpha\left(\tfrac{1}{3}\widehat{\nabla}_\alpha-\tfrac{1}{2}\frac{\partial}{\partial y^\alpha}\right)\\ &\quad+\tfrac{1}{2}A^\alpha A^\beta\left(\tfrac{1}{4}\widehat{\nabla}_\alpha\widehat{\nabla}_\beta-\frac{\partial}{\partial y^\beta}\widehat{\nabla}_\alpha+\frac{\partial^2}{\partial y^\alpha\partial y^\beta}\right)\\ &\quad+\text{terms of higher order}\end{aligned} \tag{3.10}$$

where A^α and B^α are given in appendix A.

Lastly, the first factor on the RHS of equation (3.9) is a function which involves the conformal factor $a(x)$ and its derivatives. The known formula gives the expansion:

$$\begin{aligned}A(x,y) &:= e^{(y^\alpha\widehat{\nabla}_\alpha)}a^{-1}\,e^{(-y^\alpha\widehat{\nabla}_\alpha)}\\ &= a^{-1}\left(1-y^\alpha a_\alpha+\tfrac{1}{2}(y^\alpha a_\alpha)^2-\tfrac{1}{2}y^\alpha y^\beta a_{\alpha;\beta}\right)\\ &\quad+\text{terms of higher order}\end{aligned} \tag{3.11}$$

where $a_{\alpha;\beta} := \nabla_\beta a_\alpha$. On the RHS of equations (3.10), (3.11), we have kept only those terms which yield lowest order quantum corrections to equation (3.5).

We are now prepared for writing down the transformation law for the Wigner function, correct to second adiabatic order. Before doing that we note the following identity [11]:

$$\frac{\partial}{\partial y^\alpha}\Phi(x,y)=\widehat{\nabla}_\alpha\,\Phi(x,y)+\text{terms of second or higher order.} \tag{3.12}$$

Then, with the aid of the results of appendix A, we finally obtain:

$$\begin{aligned}f(x,p) &= a^2\left(1+\tfrac{1}{4}\hbar^2 a_{\alpha;\beta}\frac{\partial^2}{\partial p_\alpha\partial p_\beta}+\tfrac{1}{8}\hbar^2A^\alpha(\partial_p)\overline{D}_\alpha-\tfrac{1}{24}\hbar^2B^\alpha(\partial_p)p_\alpha\right)\overline{f}(x,p)\\ &\quad+\text{terms of higher adiabatic order}\end{aligned} \tag{3.13}$$

where the operator $\overline{D}_\alpha$ is conformally related to (2.9), and $A^\alpha(\partial_p)$ and $B^\alpha(\partial_p)$ are the following operators (cf equations (A.1), (A.2)):

$$A^\alpha(\partial_p)=(2\,a_\mu\,\delta^\alpha_\nu-\overline{a}^\alpha\,\overline{g}_{\mu\nu})\frac{\partial^2}{\partial p_\mu\partial p_\nu} \tag{3.14}$$

$$B^\alpha(\partial_p)=(2\overline{a}_{\mu;\nu}\,\delta^\alpha_\gamma-\overline{a}^\alpha_{;\mu}\,\overline{g}_{\nu\gamma}-8a_\mu\,a_\nu\,\delta^\alpha_\gamma+2\,\overline{a}^\beta\,a_\beta\,\overline{g}_{\mu\nu}\,\delta^\alpha_\gamma+4\,\overline{a}^\alpha\,a_\mu\,\overline{g}_{\nu\gamma})\frac{\partial^3}{\partial p_\mu\partial p_\nu\partial p_\gamma}. \tag{3.15}$$

We have denoted $\overline{a}_{\mu;\nu} := \overline{\nabla}_\nu a_\mu$ and $\overline{a}^\alpha := \overline{g}^{\alpha\beta} a_\beta$.

Equations (3.13)–(3.15) give the transformation law for the Wigner function which we have been seeking, correct to second adiabatic order. It is worth noting that the transformation law has been derived irrespective of any specific form of quantum kinetic equations, and it holds independently of whether or not the Wigner functions satisfy these equations. One could follow a different way. Given equations which govern the evolution of a Wigner function, one may seek a transformation which leaves the equations invariant under the conformal transformation of the metric and the external potential. This is what we did in order to obtain the transformation law at zeroth order, equation (3.5). However, this programme meets serious problems at higher orders because the structure of the quantum kinetic equations in curved spacetime is very complicated. Instead, in appendix B we prove, to second order, the conformal invariance of the collisionless quantum kinetic equations by using the transformation law for the Wigner function.

One may make sure that equations (3.13)–(3.15) yield the right transformation law for physical observables. For example, the number-flux vector defined by [11]

$$\langle J_\alpha(x)\rangle = \int \frac{\mathrm{d}^4 p}{\sqrt{-g(x)}}\, p_\alpha\, f(x,p) \tag{3.16}$$

is transformed as follows:

$$(-g)^{1/4}\langle J_\alpha\rangle = (-\overline{g})^{1/4}\langle \overline{J}_\alpha\rangle\,. \tag{3.17}$$

The transformation law (3.17) implies that the number of 'particles' (defined as in kinetic theory) in a comoving 3-volume, $\mathrm{d}N := \langle J^\alpha\rangle \mathrm{d}\Sigma_\alpha$, is an invariant of the group of conformal transformations. Here $\mathrm{d}\Sigma_\alpha := \sqrt{-g}\,\varepsilon_{\alpha\sigma\mu\nu}\,\mathrm{d}x^\sigma\,\mathrm{d}x^\mu\,\mathrm{d}x^\nu$.

The transformation law for the stress–energy tensor depends upon an explicit form of the external potential. For a scalar field coupled to curvature one obtains (see appendix C)

$$\left(\frac{g}{\overline{g}}\right)^{1/4}\langle T_{\alpha\beta}\rangle = \langle \overline{T}_{\alpha\beta}\rangle + \tfrac{1}{2}\hbar^2(6\xi - 1)\left(2aa_{(\alpha}\overline{\nabla}_{\beta)} - a\overline{g}_{\alpha\beta}a_\nu\overline{\nabla}^\nu - \overline{g}_{\alpha\beta}\overline{a}^\nu_\nu\right)\frac{\overline{N}}{a} \tag{3.18}$$

where ξ is the non-minimal gravitational coupling constant and

$$\overline{N}(x) = \int \frac{\mathrm{d}^4 p}{\sqrt{-\overline{g}(x)}}\,\overline{f}(x,p). \tag{3.19}$$

4. Quantum corrections to isotropic distributions

Let us consider a system in external fields $g_{\alpha\beta}$ and V whose classical distribution function (which is assumed to obey the Vlasov equation) is locally isotropic in momentum space, that is $F_{\mathrm{cl}}(x,p) = F(x, u^\alpha(x)p_\alpha)$, $u^\alpha(x)$ being some world velocity field. Of course, the external fields must be consistent with the symmetry of the distribution function. From the results of [18–22] and from the conformal invariance of the Vlasov equation it follows that only two possibilities exist:

(A) the metric is conformally stationary, i.e. the line element takes the form (in preferred coordinates)

$$\mathrm{d}s^2 = a^2(\mathrm{d}t^2 + 2\overline{g}_{0i}\,\mathrm{d}x^0\,\mathrm{d}x^i + \overline{g}_{ij}\,\mathrm{d}x^i\,\mathrm{d}x^j) \tag{4.1}$$

with $\overline{g}_{0i}$ and $\overline{g}_{ij}$ being functions of the spatial coordinates x^i alone, and either the potential V equals zero or the function $\overline{V} := Va^2$ does not depend on time;

(B) the metric is conformally static, that is $\overline{g}_{0i} = 0$ in some coordinate system, and the function $\overline{V}$ depends on time alone (since the Vlasov equation is linear in V, one can conclude that in case B the function $\overline{V}$ may be of the form [19] $\overline{V} = \overline{V}_1(t) + \overline{V}_2(x^i)$; we shall not deal with this more general situation here).

In accordance with our strategy we can solve the quantum kinetic equation in an auxiliary spacetime with the line element

$$d\overline{s}^2 = dt^2 + 2\overline{g}_{0i}\, dx^0\, dx^i + \overline{g}_{ij}\, dx^i\, dx^j \tag{4.2}$$

and then use equation (3.13) to calculate the Wigner function in the physical spacetime.

In the spacetime with the metric $\overline{g}_{\alpha\beta}$ there exists a unit timelike Killing vector $\overline{\xi}^\alpha$, $\overline{\xi}^\alpha\overline{\xi}_\alpha = 1$ ($\overline{\xi}^\alpha = \delta_0^\alpha$ in the preferred coordinates), which obeys the equation [26]:

$$\overline{\nabla}_\alpha\overline{\xi}_\beta + \overline{\nabla}_\beta\overline{\xi}_\alpha = 0 \qquad \text{in case A} \tag{4.3}$$

$$\overline{\nabla}_\alpha\overline{\xi}_\beta = 0 \qquad \text{in case B.} \tag{4.4}$$

The potential must satisfy the conditions:

$$\overline{\xi}^\alpha\overline{V}_{;\alpha} = 0 = \overline{\xi}^\alpha\overline{\xi}^\beta\overline{V}_{;\alpha\beta} = \dots \qquad \text{in case A} \tag{4.5}$$

$$\overline{\Delta}^{\alpha\beta}\overline{V}_{;\alpha} = 0 = \overline{\Delta}^{\alpha\beta}\overline{V}_{;\alpha\beta} = \dots \qquad \text{in case B.} \tag{4.6}$$

Here

$$\overline{\Delta}^{\alpha\beta} := \overline{\xi}^\alpha\overline{\xi}^\beta - \overline{g}^{\alpha\beta} \tag{4.7}$$

is the spacelike projection operator and $\overline{V}_{;\alpha} := \overline{\nabla}_\alpha\overline{V}$ etc.

In the two cases considered, the locally isotropic distribution functions [19–22] take the form (one can see that in case B the distribution function below has the desired form $F(x, u^\alpha(x)p_\alpha)$ on the mass shell):

$$F_{\text{cl}} = F(\overline{\xi}^\alpha p_\alpha) \qquad \text{in case A} \tag{4.8}$$

$$F_{\text{cl}} = F\left(\sqrt{\overline{\Delta}^{\alpha\beta} p_\alpha p_\beta}\right) \qquad \text{in case B.} \tag{4.9}$$

We proceed now to the derivation of the quantum corrections to the classical distribution functions. Both the function (4.8) and the function (4.9) satisfy the off-shell Vlasov equation in the external fields $\overline{g}_{\alpha\beta}$ and $\overline{V}$, with $\overline{\Delta}_F$ being zero (see equation (2.10)). Note, however, that for systems conformally related to the ones under consideration, $\Delta_F \neq 0$, as follows from equation (3.7).

The semiclassical Wigner functions which correspond to the distribution functions (4.8) and (4.9) in the spacetime $\overline{\mathcal{M}}$ are represented by the same expansions as in equation (2.5), with the argument of the δ-function being

$$\overline{\Omega} = \overline{g}^{\alpha\beta} p_\alpha p_\beta - \overline{V}. \tag{4.10}$$

The off-shell quantum corrections are obtained by substituting the classical distribution functions (4.8) and (4.9) into equation (B.6). Both in case A and in case B, they can be written in the form:

$$\overline{F}_1 = -\tfrac{1}{6}\left(\overline{R} + 2\overline{R}^{\mu\,\nu}_{\alpha.\beta}\, p_\mu p_\nu \frac{\partial^2}{\partial p_\alpha \partial p_\beta}\right) F_{\rm cl}$$

$$\overline{F}_2 = \left(-\tfrac{1}{3}\overline{R}^{\mu\nu} p_\mu p_\nu + \tfrac{1}{4}\overline{V}^{;\alpha}_{\alpha} + \tfrac{1}{2}\overline{V}^{;\mu}_{\nu}\, p_\mu \frac{\partial}{\partial p_\nu}\right) F_{\rm cl} \tag{4.11}$$

$$\overline{F}_3 = \tfrac{1}{12}\left(2\,\overline{V}^{;\mu\nu}\, p_\mu p_\nu - \overline{V}^{;\alpha}\overline{V}_{;\alpha}\right) F_{\rm cl}\,.$$

Next, the on-shell quantum corrections are found from equation (B.7). It follows from the conditions listed in equations (4.5) and (4.6) that the equation for $\overline{F}_{\rm qu}$ can be integrated as if the potential $\overline{V}$ were constant. A solution of this equation which corresponds to the classical distribution (4.8) was found in [15]. It reads (with $\overline{\xi}^\alpha \overline{\xi}_\alpha = 1$)

$$\overline{F}_{\rm qu} = -\tfrac{1}{6}\overline{R}_{\alpha\beta} \frac{\partial^2}{\partial p_\alpha \partial p_\beta} F_{\rm cl} \qquad \text{in case A.} \tag{4.12}$$

Let us now consider case B. It follows from equation (4.4) that the following identities hold [26]:

$$\overline{\xi}^\alpha\, \overline{R}_{\alpha\beta\mu\nu} = 0 = \overline{\xi}^\sigma\, \overline{R}_{\alpha\beta\mu\nu;\sigma}\,. \tag{4.13}$$

By using equations (4.4), (4.6) and (4.13) one readily obtains that the function

$$\overline{F}_{\rm qu} = -\tfrac{1}{12}\overline{R}_{\alpha\beta} \frac{\partial^2}{\partial p_\alpha \partial p_\beta} F_{\rm cl} \qquad \text{in case B} \tag{4.14}$$

satisfies the quantum kinetic equation (B.7), with $F_{\rm cl}$ being of the form (4.9).

Equations (4.11) and (4.12) give the second-order quantum corrections to the locally isotropic distribution (4.8) in a stationary spacetime in the presence of a static potential (case A), while equations (4.11) and (4.14) give the quantum corrections to the distribution (4.9) in a static spacetime in the presence of a homogeneous potential (case B). As we have said, the Wigner functions for systems conformally related to the ones which have been considered in this section are obtained by using equations (3.1) and (3.13). In the next section we shall analyse the structure of the quantum corrected number-flux vector and stress–energy tensor in these cases.

5. Quantum corrections to a perfect fluid

The distributions (4.8) and (4.9) both imply that the classical stress–energy tensor has a perfect fluid structure [19–22]

$$\langle \overline{T}^{\alpha\beta}_{\rm cl} \rangle = (\overline{\varepsilon}_{\rm cl} + \overline{P}_{\rm cl})\overline{\xi}^\alpha\, \overline{\xi}^\beta - \overline{P}_{\rm cl}\overline{g}^{\alpha\beta} \tag{5.1}$$

the classical energy density $\overline{\varepsilon}_{\rm cl}$ and pressure $\overline{P}_{\rm cl}$ being given by

$$\begin{aligned}\overline{\varepsilon}_{\rm cl} &= 2\pi \int_0^\infty {\rm d}p\, p^2(\overline{V} + p^2)^{1/2} F_{\rm cl}(\overline{V}, p)\\ \overline{P}_{\rm cl} &= \tfrac{1}{3}2\pi \int_0^\infty {\rm d}p\, p^4(\overline{V} + p^2)^{-1/2} F_{\rm cl}(\overline{V}, p)\end{aligned} \tag{5.2}$$

with (compare equations (4.8), (4.9))

$$F_{\rm cl}(\overline{V}, p) = \begin{cases} F\left(\sqrt{\overline{V} + p^2}\right) & \text{in case A} \\ F(p) & \text{in case B.} \end{cases} \tag{5.3}$$

The classical number-flux vector associated with the locally isotropic distributions reads:

$$\langle \overline{J}^{\alpha}_{\rm cl} \rangle = \overline{n}_{\rm cl}\, \xi^\alpha \tag{5.4}$$

the classical number density being

$$\overline{n}_{\rm cl} = 2\pi \int_0^\infty {\rm d}p\, p^2 F_{\rm cl}(\overline{V}, p). \tag{5.5}$$

It should be mentioned that we have taken into account positive 'energies' only, $\overline{E} = \sqrt{\overline{V} + p^2}$, when integrating over the mass shell. This is tantamount to multiplying the distribution function by the step function [27] $\Theta(\xi^\alpha p_\alpha)$. Though $\xi^\alpha p_\alpha$ is not a constant of motion in case B, the step function is 'ignored' by the Liouville–Vlasov operator, nor does it contributes to the quantum kinetic equation, since differentiation of the step function gives $\delta(\xi^\alpha p_\alpha)$ which vanishes on the mass shell (we assume that, in the classical limit, $\overline{V}$ is non-negative everywhere in the spacetime). Note that negative 'energies', $\overline{E} = -\sqrt{\overline{V} + p^2}$, correspond to antiparticles [6].

The quantum corrections to equations (5.1), (5.4) are easily obtained by substituting the Wigner functions found in the preceding section into equations (C.4) and (3.16). We shall first consider case B. The second order quantum corrections to the number flux (5.4) can be shown to vanish. The quantum corrected stress–energy tensor can be written, after some simple algebra, in the following form:

$$\begin{aligned}\langle \overline{T}^{\alpha\beta} \rangle = \langle \overline{T}^{\alpha\beta}_{\rm cl} \rangle &+ \hbar^2 (\xi - \tfrac{1}{6})\, \overline{R}^{\alpha\beta}\, M_{1,1} + \hbar^2 \ddot{\overline{V}}\, \overline{\Delta}^{\alpha\beta} \left(\tfrac{1}{2} (\xi - \tfrac{1}{6})\, M_{1,2} - \tfrac{1}{24} \overline{V}\, M_{1,3}\right) \\ &- \hbar^2 \dot{\overline{V}}^2\, \overline{\Delta}^{\alpha\beta} \left(\tfrac{3}{4}(\xi - \tfrac{1}{6})\, M_{1,3} - \tfrac{1}{96} M_{2,4} - \tfrac{1}{16} \overline{V}\, M_{1,4}\right) \\ &+ \tfrac{1}{32} \dot{\overline{V}}^2 \xi^\alpha \xi^\beta M_{1,3}\, .\end{aligned} \tag{5.6}$$

Here $\dot{\overline{V}} := \xi^\alpha\, \overline{\nabla}_\alpha \overline{V}$, etc, $\overline{\Delta}^{\alpha\beta}$ is defined by equation (4.7), and the M denote the integrals:

$$M_{n,k} = 2\pi \int_0^\infty {\rm d}p\, p^{2n} (\overline{V} + p^2)^{1/2-k} F(p). \tag{5.7}$$

The stress–energy tensor (5.6) has an 'almost' perfect fluid form, except the second term which involves $\overline{R}^{\alpha\beta}$. The latter has in general off-diagonal components. Note, however, that the Einstein equations impose severe restrictions on the underlying spacetime. Suppose, for example, that the conformal factor in equation (4.1) depends on time only: $a = a(t)$ (recall that we are considering case B, that is $\overline{g}_{0i} = 0$ in the preferred coordinates). If the total stress–energy tensor of classical matter is assumed to have a perfect fluid form then the physical spacetime is necessarily a Robertson–Walker one [19]. Then $\overline{R}^{\alpha\beta} = 2K\,\overline{\Delta}^{\alpha\beta}$, with $K = -1, 0, 1$, that is the quantum corrected stress–energy tensor preserves the perfect fluid structure in this case. Let us write down the stress–energy tensor with the second order quantum corrections in a Robertson–Walker spacetime with the line element

$$\mathrm{d}s^2 = a^2\left(\mathrm{d}t^2 - \frac{\mathrm{d}r^2}{1-Kr^2} - r^2\,\mathrm{d}\theta^2 - r^2\sin^2\theta\,\mathrm{d}\varphi^2\right). \tag{5.8}$$

Since the Bianchi identities hold automatically for the stress–energy tensor derived in kinetic theory, it suffices to know the energy density $\varepsilon := \langle T_0^0\rangle$. Making use of equations (3.18), (5.2) and (5.6) gives

$$\varepsilon = \frac{1}{a^4}M_{1,0} + \frac{\hbar^2}{2a^4}(1-6\xi)\left(\frac{\dot a^2}{a^2} - K\right)M_{1,1} + \tfrac{1}{4}\hbar^2(1-6\xi)\frac{\dot a}{a^3}\left(\dot U + 2\frac{\dot a}{a}U\right)M_{1,2}$$
$$+ \tfrac{1}{32}\hbar^2\left(\dot U + 2\frac{\dot a}{a}U\right)^2 M_{1,3}. \tag{5.9}$$

Here dots stand for the time derivative, and the M are given by equation (5.7), with $\overline{V} = Ua^2$, U being the classical part of the external potential in the Robertson–Walker spacetime (see appendix C). Equation (5.9) extends the result of [13] to the case of a variable potential and the non-minimal coupling to curvature. Physical manifestations of the quantum kinetic corrections in the Friedmann–Robertson–Walker cosmology have been examined in [14]. It was shown there that the quantum corrected energy density causes an apparent time variation of the value of an effective gravitational constant in the early universe.

We turn now to case A of the preceding section. The structure of the quantum corrections is more complicated in this case. To avoid lengthy expressions, we shall confine our attention to systems for which $V = 0$. Substituting the distribution function (4.8) and the quantum corrections to it given by equations (4.11), (4.12) into equations (3.16) and (C.4) gives the quantum corrected number-flux vector and stress–energy tensor in a stationary spacetime (recall that $\overline{\xi}^\alpha\overline{\xi}_\alpha = 1$):

$$\langle\overline{J}^\alpha\rangle = \overline{\xi}^\alpha M_2 - \tfrac{1}{6}\hbar^2\,\overline{\Delta}^{\alpha\mu}\,\overline{\xi}^\nu\,\overline{R}_{\mu\nu}M_0 \tag{5.10}$$

$$\langle\overline{T}^{\alpha\beta}\rangle = \tfrac{1}{3}\left(4\,\overline{\xi}^\alpha\,\overline{\xi}^\beta - \overline{g}^{\alpha\beta}\right)\left(M_3 - \tfrac{1}{2}\hbar^2\,\overline{\xi}^\mu\,\overline{\xi}^\nu\,\overline{R}_{\mu\nu}M_1\right)$$
$$+ \tfrac{2}{3}\hbar^2\left(\overline{\xi}^{(\alpha}\,\overline{R}^{\beta)\mu}\,\overline{\xi}_\mu - \overline{\xi}_\mu\,\overline{\xi}_\nu\,\overline{R}^{\alpha\mu\beta\nu}\right)M_1. \tag{5.11}$$

Here $\overline{R}^{\alpha\mu\beta\nu}$ is the Riemann tensor of the stationary spacetime, and

$$M_n = 2\pi\int_0^\infty \mathrm{d}p\; p^n F(p). \tag{5.12}$$

The remarkable feature of the quantum corrections in equations (5.10), (5.11) is that they have a non-equilibrium structure. First, the quantum corrections to stress components of the stress–energy tensor keep track of anisotropic character of the gravitational background. Next, the hydrodynamical velocity associated with the number flux (5.10) is

$$\overline{v}^{\alpha} = \overline{\xi}^{\alpha} - \tfrac{1}{6}\hbar^2 \overline{\Delta}^{\alpha\mu} \overline{\xi}^{\nu} \overline{R}_{\mu\nu} \frac{M_0}{M_2} . \tag{5.13}$$

On the other hand, the eigenvector of $\langle \overline{T}^{\alpha\beta} \rangle$ is

$$\overline{u}^{\alpha} = \overline{\xi}^{\alpha} - \tfrac{1}{4}\hbar^2 \overline{\Delta}^{\alpha\mu} \overline{\xi}^{\nu} \overline{R}_{\mu\nu} \frac{M_1}{M_3} . \tag{5.14}$$

In general, $\overline{u}^{\alpha}$ does not coincide with $\overline{v}^{\alpha}$. This implies that there is a non-vanishing energy flux relative to the particle stream†:

$$\overline{q}^{\alpha} = \tfrac{1}{9}\hbar^2 \left(\frac{2M_0 M_3}{M_2} - M_1 \right) \overline{\Delta}^{\alpha\mu} \overline{\xi}^{\nu} \overline{R}_{\mu\nu} . \tag{5.15}$$

The combination of the moments on the RHS of equation (5.15) can be zero for a very special choice of the function $F(p)$. It is surprising that $\overline{q}^{\alpha}$ does not vanish even when the classical distribution function describes local thermal equilibrium [28]. Though the effect is purely quantum, it could play an important role in the early universe and in the evolution of massive stars.

If (and only if) the manifold $\overline{\mathcal{M}}$ is static, that is the classical fluid motion is irrotational, the quantum corrections in equations (5.10), (5.11) vanish, which is consistent with case B in the limit $\overline{V} \to 0$, $\xi = \frac{1}{6}$ (compare equation (5.6)).

Finally, the quantum corrected number-flux vector and stress–energy tensor in a conformally stationary spacetime can be easily found with the aid of equations (3.17), (3.18). For the conformal coupling ($\xi = \frac{1}{6}$) they are obtained by simply multiplying equations (5.10) and (5.11), respectively, by a^{-4} and a^{-6}.

6. Concluding remarks

In summary, we have considered conformal properties of the covariant Wigner function. The transformation law (3.13) relates Wigner functions in manifolds conformally related to each other, correct to second adiabatic order. Given a solution of quantum kinetic equations which govern the evolution of a quantum distribution function (a Wigner function) of a specific system, one is able to evaluate Wigner functions for a wide class of systems conformally related to the one under consideration.

There exist two possible applications of the result. First, our method allows one to simplify an original system by reducing it to a simpler one for which the evolution can be easily solved. As an example, we have found explicit solutions of the collisionless quantum

† Usually an energy flux defined in such a way is interpreted as the heat flux caused by small deviations from thermal equilibrium [25, 27, 6] and is proportional to a gradient of the temperature. In our case, however, the temperature ($\propto 1/\overline{\xi}^{\alpha}\overline{\xi}_{\alpha}$) is constant, so there is no heat flux as is understood in transport theory. Likewise, there is no viscosity in the system, though the quantum corrected stress–energy tensor is non-diagonal in the fluid's rest frame.

kinetic equations in conformally stationary/static spacetimes and analysed the structure of the quantum corrected stress–energy tensor and number-flux vector.

Second, in some theories of gravity, like ones of Brans–Dicke–Jordan type which have been attracting a great deal of attention in recent years in connection with the low energy limit of string theory (see, for instance, [29]), two metrics are involved: a physical metric which enters matter equations, and the so called Einstein metric which satisfies Einstein-like equations. The two metrics are usually conformally related to each other (though more general relations have been suggested, see [30]), and one might want to solve matter equations on the physical manifold and then express the solutions in terms of the metric associated with gravity. Our method described in the present paper is especially suitable for such theories.

We would like to finish our discussion by the following remark. It is well known [18, 19, 25, 27] that local thermal equilibrium in a relativistic gas is possible in a very restrictive class of physical spacetimes. As we have shown, in a quantum system thermal equilibrium is even more exceptional than in a classical one. Spacetimes in which quantum fields can be in local thermal equilibrium are restricted to conformally static ones. Although we have analysed collisionless equations only, this is also correct for collision-dominated systems [27], because quantum corrections to the collision integral are supposed to be small compare with those for the Liouville–Vlasov operator when a system is nearly in equilibrium. Thus our results provide further motivation for studying non-equilibrium processes in quantum systems in curved spacetime.

Note added in proof. In [33, 34] another approach to treating conformal properties of the Wigner function was developed. It would be useful to see how that approach is related to ours.

Acknowledgments

I am grateful to Professors L P Horwitz and J D Bekenstein for helpful discussions about this work.

Appendix A. Evaluation of the transformation law for the Wigner function

To derive equation (3.13) we use essentially the same trick as in appendix A of [11]. First, making use of the Campbell–Baker–Hausdorff formula [32] together with equation (3.8), and taking into account the fact [11] that the operator ∇_α commutes with $\partial/\partial y^\alpha$ and annihilates y^α, immediately give equation (3.10) with

$$A^\alpha = 2y^\alpha y^\nu a_\nu - y^\nu y_\nu a^\alpha \tag{A.1}$$

$$\begin{aligned} B^\alpha &= y^\mu \widehat{\overline{\nabla}}_\mu A^\alpha - A^\mu \frac{\partial}{\partial y^\mu} A^\alpha \\ &= 2y^\alpha y^\mu y^\nu (\overline{\nabla}_\mu a_\nu) - y^\mu y^\nu y_\nu (\overline{\nabla}_\mu a^\alpha) - 8y^\alpha (y^\nu a_\nu)^2 + 2y^\alpha y^\nu y_\nu a^\mu a_\mu \\ &\quad + 4y^\nu y_\nu y^\mu a_\mu a^\alpha . \end{aligned} \tag{A.2}$$

Next, with the aid of equations (3.10) and (3.12) we obtain:

$$\widehat{\overline{Z}}(x, y)\overline{\Phi}(x, y) = \left(1 - \tfrac{1}{2}A^{\alpha}\widehat{\overline{\nabla}}_{\alpha} + \tfrac{1}{6}B^{\alpha}\frac{\partial}{\partial y^{\alpha}} + \tfrac{1}{8}A^{\alpha}A^{\mu}\widehat{\overline{\nabla}}_{\alpha}\widehat{\overline{\nabla}}_{\mu}\right)\overline{\Phi}(x, y)$$

$$+ \text{terms of third or higher orders.} \tag{A.3}$$

Using equations (3.9), (3.11), (A.3) and analogous identities for the conjugate field, we arrive at

$$\Phi(x, -y)\Phi^{\dagger}(x, y) = a^{-2}\left(1 - y^{\alpha}y^{\beta}a_{\alpha;\beta} - \tfrac{1}{2}A^{\alpha}\widehat{\overline{\nabla}}_{\alpha} - \tfrac{1}{6}B^{\alpha}\frac{\partial}{\partial y^{\alpha}} + \tfrac{1}{8}A^{\alpha}A^{\mu}\widehat{\overline{\nabla}}_{\alpha}\widehat{\overline{\nabla}}_{\mu}\right)$$

$$\times \left(\overline{\Phi}(x, -y)\overline{\Phi}^{\dagger}(x, y)\right)$$

$$+ \text{terms of higher orders.} \tag{A.4}$$

Finally, substituting equation (A.4) into the definition of the Wigner function (2.3) and integrating by parts give equation (3.13) (the last term in the bracket on the RHS of equation (A.4) leads to terms of fourth adiabatic order, which we do not keep).

Appendix B. Conformal invariance of the collisionless quantum kinetic equations

The equations for a Wigner function of a scalar field obeying the generalized Klein–Gordon equation (2.4) read [8, 9, 11], to second adiabatic order:

$$\widehat{\mathcal{L}}f(x, p) = \hbar^2\widehat{\Lambda}f(x, p) \tag{B.1}$$

$$\Omega f(x, p) = -\hbar^2\widehat{\Pi}f(x, p) \tag{B.2}$$

where Ω and $\widehat{\mathcal{L}}$ are respectively given by equations (2.6) and (2.8), and the operators on the RHS of equations (B.1) and (B.2) are

$$\widehat{\Lambda} = \tfrac{1}{6}R_{\nu\beta\mu\alpha}p^{\mu}\frac{\partial^2}{\partial p_{\alpha}\partial p_{\beta}}D^{\nu} - \tfrac{1}{24}R_{\alpha\mu\beta\nu;\sigma}p^{\mu}p^{\nu}\frac{\partial^3}{\partial p_{\alpha}\partial p_{\beta}\partial p_{\sigma}} + \tfrac{1}{12}R^{\nu}_{\alpha}\frac{\partial}{\partial p_{\alpha}}D_{\nu}$$

$$- \tfrac{1}{24}R_{\alpha\beta;\nu}p^{\nu}\frac{\partial^2}{\partial p_{\alpha}\partial p_{\beta}} - \tfrac{1}{24}R_{;\alpha}\frac{\partial}{\partial p_{\alpha}} + \tfrac{1}{48}V_{;\alpha\beta\sigma}\frac{\partial^3}{\partial p_{\alpha}\partial p_{\beta}\partial p_{\sigma}} \tag{B.3}$$

$$\widehat{\Pi} = -\tfrac{1}{4}D^{\alpha}D_{\alpha} - \tfrac{1}{6}R - \tfrac{1}{12}R_{\alpha\mu\beta\nu}p^{\mu}p^{\nu}\frac{\partial^2}{\partial p_{\alpha}\partial p_{\beta}} - \tfrac{1}{4}R_{\mu\nu}p^{\mu}\frac{\partial}{\partial p_{\nu}} + \tfrac{1}{8}V_{;\alpha\beta}\frac{\partial^2}{\partial p_{\alpha}\partial p_{\beta}}. \tag{B.4}$$

Here D_{α} is defined by equation (2.9), semicolons signify the covariant differentiation associated with the metric $g_{\mu\nu}$.

The operators $\widehat{\Lambda}$ and $\widehat{\Pi}$ satisfy the following identity [11]:

$$[\widehat{\Lambda}, \Omega] = [\widehat{\mathcal{L}}, \widehat{\Pi}] \tag{B.5}$$

which allows one to look for solutions to equations (B.1), (B.2) of the form (2.5) [9, 11]. The F in equation (2.5) are given by

$$
\begin{aligned}
F_1(x,p) &= \widehat{\Pi} F_{\rm cl}(x,p)\\
F_2(x,p) &= \tfrac{1}{2}[\widehat{\Pi},\Omega] F_{\rm cl}(x,p)\\
F_3(x,p) &= \tfrac{1}{6}[[\widehat{\Pi},\Omega],\Omega]\, F_{\rm cl}(x,p)
\end{aligned}
\tag{B.6}
$$

while the function $F_{\rm qu}(x,p)$ satisfies the equation [11]:

$$
\widehat{\mathcal{L}}\, F_{\rm qu}(x,p) = \widehat{\Lambda}\, F_{\rm cl}(x,p) + \widehat{\Pi}\,\Delta_F(x,p) \tag{B.7}
$$

$\Delta_F(x,p)$ being defined by equation (2.10).

Consider now a conformally related spacetime with the metric $\overline{g}_{\alpha\beta} = g_{\alpha\beta}/a^2$ and suppose that $\overline{f}(x,p)$ satisfies the quantum corrected Vlasov equation and mass-shell constraint in the presence of the external fields $\overline{g}_{\alpha\beta}$ and $\overline{V} = Va^2$. As we have shown in section 3, the conformal transformation of the spacetime induces the following transformation of the Wigner function defined in the cotangent bundle:

$$
f(x,p) = a^2(1+\hbar^2\widehat{C})\,\overline{f}(x,p) \tag{B.8}
$$

where the operator $\widehat{C}$ is, to second adiabatic order (cf equation (3.13),

$$
\begin{aligned}
\widehat{C} = {}& \tfrac{1}{24}\left(-4\overline{a}_{\mu;\nu} + \overline{a}^{\alpha}_{;\alpha}\overline{g}_{\mu\nu} + 28a_\mu a_\nu - 10\overline{a}^{\alpha}a_\alpha\overline{g}_{\mu\nu} + 6a_\mu\overline{D}_\nu - 3\overline{g}_{\mu\nu}\overline{a}^{\alpha}\overline{D}_\alpha\right)\frac{\partial^2}{\partial p_\mu\partial p_\nu}\\
&+\tfrac{1}{24}\left(-2\overline{a}_{\mu;\nu}p_\sigma + \overline{a}^{\alpha}_{;\sigma}\overline{g}_{\mu\nu}p_\alpha + 8a_\mu a_\nu p_\sigma - 2\overline{a}^{\alpha}a_\alpha\overline{g}_{\mu\nu}p_\sigma - 4\overline{g}_{\mu\nu}a_\sigma\overline{a}^{\alpha}p_\alpha\right)\\
&\times\frac{\partial^3}{\partial p_\mu\partial p_\nu\partial p_\sigma}.
\end{aligned}
\tag{B.9}
$$

If one takes into account the transformation laws for the Wigner function, equation (B.8), and for the Liouville–Vlasov operator, equation (3.3), and uses the equations for the function $\overline{f}(x,p)$, one arrives at the following equations:

$$
\begin{aligned}
\widehat{\mathcal{L}}f(x,p) = {}& \hbar^2\left(\widehat{\overline{\Lambda}} + [\widehat{\overline{\mathcal{L}}},\widehat{C}] - a_\alpha\frac{\partial}{\partial p_\alpha}(\widehat{\overline{\Pi}} + [\widehat{C},\overline{\Omega}])\right)\overline{f}(x,p)\\
&+\text{terms of higher order}
\end{aligned}
\tag{B.10}
$$

$$
\Omega f(x,p) = -\hbar^2\left(\widehat{\overline{\Pi}} + [\widehat{C},\overline{\Omega}]\right)\overline{f}(x,p) + \text{terms of higher order.} \tag{B.11}
$$

Now, our task is to prove that these equations are tantamount, to second adiabatic order, to equations (B.1), (B.2). Tedious but straightforward calculations give the following results:

$$
\widehat{\overline{\Pi}} + [\widehat{C},\overline{\Omega}] = \widehat{\Pi}\,a^2 + a_\alpha\frac{\partial}{\partial p_\alpha}\widehat{\overline{\mathcal{L}}} + \tfrac{1}{4}(4a_\mu a_\nu - \overline{a}^{\alpha}a_\alpha\overline{g}_{\mu\nu} - \overline{a}_{\mu;\nu})\frac{\partial^2}{\partial p_\mu\partial p_\nu}\overline{\Omega} \tag{B.12}
$$

$$
\begin{aligned}
\widehat{\overline{\Lambda}} + [\widehat{\overline{\mathcal{L}}},\widehat{C}] = {}& \left(\widehat{\Lambda} + a_\alpha\frac{\partial}{\partial p_\alpha}\widehat{\Pi}\right)a^2 + \tfrac{1}{4}(\overline{a}_{\mu;\nu} + \overline{a}^{\alpha}a_\alpha\overline{g}_{\mu\nu})\frac{\partial^2}{\partial p_\mu\partial p_\nu}\widehat{\overline{\mathcal{L}}}\\
&-\tfrac{1}{24}(\overline{a}_{\mu;\nu\sigma} - 8\overline{a}_{\mu;\nu}a_\sigma + 4\overline{a}^{\alpha}\overline{a}_{\mu;\alpha} - 6\overline{a}^{\alpha}a_\alpha a_\mu\overline{g}_{\nu\sigma})\frac{\partial^3}{\partial p_\mu\partial p_\nu\partial p_\sigma}\overline{\Omega}.
\end{aligned}
\tag{B.13}
$$

When deriving equations (B.12), (B.13) we have used the transformation laws for the Christoffel symbols and Riemann tensor [26]:

$$\Gamma^{\alpha}_{\mu\nu} - \overline{\Gamma}^{\alpha}_{\mu\nu} = \delta^{\alpha}_{\mu} a_{\nu} + \delta^{\alpha}_{\nu} a_{\mu} - \overline{g}_{\mu\nu}\overline{a}^{\alpha} \tag{B.14}$$

$$a^{-2} R_{\alpha\mu\beta\nu} - \overline{R}_{\alpha\mu\beta\nu} = 2a_{\nu[\alpha}\overline{g}_{\mu]\beta} - 2a_{\beta[\alpha}\overline{g}_{\mu]\nu} \tag{B.15}$$

where $a_{\alpha\beta} := \overline{a}_{\alpha;\beta} - a_{\alpha}a_{\beta} + \frac{1}{2}\overline{g}_{\alpha\beta}\overline{a}^{\nu}a_{\nu}$. Recall also that $a_{\alpha} := \partial_{\alpha}\ln a$, $\overline{a}^{\alpha} := \overline{g}^{\alpha\beta}a_{\beta}$ and semicolons in equations (B.9), (B.12)–(B.15) signify covariant differentiation associated with the metric $\overline{g}_{\alpha\beta}$.

Substituting equations (B.12), (B.13) into equations (B.10), (B.11) completes the proof.

Appendix C. The stress–energy tensor

Consider, for simplicity, the theory of a real scalar field described by the action:

$$S = \frac{1}{2}\int \mathrm{d}^4 x\,\sqrt{-g}\,(\hbar^2\nabla^{\alpha}\varphi\nabla_{\alpha}\varphi + \hbar^2\xi R\varphi^2 - U\varphi^2) \tag{C.1}$$

R being the Ricci scalar, U an external potential not involving a metric dependence.

The variation of the action with respect to the metric tensor yields the stress–energy tensor [1]:

$$\begin{aligned} T_{\mu\nu}(\xi, V) = {} & \hbar^2\nabla_{\mu}\varphi\nabla_{\nu}\varphi + \hbar^2\xi R_{\mu\nu}\varphi^2 - \tfrac{1}{2}g_{\mu\nu}(\hbar^2\nabla^{\alpha}\varphi\nabla_{\alpha}\varphi + \tfrac{1}{6}\hbar^2 R\varphi^2 - V\varphi^2) \\ & - \hbar^2\xi(\nabla_{\mu}\nabla_{\nu} - g_{\mu\nu}\nabla^{\alpha}\nabla_{\alpha})\varphi^2. \end{aligned} \tag{C.2}$$

We have introduced, for convenience, the generalized external potential [31]

$$V = U - \hbar^2(\xi - \tfrac{1}{6})R. \tag{C.3}$$

The expectation value of the stress–energy tensor can be expressed in terms of the covariant Wigner function (2.3) as follows [11]:

$$\begin{aligned} \langle T_{\mu\nu}(\xi, V)\rangle = {} & \int \frac{\mathrm{d}^4 p}{\sqrt{-g(x)}}\,(p_{\mu}p_{\nu} + \hbar^2\xi R_{\mu\nu})f(x,p) + \hbar^2(\tfrac{1}{4} - \xi)(\nabla_{\mu}\nabla_{\nu} - g_{\mu\nu}\nabla^{\alpha}\nabla_{\alpha}) \\ & \times \int \frac{\mathrm{d}^4 p}{\sqrt{-g(x)}}\,f(x,p). \end{aligned} \tag{C.4}$$

The action (C.1) is invariant under the conformal transformation (3.1). Notice that the potential U is transformed as the following:

$$U = \frac{\overline{U}}{a^2} + \hbar^2(1 - 6\xi)a^{-3}\overline{\nabla}^{\alpha}\overline{\nabla}_{\alpha}a. \tag{C.5}$$

Since the transformation law for the potential explicitly involves the dependence on the metric, the stress–energy tensor is not conformally invariant. With the aid of equation (3.13) we obtain the following relation:

$$\begin{aligned} a^2\langle T_{\mu\nu}\rangle - \langle\overline{T}_{\mu\nu}\rangle = {} & \hbar^2(6\xi - 1)(a_{(\mu}\overline{\nabla}_{\nu)} - \tfrac{1}{2}\overline{g}_{\mu\nu}a_{\alpha}\overline{\nabla}^{\alpha} - a_{\mu}a_{\nu} - \tfrac{1}{2}\overline{g}_{\mu\nu}(\overline{\nabla}^{\alpha}a_{\alpha})) \\ & \times \int \frac{d^4 p}{\sqrt{-\overline{g}(x)}}\,\overline{f}(x,p). \end{aligned} \tag{C.6}$$

Simple algebra then leads to equation (3.18). When deriving equation (C.6) we have used the identity (B.15) and the following property of the operator (2.9) (see [11]):

$$\int \frac{d^4 p}{\sqrt{-g}} D_\alpha f(x, p) = \nabla_\alpha \int \frac{d^4 p}{\sqrt{-g}} f(x, p). \qquad \text{(C.7)}$$

From equation (C.6) it follows that the stress–energy tensor (multiplied by the density $(-g)^{1/4}$) is conformally invariant, but not trace-free, if $\xi = \frac{1}{6}$. T^α_α vanishes only if $V = 0$.

References

[1] Birrell N D and Davies P C W 1982 *Quantum Fields in Curved Space* (Cambridge: Cambridge University Press)

[2] Parker L and Simon J Z 1993 *Phys. Rev.* D **47** 1339 (1993).

[3] Ford L H 1982 *Ann. Phys., NY* **144** 238
Kuo Ch–I and Ford L H 1993 *Preprint* gr–qc/9304008

[4] Wigner E P 1932 *Phys. Rev.* **40** 749

[5] Carruthers P and Zachariasen F 1983 *Rev. Mod. Phys.* **55** 245

[6] de Groot S R, van Leeuwen W A and van Weert Ch G 1980 *Relativistic Kinetic Theory* (Amsterdam: North-Holland)

[7] Habib S and Kandrup H E 1989 *Ann. Phys., NY* **191** 335

[8] Winter J 1985 *Phys. Rev.* D **32** 1871

[9] Calzetta E, Habib S and Hu B L 1988 *Phys. Rev.* D **37** 2901

[10] Fonarev O A 1990 *Izv. VUZov. Fiz.* **9** 47 (*Sov. Phys.–JETP* **33** 759)

[11] Fonarev O A 1994 *J. Math. Phys.* **35** 2105

[12] Bunch T S and Parker L 1979 *Phys. Rev.* D **20** 2499

[13] Pirk K and Börner G 1989 *Class. Quantum Grav.* **6** 1855

[14] Fonarev O A 1994 *Phys. Lett.* **190A** 29

[15] Fonarev O A 1991 *Phys. Lett.* **152A** 153

[16] Fonarev O A 1991 *PhD Thesis* (Tomsk, Russia: Tomsk University) unpublished

[17] Bekenstein J D 1974 *Ann. Phys., NY* **82** 535
Schmidt H-J 1988 *Phys. Lett.* **214B** 519
Futamase T and Maeda K 1989 *Phys. Rev.* D **39** 399

[18] Tauber G E and Weinberg J W 1961 *Phys. Rev.* **122** 1342

[19] Ehlers J, Geren P and Sachs R K 1968 *J. Math. Phys.* **9** 1344

[20] Chernikov N A 1962 *Sov. Phys.–Dokl.* **7** 428

[21] Hakim R 1968 *Phys. Rev.* **173** 1235

[22] Bel L 1969 *Astrophys. J.* **155** 83

[23] Yano K and Ishihara Sh 1973 *Tangent and Cotangent Bundles: Differential Geometry* (Pure and Applied Mathematics: A Series of Monographs and Textbooks) (New York: Dekker)

[24] Vlasov A A 1966 *Statistical Distribution Functions* (Moscow: Nauka) in Russian

[25] Israel W 1972 *General Relativity: Papers in Honour of J L Synge* ed L O'Raifeartaigh (Oxford: Clarendon) p 201

[26] Yano K 1970 *Integral Formulas in Riemannian Geometry* (Pure and Applied Mathematics: A Series of Monographs and Textbooks) (New York: Dekker)

[27] Stewart J M 1971 *Non–Equilibrium Relativistic Kinetic Theory* (Lecture Notes in Physics **10**) (Berlin: Springer)

[28] Fonarev O A 1993 *Proc. 3rd Int. Wigner Symp.* (Oxford) (to appear); 1993 *Preprint* gr–qc/9311018

[29] Tseytlin A A 1992 *Class. Quantum Grav.* **9** 979
Goldwirth D S and Perry M J 1993 *Preprint* hep–th/9308023
Damour T and Polyakov A M 1994 *Preprint* hep–th/9401069

[30] Bekenstein J D 1993 *Phys. Rev.* D **48** 3641

[31] Parker L and Toms D J 1984 *Phys. Rev.* D **29** 1584

[32] Serre J-P 1965 *Lie algebras and Lie groups* (New York: Benjamin)

[33] Calzetta E and Hu B L 1989 *Phys. Rev.* D **40** 380

[34] Calzetta E 1989 *Physica* A **158** 261

Class. Quantum Grav. **11** (1994) 2615–2618. Printed in the UK

Book reviews

Introduction to Cosmology
Matts Roos
New York: Wiley 1994
206 pp price £19.95
ISBN 0-471-94298-7

Particle physics and traditional cosmology have come together in the past twenty years to create an exciting new subject. The Big Bang has caught the public imagination, as witness the phenomenal sales of Steven Weinberg's *The First Three Minutes* and Stephen Hawking's *A Brief History of Time*, to mention but two. Physics departments everywhere have discovered that this is their best selling point. It is always what prospective students ask about. Many departments are responding by laying on courses on 'Cosmology' or 'The Universe' for first-year students. We certainly need a good introduction to cosmology for first-year students. But is this it?

Our new subject does not yet have an accepted name. 'Particle cosmology' is perhaps the best, though there are many weirder coinages, for example 'astrofundamental physics' for the Erice summer school. Roos seems to regard it as synonymous with cosmology: he talks of astronomy and particle physics merging in the common field of cosmology, though astronomers may object that they were studying cosmology long before particle physicists got in on the act!

Whatever name we give it, there are good reasons for writing an introductory text on the subject, but it is hard to decide just what to include. One of the things that makes particle cosmology such fun for the professionals is that it is among the most interdisciplinary of subjects. To follow what is going on, you need at least some knowledge of general relativity, quantum mechanics, astronomy, particle physics, statistical mechanics, nuclear physics and quantum field theory. How on earth can you convey all this to a first-year audience?

Roos takes a very ambitious line. He tries to cover all the current issues—dark matter, baryogenesis, perturbations in the microwave background, galaxy formation, and so on. He acknowledges that in many subjects 'the explanations have to be qualitative and the derivations meagre', but despite that he goes into considerable mathematical detail. There is a lot of excellent stuff here, but I kept wondering as I read it, 'What would a first-year student make of all this?'

In Chapter 2, Roos tells us, 'the main purpose of this chapter is to derive Einstein's law of gravitation using as few mathematical tools as possible'. A bold aim no doubt, but of course it cannot be done. Roos reaches the point of writing down Einstein's equations, complete with all the paraphernalia of tensor algebra, with the summation convention and upper and lower indices, but he is not in a position to explain what any of these things mean, and there are huge gaps in the 'derivation'. So what is the point? I am afraid this sort of treatment will only serve to reinforce the most serious stumbling block that many students face, that they cease to expect physics to make sense.

Of course, we should try to explain to them what the current debates are all about, why the results from COBE are so exciting, or what the dark matter might be. A great deal of this *can* be explained without going beyond first-year physics—with thought and imagination a lot more than one might initially suppose. But a lot of it *cannot* be explained, and to try to pretend that it can is to do the students a disservice.

There are a great many things here that will puzzle or confuse the best of students. They are told that the affine connections are not tensors. But this is meaningless since they have not been told what tensors *are*. The only definition is 'In general a tensor is a quantity with two or more indices running over the dimensions of a manifold'. What, they might then ask, is a manifold? The term does not appear in the index, and so far as I can see is never explained. This is, unfortunately, not an isolated example. Because of oversimplification, many definitions and mathematical derivations are misleading: the definition of Gaussian curvature is wrong; a spacetime diagram illustrating a black hole has the world line of a spacecraft outside the light cone; the 'proof' that if $\hat{U}$ is a unitary operator and $\hat{U} = \exp \mathrm{i}\hat{H}$, then

$\hat{H}^\dagger = \hat{H}$ is invalid. Terms ranging from 'the local supergalaxy' to 'adiabatic fluctuations' are used without explanation. There is considerable confusion over units. Some data are in SI units, some of course in the peculiar units of astronomers, other in CGS—a serious problem since many students have no idea what an erg is. Abbreviations like Gyr are used without explanation.

For all these faults, this is a bold and imaginative book. Once they have learned enough basic physics to make sense of it, students will find a lot here to excite their interest. But I cannot see it as a successful textbook for its intended audience.

T Kibble

Effective Action in Quantum Gravity
I L Buchbinder, S D Odintsov and I L Shapiro
Bristol: Adam Hilger 1992
424 pp price £51.00 $102.00
ISBN 0-7503-0122-8

This book gives a very good picture of the present status of field theoretic approaches to quantum gravity. It is divided into three parts. Part one is a preliminary review of quantum field theory in flat space, with particular emphasis on the method of the effective action. Part two deals with interacting quantum field theory in a fixed curved background, both with and without torsion. It includes a treatment of the Vilkovisky–de Witt effective action, and of the effective action for composite operators. Particular emphasis is placed on renormalization group effects, for example the possibility that strong curvature induces phase transitions. Part three deals with quantum gravity proper. By this is meant the theory whose Lagrangian is the sum of a cosmological term, an Einstein term and higher derivative terms quadratic in curvature. As far as we know this theory is not completely consistent: it is perturbatively renormalizable but not unitary. The point of view of the authors on this issue is pragmatic: they regard this theory as an effective one, probably with a limited range of validity. Again much emphasis is put on the calculation of beta functions, but there is also a treatment of canonical quantization. There is then a discussion of the coupling of grand unified theories to quantum gravity, which is very relevant for example in early cosmology, and of the effective action in higher dimensional theories. The last chapter deals, somewhat spuriously, with membranes.

This book is in a certain sense an ideal continuation of Birrell and Davies' *Quantum Fields in Curved Space*. The book by Birrell and Davies contains the results of research done during the seventies, dealing almost exclusively with free fields in curved space. Buchbinder *et al* begin where Birrell and Davies end. They give a survey of work done mostly during the eighties, and the emphasis is on interacting theories. As a result the style is also quite different: methods of quantum field theory play a very prominent role here. The overall level of technical sophistication is quite high, but sufficient details are given so that the reader can with some patience reproduce all the results.

This is the only book where one can find a complete and coherent discussion of the topics listed above. It is therefore an indispensable reference for all those who want to work in this field. Apart from quantizers of gravity, I have in mind astroparticle physicists and 'early cosmologists', who will find here all the tools that are needed to go beyond the semiclassical approximation.

The failure of perturbative Einstein theory and the unitarity problems of higher derivative theories seem to have discouraged most researchers from pursuing this line of thought. The prevalent attitude nowadays seems to be that either quantum field theory is not enough (strings) or at least that a fully fledged non-perturbative approach is necessary (canonical methods triangulations etc). Hopefully this book will stimulate some people to reconsider this issue. It is possible that with some fresh ideas and a clever use of the renormalization group, further progress along these lines is possible.

R Percacci

Approaches to Numerical Relativity
R d'Inverno (ed)
Cambridge: Cambridge University Press 1992
378+xx pp price £40.00
ISBN 0-521-43976-0

'General relativity was for too long the ugly duckling of science'. So says Chris Clarke in the preface of *Approaches to Numerical Relativity* edited by R A d'Inverno. To paraphrase Clarke, one could also say that for too long numerical relativity—the simulation of highly relativistic systems using numerical computations—has been the runt of the general relativistic litter. However developments in high performance computing have created an opportunity for numerical relativity to make definite predictions concerning realistic astrophysical systems. In turn this has made general relativity relevant to understanding the nature of the physical universe.

This new volume, in a series of numerical relativity workshop proceedings published by Cambridge University Press, documents the presentations and formal discussions that took place during an international meeting held in Southampton, UK, late in 1991. In the past, numerical relativity has stood apart from the formal mathematical investigations of the Einstein equations that were in part responsible for placing general relativity in a position where there was too much theory following upon too little experimental evidence. The primary motivation for constructing numerical solutions to the Einstein equations was to remedy this situation by simulating realistic physical systems whose dynamics are governed by highly non-linear gravitational fields. This meant giving up many of the simplifying symmetries, gauges and formalisms that often formed the basis of a purely analytical approach to general relativity for a direct attack on the Einstein equations in their most general form.

In order that numerical relativity could employ the methods developed in more mature numerically oriented disciplines (e.g. computational hydrodynamics) the inherently covariant four-dimensional spacetime approach had to be abandoned for a formalism that split spacetime into space and time. The full four-dimensional manifold could then be constructed spatial slice by spatial slice from self-consistent initial conditions. Since a great deal of theoretical work had already been accomplished in the '3+1' decomposition of spacetime into space and time (more for the purposes of quantizing the gravitational field), it was the natural choice for most early work in numerical relativity have reported on the results obtained from this approach.

This volume is encouraging in that a large percentage of the papers present alternative ways of foliating spacetime in terms of its null cone structure. Two alternatives arise where the spacetime can be split into double-null foliations (the '2+2' formalism developed by d'Inverno and his collaborators) or a combination of null and timelike foliations (a method that has been used to a large extent by Winicour, Issacson and their collaborators).

The book itself is foliated into two major sections; Part A for the theoretical approaches and Part B for the more practical approaches. The first part includes not only more formal methods for studying various null, hyperbolic (spacelike to null), and pure spacelike foliations (represented respectively by articles from Hayward, Friedrich and Ó Murchadha) but also some more 'practical' methods described by d'Inverno, Bishop and Stewart. These authors have taken up the computational challenge by applying their proposals on non-trivial 'test-cases'. Also included in this section are theoretical discussions of colliding plane waves, rotating sources, boson stars and even a new-finite element approach to constructing spacetimes.

The second part of the volume consists of contributions from researchers who have implemented practical computational algorithms on different computer systems and who subsequently report on their results. If one defines numerical relativity as the use of numerical methods to solve the Einstein equations then the majority of the contributions in this section lie outside of this strict definition. Only four of the fifteen articles (those by Choptuik, Bona and Massó, Gómez and Winicour, and Bonnazola and Marck) actually discuss methods and results that involve solving the time-dependent Einstein equations. Two more contributions (Dubal *et al* and Nakao *et al*) discuss solutions to the Cauchy (initial value) problem in general relativity where one solves a subset of the Einstein equations (the elliptic constraint equations) that relate the dynamic and kinematic variables on a single t=const (3+1) hypersurface.

The remaining contributions to the second part deal with the dynamics of matter on fixed background spacetimes, or employ post-Newtonian or Newtonian approximations to calculate gravitational effects. Such problems are still of interest since the systems studied by such methods (stellar collapse, the coalescence of neutron stars, the formation of accretion discs around black holes or the growth of topological defects in early cosmologies, etc) may well be important sources of gravitational radiation where the complete solution to the Einstein equations is not necessary for the dynamical evolution.

That numerical relativity (in its strict definition) can provide us with a better understanding of the nature of Einstein's equations is well represented by Choptuik's article that discusses a newly discovered critical behaviour appearing in the gravitational collapse of scalar fields. While most of the dynamical fully relativistic systems discussed involve only one spatial dimension, the article by Bona and Massó describes the development of a fully relativistic three-dimensional code. This paper represents one of the first attempts to move towards the ultimate goal of numerical relativity: a computer code capable of providing a solution to the Einstein equations under completely general conditions.

Unfortunately this part of the volume fails to provide an update on the interesting (and sometimes controversial) results obtained through the use of highly accurate and stable fully relativistic two-dimensional codes that now simulate gravitational collapse, naked singularity formation, inhomogeneous cosmologies, dynamical black holes and the generation and propagation of non-linear gravitational waves. These codes can now construct substantial portions of fully dynamical spacetimes and much work has been done to provide tools capable of analysing the numerically constructed spacetimes using methods that have already proven successful for the analysis of analytic spacetimes. In particular, numerical methods exist for locating horizons, extracting the gravitational waves, analysing the Newman–Penrose quantities, integrating geodesics, and performing other procedures that allow one to understand the nature of a tabulated spacetime.

Finally the volume concludes with a substantially documented panel discussion on the current state and future goals of numerical relativity. This is interesting from the point of view that it allows one to see the individual concerns and personal views of researchers in numerical relativity that often do not appear in the more formal reports of their methods and results.

In spite of the somewhat biased view of the state of numerical relativity provided by this volume, it is an important contribution to the discipline. It demonstrates that the input of ideas to numerical relativity is coming from many different directions and that there is even a growing interest in the field on the part of those with traditionally theoretical interests. If anything this volume clearly shows that a significant amount of genetic diversity is beginning to appear in numerical relativity. All this bodes well for a fledgling discipline that is just beginning to make significant contributions to our understanding of general relativity as a physical theory.

D Hobill

Origin and Evolution of the Elements
N Prantzos, E Vangioni-Flam and M Cassé (ed)

Cambridge: Cambridge University Press 1993
545+xvi pp price £45.00 $69.95
ISBN 0-521-43428-9

This book contains the proceedings of a conference held in Paris in June 1992 to honour Hubert Reeves on his 60th birthday. Sandwiched between E Salpeter's historical introduction and his closing remarks are over 500 pages of reviews and contributions that offer state-of-the-art interpretations of element abundances observed in stars, interstellar gas, and meteorites.

A feature of the book is the provision of major reviews by international experts. These range from the general introductory review of cosmic abundances by N Grevasse and A Noels, through primordial abundances of H and He isotopes (J Geiss), Li abundance problems (F and M Spite), primordial nucleosynthesis (D Schramm), light elements in general (H Reeves), stellar nucleosynthesis up to Fe (D Lambert), heavy elements (M Arnould and K Takahashi), the first population of stars in the Galaxy (R Cayrel) element fractionation in the Sun (J P Meyer) and isotope abundance anomalies in the early solar system (F Begemann).

In addition there is the usual conference fare of short contributions containing new results. Some of these are substantial research articles in their own right—in fact, the published proceedings of this conference are noteworthy for excluding the category of abstracts of longer articles published elsewhere.

By reading first the reviews, which are all well written, and then the short contributions the reader can quickly reach the forefront of abundance research. There is no general text available that is as up to date as the contents of this proceedings; it is strongly recommended as both an introduction and a research monograph.

B Warner